TRENDS IN MATHEMATICS

Trends in Mathematics is a series devoted to the publication of volumes arising from conferences and lecture series focusing on a particular topic from any area of mathematics. Its aim is to make current developments available to the community as rapidly as possible without compromise to quality and to archive these for reference.

Proposals for volumes can be sent to the Mathematics Editor at either

Birkhäuser Verlag
P.O. Box 133
CH-4010 Basel
Switzerland

or

Birkhäuser Boston Inc.
675 Massachusetts Avenue
Cambridge, MA 02139
USA

Material submitted for publication must be screened and prepared as follows:

All contributions should undergo a reviewing process similar to that carried out by journals and be checked for correct use of language which, as a rule, is English. Articles without proofs, or which do not contain any significantly new results, should be rejected. High quality survey papers, however, are welcome.

We expect the organizers to deliver manuscripts in a form that is essentially ready for direct reproduction. Any version of $\TeX$ is acceptable, but the entire collection of files must be in one particular dialect of $\TeX$ and unified according to simple instructions available from Birkhäuser.

Furthermore, in order to guarantee the timely appearance of the proceedings it is essential that the final version of the entire material be submitted no later than one year after the conference. The total number of pages should not exceed 350. The first-mentioned author of each article will receive 25 free offprints. To the participants of the congress the book will be offered at a special rate.

Topics in Cohomological Studies of Algebraic Varieties

Impanga Lecture Notes

Piotr Pragacz
Editor

Birkhäuser Verlag
Basel · Boston · Berlin

Editor:

Piotr Pragacz
Institute of Mathematics of the
Polish Academy of Sciences
ul. Sniadeckich 8
P.O. Box 21
00-956 Warszawa
Poland

e-mail: p.pragacz@impan.gov.pl

2000 Mathematical Subject Classification 05E15, 14C17, 14C20, 14C22, 14C35, 14L35, 14M10, 14N15, 14N35, 15A52, 19Axx, 19Dxx, 19Exx, 32Q35, 32Sxx, 51P05, 58K30, 70H06

A CIP catalogue record for this book is available from the
Library of Congress, Washington D.C., USA

Bibliographic information published by Die Deutsche Bibliothek
Die Deutsche Bibliothek lists this publication in the Deutsche Nationalbibliografie; detailed bibliographic data is available in the Internet at http://dnb.ddb.de

ISBN 3-7643-7214-1 Birkhäuser Verlag, Basel – Boston – Berlin

Part of Springer Science+Business Media
Printed on acid-free paper produced from chlorine-free pulp. TCF ∞
Printed in Germany
ISBN-10: 3-7643-7214-1
ISBN-13: 976-3-7643-7214-9

9 8 7 6 5 4 3 2 1 www.birkhauser.ch

Contents

Dédié à Alexandre Grothendieck

Preface

The articles in this volume[1] are an outgrowth of seminars and schools of Impanga. Impanga is an algebraic geometry group operating since 2000 at the Institute of Mathematics of Polish Academy of Sciences in Warsaw. Besides seminars Impanga organized the following schools at the Banach Center in Warsaw:

- *Characteristic classes of singular varieties*, April 2002,
- *Stratifications of moduli spaces*, May 2002,
- *Schubert varieties*, May 2003, and
- *Hommage à Grothendieck*, January 2004.

More information about the Impanga seminars and schools[2] can be found on:

`http://www.impan.gov.pl/~pragacz/impanga.htm`

Impanga also co-organized the school *Algebraic geometry, algebra, and applications* in Borovetz, Bulgaria (September 2003).

Let us describe briefly the contents of the lecture notes in this volume[3].

Paolo Aluffi discusses various characteristic classes generalizing classical Chern classes for nonsingular varieties: the classes of Mather, Schwartz-MacPherson, and Fulton. A particular emphasis is put on concrete computations of these classes, often with the help of Segre classes.

Michel Brion gives a comprehensive introduction to the geometry of flag varieties and Grassmannians. A special emphasis is put on geometric properties of Schubert varieties and their resolutions: Bott-Samelson schemes. Vanishing properties of line bundles over Schubert varieties are studied. Also Richardson varieties together with their applications are discussed. One of the main goals is to present a proof of Buch's conjecture on the structure constants in the Grothendieck ring of a flag variety.

---

[1]During the preparation of this volume, the Editor was partially supported by KBN grant No 2P03A 024 23.

[2]The first three schools were partially supported by EAGER.

[3]The lecture notes by Aluffi and Schürmann stem from the first school, the notes by Brion, Buch, and Tamvakis from the third school, the article by Pragacz opening the volume from the fourth school, and the notes by Szyjewski from the school in Borovetz. The remaining contributions by Duan, Kisisel, and Pragacz come from the seminars of Impanga.

Anders Buch studies Grothendieck polynomials and their properties (following the work of Lascoux-Schützenberger and Fomin-Kirillov). They are used to show various combinatorial aspects of the Grothendieck ring of a Grassmannian and K-theoretic formulas for "quiver degeneracy loci".

Haibao Duan presents an elementary introduction to classical Morse theory, and shows its applications to homogeneous spaces. In particular, "Morse functions of Bott-Samelson type" are discussed together with Bott-Samelson cycles. Some applications to Schubert calculus are also mentioned.

The starting point of the article by Ali Kisisel are completely integrable systems, in particular: KP and KdV hierarchies. Then the infinite Grassmannian of Sato and Toda hierarchy are described. Finally, through the studies of random matrices and enumeration of graphs, the article approaches the solution by Kontsevich of Witten's conjecture, as well as some aspects of Gromov-Witten theory.

Piotr Pragacz shows a way of computing the structure constants for multiplication of Schubert classes in the cohomology rings of generalized flag varieties G/P.

Jörg Schürmann presents stratified Morse theory for constructible functions and its applications to characteristic classes of singular varieties. The point of view of characteristic classes of Lagrangian cycles is emphasized and a Verdier-type Riemann-Roch theorem is discussed.

Marek Szyjewski gives an introduction to higher K-groups of Quillen (and also to those of Milnor). Computations of higher K-groups of fields and of projective bundles, quadrics, and Severi varieties are presented. Some arithmetical aspects of the theory are also discussed.

Harry Tamvakis studies quantum cohomology of homogeneous spaces, notably of various Grassmannians. It is shown that three-point genus zero Gromov-Witten invariants are equal to classical triple intersection numbers on homogeneous spaces of the same Lie type. Quantum analogs of the Pieri and Giambelli formulas are also presented.

We dedicate the whole volume to **Alexander Grothendieck** who remains for us an unsurpassed master in cohomological studies of algebraic varieties. The opening article by Piotr Pragacz discusses some aspects of his life and work.

Acknowledgments. The Editor wishes to thank the authors for their scientific contributions. He is grateful to Andrzej Weber for his help with the Impanga schools. Warm thanks go also to Dr. Thomas Hempfling from Birkhäuser-Verlag for his invitation to publish this material in *Trends in Mathematics* and for a pleasant cooperation during the preparation of this volume.

Warszawa, October 2004 *The Editor*

Trends in Mathematics:
Topics in Cohomological Studies of Algebraic Varieties, xi–xxviii

Notes on the Life and Work of Alexander Grothendieck*

Piotr Pragacz**

When I was a child I loved going to school. The same instructor taught us reading, writing and arithmetic, singing (he played upon a little violin to accompany us), the archaeology of prehistoric man and the discovery of fire. I don't recall anyone ever being bored at school. There was the magic of numbers and the magic of words, signs, and sounds ...

A. Grothendieck: *Récoltes et Semailles*

Abstract. This is a story of Alexander Grothendieck – a man who has changed the face of mathematics during some 20 years of his work on functional analysis and algebraic geometry. Last year he turned 75. This paper, written in April 2004, is based on a talk presented at the *Hommage à Grothendieck* session of Impanga[1], held at the Banach Centre in Warsaw (January 2004).

Alexander Grothendieck was born in Berlin in 1928. His father, Alexander Shapiro (1890–1942) was a Russian Jew from a Hassidic town on a now Russian-Ukrainian-Belorussian border. He was a political activist – an anarchist involved in all the major European revolutions during 1905–1939. In the 20's and 30's he lived mostly in Germany, operating in the left-wing movements against more and more powerful Nazis, and working as a street photographer. In Germany, he met Hamburg-born Hanka Grothendieck (1900–1957). (The name *Grothendieck* comes from *plattdeutsch*, a Northern German dialect.) Hanka Grothendieck worked on and off as a

---

*Translated from the Polish by Janusz Adamus. This paper was originally published in *Wiadomości Matematyczne (Ann. Soc. Math. Pol.)* vol. 40 (2004). We thank the Editors of this journal for permission to reprint the paper.
**Partially supported by Polish KBN grant No. 2 P03A 024 23.
[1] *Impanga* is an algebraic geometry group, operating since 2000 at the Institute of Mathematics of the Polish Academy of Sciences. This session hosted the talks of: M. Chałupnik, *Grothendieck topologies and étale cohomology*, T. Maszczyk, *Toposes and the unity of mathematics*, J. Gorski, *Grothendieck stacks on Mazovia plains*, O. Kędzierski, *Why the derived categories?*, A. Weber, *The Weil conjectures*, G. Banaszak, *l-adic representations*, P. Krasoń, *Mordell-Weil groups of Abelian varieties*.

journalist, but her true passion was writing. On March 28, 1928 she gave birth to their son Alexander.

During 1928–1933 Alexander lived with his parents in Berlin. After Hitler's rise to power, Alexander's parents immigrated to France, leaving their son (for about 5 years) with the Heydorns, a surrogate family in Hamburg, where he went to a primary and secondary school. In 1939 Alexander joined his parents in France. His father was soon interned by the French Vichy police in the Vernet camp in the Pyrenees, and then handed out to the Nazis occupiers. He was murdered in the German concentration camp Auschwitz-Birkenau in 1942.

Very young Alexander Grothendieck

Hanka and Alexander Grothendieck did not survive the occupation without problems. In the years 1940–1942 they were interned – as "undesirable dangerous foreigners" – in the Rieucros camp near Mende in southern France. Hanka was later transferred to the Gurs camp in the Pyrenees, whilst Alexander was allowed to continue his education in Collège Cévenol in a Cévennes Mountains town of Chambon-sur-Lignon in the southern Massif Central. The college, run by local Protestants under the leadership of Pastor Trocmé, was a sanctuary to many children (mainly Jews) whose lives were endangered during the war.

Already then, Alexander asked *himself* a question that showed the uniqueness of his mind: How to accurately measure the length of a curve, area of a surface, or volume of a solid? Continuing the reflection on these problems during his university studies in Montpellier (1945–1948), he independently obtained results equivalent to Lebesgue's measure and integration theory. As expressed by J. Dieudonné in [D], the university in Montpellier – in Grothendieck's days – wasn't a "proper place" for studying great mathematical problems .... In the fall of 1948 Grothendieck arrived in Paris, where he spent a year attending courses in the famous École Normale Supérieure (ENS), the "birthplace" of most of the French mathematical elite. In particular, he took part in Cartan's legendary seminar, that year devoted to algebraic topology. (More information about this period of Grothendieck's life, his parents, and France of those days, can be found in [C2].)

Grothendieck's interests, however, began focusing on functional analysis. Following Cartan's advice, in October 1949 he comes to Nancy, a centre of functional analysis studies, where J. Dieudonné, L. Schwartz, and others run a seminar on Fréchet spaces and their direct limits. They encounter a number of problems which they are unable to solve, and suggest Grothendieck to try and attack them. The result surpasses all expectations. In less than a year, Grothendieck manages to solve all the problems by means of some very ingenious constructions. By the time of his doctorate, Grothendieck holds 6 papers, each of which could make a very good doctoral thesis. The thesis, dedicated to his mother[2]:

Produits tensoriels topologiques et espaces nucléaires

HANKA GROTHENDIECK in Verehrung und Dankbarkeit gewidmet

is ready in 1953. This dissertation, published in 1955 in the *Memoirs of the Amer. Math. Soc.* [18] [3], is generally considered one of the most important events in the post-war functional analysis[4]. The years 1950–1955 mark the period of Grothendieck's most intensive work on functional analysis. In his early papers (written at the age of about 22) Grothendieck poses many questions concerning the structure of locally convex linear topological spaces, particularly the complete linear metric spaces. Some of them are related to the theory of linear partial differential equations and analytic function spaces. The Schwartz kernel theorem leads Grothendieck to distinguishing the class of *nuclear spaces*[5]. Roughly speaking, the kernel theorem asserts that "decent" operators on distributions are distributions themselves, which Grothendieck expressed abstractly as an isomorphism of certain injective and projective tensor products. The main difficulty in introducing the theory of nuclear spaces is the problem of equivalence of two interpretations of kernels: as elements of tensor products, and as linear operators (in the case of finite-dimensional spaces, matrices are in one-to-one correspondence with linear transformations). This leads to the so-called *approximation problem* (a version of which was first posed in S. Banach's famous monograph [B]), whose deep study takes a considerable part of the Red Book. Grothendieck discovers many beautiful

---

[2]Grothendieck was exceptionally attached to his mother, with whom he spoke in German. She wrote poems and novels (presumably her best-known work is an autobiographical novel *Eine Frau*).

[3]A complete list of Grothendieck's mathematical publications is contained in [C R], vol. 1, pp. xiii–xx. When citing a Grothendieck's publication here, we refer to an item on that list.

[4]And called *Grothendieck's (little) Red Book.*

[5]All his life Grothendieck has been a fervent pacifist. He believed that the term "nuclear" should only be used to describe abstract mathematical objects. During the Vietnam war he taught a course on the theory of categories in a forest near Hanoi the same time that Americans were bombarding the city.

equivalences (some of the implications were earlier known to S. Banach and S. Mazur); in particular, he shows that the approximation problem is equivalent to problem 153 from the *Scottish Book* [Ma] posed by Mazur, and that, for reflexive spaces, the approximation property is equivalent to the so-called metric approximation property. Nuclear spaces are also related to the following 1950 Dvoretzky-Rogers theorem (solving problem 122 from [Ma]): In every infinite-dimensional Banach space, there exists an unconditionally convergent series that is not absolutely convergent. Grothendieck showed that the nuclear spaces are precisely those for which unconditional convergence is equivalent to the absolute convergence of a series (see [Ma, problem 122 and remarks]). The fundamental importance of nuclear spaces comes from the fact that almost all non-Banach locally convex spaces naturally occurring in analysis are nuclear. We mean here various spaces of smooth functions, distributions, or holomorphic functions with their natural topologies – in many cases their nuclearity was shown by Grothendieck himself.

Another important result of the Red Book is the equivalence of the product definition of nuclear spaces with their realization as inverse limits of Banach spaces with morphisms being nuclear or absolutely summable operators (which Grothendieck calls *left semi-integral* operators). His study of various classes of operators (Grothendieck has been the first to define them in a functorial way, in the spirit of the theory of categories) yields deep results that gave rise to the modern, so-called *local theory* of Banach spaces. The results are published in two important papers [22, 26] in *Bol. Soc. Mat. São Paulo*, during his stay in that city (1953–1955). He shows there, in particular, that operators from a measure space into a Hilbert space are absolutely summable (a fact analytically equivalent to the so-called *Grothendieck inequality*), and makes a conjecture concerning a central problem in the theory of convex bodies, solved by A. Dvoretzky in 1959. Many very difficult questions posed in those papers were later solved by: P. Enflo (negative resolution of the approximation problem, in 1972), B. Maurey, G. Pisier, J. Taskinen ("problème des topologies" on bounded sets in tensor products), U. Haagerup (non-commutative analogue of Grothendieck's inequality for C^*-algebras), J. Bourgain – a Fields medalist, and indirectly influenced the results of another "Banach" Fields medalist, T. Gowers. Supposedly, of all the problems posed by Grothendieck in functional analysis, there is only one left open to these days, see [PB, 8.5.19].

To sum up, Grothendieck's contributions to functional analysis include: nuclear spaces, topological tensor products, Grothendieck inequality, relations with absolutely summable operators, and ... many other dispersed results.[6]

In 1955 Grothendieck's mathematical interests shift to homological algebra. This is a time of the triumph of homological algebra as a powerful tool in algebraic topology, due to the work of H. Cartan and S. Eilenberg. During his stay at the University of Kansas in 1955, Grothendieck constructs his axiomatic theory of

[6]The above information about Grothendieck's contribution to functional analysis comes mostly from [P].

Abelian categories. His main result asserts that the sheaves of modules form an Abelian category with sufficiently many injective objects, which allows one to define cohomology with values in such a sheaf without any constraints on the sheaf or the base space (the theory appears in [28]).

After homological algebra, Grothendieck's curiosity directs towards algebraic geometry – to a large extent due to the influence of C. Chevalley and J.-P. Serre. Grothendieck considers the former a great friend of his, and in later years participates in his famous seminar in the ENS, giving a number of talks on algebraic groups and intersection theory [81–86]. He also exploits J.-P. Serre's extensive knowledge of algebraic geometry, asking him numerous questions (recently, the French Mathematical Society published an extensive selection of their correspondence [CS]; this book can teach more algebraic geometry than many monographs). Serre's paper [S1], building the foundations of the theory of sheaves and their cohomology in algebraic geometry, is of key importance to Grothendieck.

One of Grothendieck's first results in algebraic geometry is a classification of holomorphic bundles over the Riemann sphere [25]. It says that every such bundle is the direct sum of a certain number of tensor powers of the tautological line bundle. Some time after this publication it turned out that other "incarnations" of this result were much earlier known to mathematicians such as G. Birkhoff, D. Hilbert, as well as R. Dedekind and H. Weber (1892). This story shows, on the one hand, Grothendieck's enormous intuition for important problems in mathematics, but on the other hand, also his lack of knowledge of the classical literature. Indeed, Grothendieck wasn't a bookworm; he preferred to learn mathematics through discussions with other mathematicians. Nonetheless, this work of Grothendieck initiated systematic studies on the classification of bundles over projective spaces and other varieties.

Algebraic geometry absorbs Grothendieck throughout the years 1956–1970. His main motive at the beginning of this period is transformation of "absolute" theorems (about varieties) into "relative" results (about morphisms). Here is an example of an absolute theorem[7]:

> *If X is a complete variety and $\mathcal{F}$ is a coherent sheaf on X, then* $\dim H^j(X, \mathcal{F}) < \infty$.

And this is its relative version:

> *If $f\colon X \to Y$ is a proper morphism, and $\mathcal{F}$ is a coherent sheaf on X, then $\mathcal{R}^j f_* \mathcal{F}$ is coherent on Y.*

Grothendieck's main accomplishment of that period is concerned with the *relative Hirzebruch-Riemann-Roch* theorem. The original problem motivating the work on this topic can be formulated as follows: given a connected smooth projective variety X and a vector bundle E over X, calculate the dimension $\dim H^0(X, E)$

[7] In the rest of this paper we will use some standard algebraic geometry notions and notation (see [H]). Unless otherwise implied, by a *variety* we will mean a complex algebraic variety. Cohomology groups of such a variety – unless otherwise specified – will have coefficients in the field of rational numbers.

of the space of global sections of E. The great intuition of Serre told him that the problem should be reformulated using higher cohomology groups as well. Namely, Serre conjectured that the number

$$\sum(-1)^i \dim H^i(X,E)$$

could be expressed in terms of topological invariants related to X and E. Naturally, Serre's point of departure was a reformulation of the classical Riemann-Roch theorem for a curve X: given a divisor D and its associated line bundle $\mathcal{L}(D)$,

$$\dim H^0(X,\mathcal{L}(D)) - \dim H^1(X,\mathcal{L}(D)) = \deg D + \frac{1}{2}\chi(X)\,.$$

(An analogous formula for surfaces was also known.)

The conjecture was proved in 1953 by F. Hirzebruch, inspired by earlier ingenious calculations of J.A. Todd. Here is the formula discovered by Hirzebruch for an n-dimensional variety X:

$$\sum(-1)^i \dim H^i(X,E) = \deg(\mathrm{ch}(E)\mathrm{td}\,X)_{2n}\,, \qquad (*)$$

where $(-)_{2n}$ denotes the degree $2n$ component of an element of the cohomology ring $H^*(X)$, and

$$\mathrm{ch}(E) = \sum e^{a_i} \quad , \quad \mathrm{td}\,X = \prod \frac{x_j}{1-e^{-x_j}}$$

(where the a_i are the Chern roots of E [8], and the x_j are the Chern roots of the tangent bundle TX).

To formulate a relative version of this result, let a proper morphism $f\colon X \to Y$ between smooth varieties be given. We want to understand the relationship between

$$\mathrm{ch}_X(-)\mathrm{td}X \quad \text{and} \quad \mathrm{ch}_Y(-)\mathrm{td}Y,$$

"induced" by f. In the case of $f\colon X \to Y = \textit{point}$, we should obtain the Hirzebruch-Riemann-Roch theorem. The relativization of the right-hand side of $(*)$ is easy: there exists a well defined additive mapping of cohomology groups $f_*\colon H(X) \to H(Y)$, and $\deg(z)_{2n}$ corresponds to $f_*(z)$ for $z \in H(X)$. What about the left-hand side of $(*)$? The relative version of the $H^j(X,\mathcal{F})$ are the coherent modules $\mathcal{R}^j f_*\mathcal{F}$, vanishing for $j \gg 0$. In order to construct a relative version of the alternating sum, Grothendieck defines the following group $K(Y)$ (now called the *Grothendieck group*): It is the quotient group of a "very large" free Abelian group generated by the isomorphism classes $[\mathcal{F}]$ of coherent sheaves on Y, modulo the relation

$$[\mathcal{F}] = [\mathcal{F}'] + [\mathcal{F}'']$$

for each exact sequence

$$0 \to \mathcal{F}' \to \mathcal{F} \to \mathcal{F}'' \to 0. \qquad (**)$$

[8]These are the classes of divisors associated with line bundles, splitting E (see [H, p. 430]).

The group $K(Y)$ has the following universal property: every mapping φ from $\bigoplus \mathbb{Z}[\mathcal{F}]$ to an Abelian group, satisfying

$$\varphi([\mathcal{F}]) = \varphi([\mathcal{F}']) + \varphi([\mathcal{F}'']), \qquad (***)$$

factors through $K(Y)$. In our situation, we define

$$\varphi([\mathcal{F}]) := \sum (-1)^j [\mathcal{R}^j f_* \mathcal{F}] \in K(Y).$$

Observe that $(***)$ follows from the long exact sequence of derived functors

$$\cdots \longrightarrow \mathcal{R}^j f_* \mathcal{F}' \longrightarrow \mathcal{R}^j f_* \mathcal{F} \longrightarrow \mathcal{R}^j f_* \mathcal{F}'' \longrightarrow \mathcal{R}^{j+1} f_* \mathcal{F}' \longrightarrow \cdots,$$

associated with the short exact sequence $(**)$ (see [H, Chap. III]). Thus, we obtain an additive mapping

$$f_! : K(X) \to K(Y).$$

Now the *relative* Hirzebruch-Riemann-Roch theorem, discovered by Grothendieck ([102], [BS]) and being a sign of his genius, asserts the commutativity of the diagram

$$\begin{array}{ccc} K(X) & \xrightarrow{\quad f_! \quad} & K(Y) \\ {\scriptstyle \mathrm{ch}_X(-)\mathrm{td}\,X}\downarrow & & \downarrow{\scriptstyle \mathrm{ch}_Y(-)\mathrm{td}\,Y} \\ H(X) & \xrightarrow{\quad f_* \quad} & H(Y). \end{array}$$

(Note that due to its additivity, the Chern character ch$(-)$ is well defined in K-theory.) More information about various aspects of the intersection theory, of which the ultimate result is the above Grothendieck-Riemann-Roch theorem, can be found in [H, Appendix A][9]. The theorem has been applied in many specific calculations of characteristic classes.

Grothendieck's group K spurred the development of K-theory, marked with the works of D. Quillen and many others. Note that K-theory plays an important role in many areas of mathematics, from the theory of differential operators (the Atiyah-Singer theorem) to the modular representation theory of finite groups (the Brauer theorem).[10]

Following this spectacular result, Grothendieck is proclaimed a "superstar" of algebraic geometry, and invited to the International Congress of Mathematicians in Edinburgh in 1958, where he sketches a program to define a cohomology theory for positive characteristics that should lead to a proof of the Weil conjectures, see [32]. The Weil conjectures [W] suggested deep relations between the arithmetic of algebraic varieties over finite fields, and the topology of complex algebraic varieties. Let $k = \mathbb{F}_q$ be a finite field with q elements, and let $\bar{k}$ be its algebraic closure.

---

[9] In fact, the Grothendieck-Riemann-Roch theorem was proved for varieties over any algebraically closed field (of arbitrary characteristic) by taking the values of the Chern character in the Chow rings (cf. [102], [BS]).

[10] Three contributions in the present volume: by M. Brion, A.S. Buch, and M. Szyjewski present various developments of K-theory initiated by Grothendieck.

Fix a finite collection of homogeneous polynomials in $n+1$ variables with coefficients in k. Let X (resp. $\bar{X}$) be the zero-set of this collection in the n-dimensional projective space over k (resp. $\bar{k}$). Denote by N_r the number of points in $\bar{X}$ whose coordinates lie in the field $\mathbb{F}_{q^r}$ with q^r elements, $r = 1, 2, \ldots$. "Organize" the N_r into a generating function, called the *zeta function of* X:

$$Z(t) := \exp\Big(\sum_{r=1}^{\infty} N_r \frac{t^r}{r}\Big).$$

The Weil conjectures, for a smooth variety X, concern the properties of $Z(t)$, as well as the relations with the classical Betti numbers of the complex variety "associated" with X. The formulation of the Weil conjectures can be found in 1.1–1.4 of [H, Appendix C], or W1–W5 of [M, Chap. VI, § 12] (both lists begin with the conjecture on *rationality* of the zeta function $Z(t)$). The above sources also contain some introductory information about the Weil conjectures, as well as an account of the struggle for their proof, which (besides Weil and the Grothendieck school) involved mathematicians such as B. Dwork, J.-P. Serre, S. Lubkin, S. Lang, Yu. Manin, and many others.

The Weil conjectures become the main motivation for Grothendieck's work in algebraic geometry during his stay at the IHES[11]. He begins working at the IHES in 1959, and soon under his charismatic leadership, emerges the *Séminaire de Géométrie Algébrique du Bois-Marie* (after the wood surrounding the IHES). For the next decade, the seminar will become the world's "capital" of algebraic geometry. Working on mathematics 12 hours a day, Grothendieck generously shares his ideas with his co-workers. The atmosphere of this exceptional seminar has been captured in an interview [Du] with one of Grothendieck's students, J. Giraud. Let us concentrate now on the main ideas explored by Grothendieck at the IHES[12].

Schemes are objects that allow for unification of geometry, commutative algebra, and number theory. Let X be a set, and let F be a field. Consider the ring

$$F^X = \{\text{functions } f : X \to F\}$$

with multiplication defined pointwise. For $x \in X$, define $\alpha_x \colon F^X \to F$ by $f \mapsto f(x)$. The kernel of α_x being a maximal ideal, we can identify X with the set of all maximal ideals in F^X. Thus, we replace a simpler object, X by a more complicated one, which is the set of all maximal ideals in F^X. Variants of this idea appeared in the work of M. Stone on the theory of Boolean lattices, as well as in papers of I.M. Gelfand on commutative Banach algebras. In commutative algebra, similar ideas were first exploited by M. Nagata and E. Kähler. In the late 50's, many mathematicians in Paris (Cartan, Chevalley, Weil, ...) intensively searched for a generalization of the concept of variety over an algebraically closed field.

---

[11]IHES = Institut des Hautes Études Scientifiques: mathematical research institute in Bures-sur-Yvette near Paris – a fantastic location for doing mathematics, also thanks to its lovely canteen that will probably never run out of bread and wine.

[12]See also [D] for a more detailed account of the theory of schemes.

Serre showed that the notion of localization of a commutative ring leads to a sheaf over the maximal spectrum Specm of an (arbitrary) commutative ring. Note that the mapping $A \to \mathrm{Specm}(A)$ is not a functor (the inverse image of a maximal ideal need not be maximal). On the other hand,

$$A \to \mathrm{Spec}(A) := \{\text{prime ideals in } A\}$$

is a functor. It seems that it was P. Cartier who in 1957 first proposed the following: *a proper generalization of the classical algebraic variety is a ringed space* $(X, \mathcal{O}_X)$ *locally isomorphic to* $\mathrm{Spec}(A)$ (although it was a result of speculations of many algebraic geometers). Such an object was called a *scheme*.

The music pavilion of the IHES, Bures-sur-Yvette;
venue of the first algebraic geometry seminars.

Grothendieck was planning to write a 13-volume course in algebraic geometry EGA[13] based on the concept of schemes and culminating in the proof of the Weil conjectures. He managed to publish 4 volumes, written together with Dieudonné. But in fact, most of the material to appear in the later volumes was covered by SGA[14] – publications of the algebraic geometry seminar at the IHES. (The text [H], to which we often refer here, is a didactic recapitulation of the most useful parts of EGA concerning schemes and cohomology.)

Let us now turn to constructions in algebraic geometry that make use of representable functors. Fix an object X in the category $\mathcal{C}$. We associate with it a contravariant functor from $\mathcal{C}$ to the category of sets,

$$h_X(Y) := \mathrm{Mor}_{\mathcal{C}}(Y, X).$$

[13]EGA – *Éléments de Géométrie Algébrique*, published by the Publ. IHES and Springer Verlag [57–64].
[14]SGA – *Séminaire de Géometrie Algébrique*, published by the Springer Lecture Notes in Mathematics and (SGA 2) by North-Holland [97–103].

At first sight, it is hard to see any use of such a simple assignment. However, the knowledge of this functor gives us a unique (up to isomorphism) object X that "represents" it (a fact known as the *Yoneda Lemma*). It is thus natural to make the following definition: A contravariant functor from $\mathcal{C}$ to the category of sets is called *representable* (by X) if it is of the form h_X for some object X in $\mathcal{C}$. Grothendieck masterfully exploits the properties of representable functors to construct various *parameter spaces.* Such spaces are often encountered in algebraic geometry, a key example being the *Grassmannian* parametrizing linear subspaces of a given dimension in a given projective space. A natural question is whether there exist more general schemes parametrizing subvarieties of a given projective space, and having certain fixed numerical invariants.

Let S be a scheme over a field k. A *family of closed subschemes of* $\mathbb{P}^n$ *with the base* S is a closed subscheme $X \subset \mathbb{P}^n \times_k S$ together with the natural morphism $X \to S$. Fix a numerical polynomial P. Grothendieck considers the functor Ψ^P from the category of schemes to the category of sets, that assigns to S the set $\Psi^P(S)$ of flat families of closed subschemes of $\mathbb{P}^n$ with base S and Hilbert polynomial P. If $f\colon S' \to S$ is a morphism, then

$$\Psi^P(f)\colon \Psi^P(S) \to \Psi^P(S')$$

assigns to a family $X \to S$ the family $X' = X \times_S S' \to S'$. Grothendieck proves that the functor Ψ^P is representable by a scheme (called a *Hilbert scheme*) that is projective [74][15]. This is a (very) ineffective result – for example, estimating the number of irreducible components of the Hilbert scheme of curves in three-dimensional projective space, with a given genus and degree, is still an open problem. Nonetheless, in numerous geometric considerations it suffices to know that such an object exists, which makes this theorem of Grothendieck useful in many applications. More generally, Grothendieck constructs a so-called *Quot-scheme* parametrizing (flat) quotient sheaves of a given coherent sheaf, with a fixed Hilbert polynomial [73]. Quot-schemes enjoy many applications in constructions of moduli spaces of vector bundles. Yet another scheme, constructed by Grothendieck in the same spirit, is the *Picard scheme* [75, 76].

In 1966 Grothendieck receives the Fields Medal for his contributions to functional analysis, the Grothendieck-Riemann-Roch theorem, and the work on the theory of schemes (see [S2]).

The most important subject of Grothendieck's research at the IHES is, however, the theory of *étale cohomology.* Recall that, for the purpose of the Weil conjectures, the issue is to construct an analogue of the cohomology theory of complex varieties for algebraic varieties over a field of positive characteristic (but with coefficients in a field of characteristic zero, so that one could count the fixed points of a morphism as a sum of traces in cohomology groups, à la Lefschetz). Earlier efforts to exploit the classical topology used in algebraic geometry – the

[15] In fact, Grothendieck proves a much more general result for projective schemes over a base Noetherian scheme.

Zariski topology (closed subsets = algebraic subvarieties), turned out unsuccessful, the topology being "too poor" for homological needs. Grothendieck observes that a "good" cohomology theory can be built by considering a *variety together with all its unramified coverings* (see [32] for details on the context of this discovery). This is the beginning of the theory of étale topology, developed together with M. Artin and J.-L. Verdier. Grothendieck's brilliant idea was the revolutionary generalization of the notion of topology, differing from the classical topological space in that the "open sets" need not be all contained in the same set, but do have some basic properties that allow one to build a "satisfactory" cohomology theory of sheaves.

Alexander Grothendieck

The origins of these ideas are sketched in the following discussion of Cartier [C1]. When using sheaves on a variety X or studying cohomology of X with coefficients in sheaves, the key role is played by the lattice of open subsets of X (the points of X being of secondary importance). In our considerations, we can thus, without any harm, "replace" the variety by the lattice of its open subsets. Grothendieck's idea was to adapt B. Riemann's concept of multivalued holomorphic functions that actually "live" not on open subsets of the complex plane, but rather on suitable Riemann surfaces that cover it (Cartier uses a suggestive term "les surfaces de Riemann étalées"). Between these Riemann surfaces there are projections, and hence they form objects of a certain category. A lattice is an example of a category in which between any two objects there is at most one morphism. Grothendieck suggests then to replace the lattice of open sets with the category of open étale sets. Adapted to algebraic geometry, this concept allows one to resolve the fundamental difficulty of the lack of an implicit function theorem for algebraic functions. Also, it allows us to view the étale sheaves in a functorial way.

To continue our discussion in a more formal way, suppose that a category $\mathcal{C}$ is given, which admits fibre products. A *Grothendieck topology* on $\mathcal{C}$ is an assignment to every object $X \in C$ of a set $\mathrm{Cov}(X)$ of a families of morphisms $\{f_i\colon X_i \to X\}_{i\in I}$, called the *coverings* of X, satisfying the following conditions:

1) $\{id\colon X \to X\} \in \mathrm{Cov}(X)$;
2) if $\{f_i\colon X_i \to X\} \in \mathrm{Cov}(X)$, then, induced by a base change $Y \to X$, the family $\{X_i \times_X Y \to Y\}$ belongs to $\mathrm{Cov}(Y)$;
3) if $\{X_i \to X\} \in \mathrm{Cov}(X)$ and, for all i, $\{X_{ij} \to X_i\} \in \mathrm{Cov}(X_i)$, then the bi-indexed family $\{X_{ij} \to X\}$ belongs to $\mathrm{Cov}(X)$.

If $\mathcal{C}$ admits direct sums – and let us suppose so – then a family $\{X_i \to X\}$ can be replaced with a single morphism

$$X' = \coprod_i X_i \to X\,.$$

Having coverings, one can consider sheaves and their cohomology. A contravariant functor F from $\mathcal{C}$ to the category of sets is called a *sheaf of sets* if, for every covering $X' \to X$, have

$$F(X) = \{s' \in F(X') : p_1^*(s') = p_2^*(s')\}\,,$$

where p_1, p_2 are the two projections from $X' \times_X X'$ onto X'. A *canonical topology* in the category $\mathcal{C}$ is the topology "richest in coverings" in which all the representable functors are sheaves. If in turn, every sheaf in a canonical topology is a representable functor, then the category $\mathcal{C}$ is called a *topos*. More information about the Grothendieck topologies can be found for instance in [BD].

Let us return to geometry. Very importantly: the above f_i need not be embeddings! The most significant example of a Grothendieck topology is the *étale topology*, where the $f_i\colon X_i \to X$ are étale morphisms[16] that induce a surjection $\coprod_i X_i \to X$. Grothendieck's cohomological machinery applied to this topology yields the construction of the étale cohomology $H^i_{\text{ét}}(X,-)$. Although the basic ideas are relatively simple, the verification of many technical details regarding the properties of étale cohomology required many years of hard work, which involved the "cohomological" students of Grothendieck: P. Berthelot, P. Deligne, L. Illusie, J.-P. Jouanolou, J.-L. Verdier, and others, successively filling up the details of new results sketched by Grothendieck. The results of the Grothendieck school's work on étale cohomology are published in [100][17].

The proof of the Weil conjectures required a certain variant of étale cohomology – the *l-adic cohomology*. Its basic properties, particularly a *Lefschetz-type formula*, allowed Grothendieck to prove some of the Weil conjectures, but the most difficult one – the analogue of the Riemann Hypothesis – remained unsolved. In

[16]These are smooth morphisms of relative dimension zero. For smooth varieties, étale morphisms are precisely those that induce isomorphisms of the tangent spaces at all points – naturally, such morphisms need not be injective. A general discussion of étale morphisms can be found in [M].
[17]A didactic exposition of étale cohomology can be found in [M].

the process of proving the conjecture, Grothendieck has played a role similar to that of the biblical Moses, who led the Israelis off Egypt and towards the Promised Land, was their guide for the most part of the trip, but was not supposed to reach the goal himself. In the case of the Weil-Riemann conjecture, the goal was reached by Grothendieck's most brilliant student – Deligne [De]. (Grothendieck's plan to prove the Weil-Riemann conjecture by first proving the so-called *standard conjectures* has not been realized to these days – the conjectures are discussed in [44].)

In 1970 Grothendieck accidentally discovers that the IHES finances are partially supported by military sources, and leaves the IHES instantly. He receives a prestigious position at the Collège de France, however by that time (Grothendieck is about 42) there are things that interest him more than mathematics: one has to save the endangered world! Grothendieck cofounds an ecological group called *Survivre et Vivre (Survive and Live)*. In this group he is accompanied by two outstanding mathematicians and friends: C. Chevalley and P. Samuel. The group publishes in 1970–1975 a magazine under the same name. Typically for his temperament, Grothendieck engages wholly in this activity, and soon his lectures at the Collège de France have little to do with mathematics, concerning instead the issues like ... how to avoid the world war and live ecologically. Consequently, Grothendieck needs to find himself a new job. He receives an offer from his "home" university in Montpellier, and soon settles down on a farm near the city and works as an "ordinary" professor (with teaching duties) at the university. Working in Montpellier, Grothendieck writes a number of (long) sketches of new mathematical theories in an effort to obtain a position in the CNRS[18] and talented students from the ENS to work with. He "receives" no students, but for the last four years before retirement (at the age of 60) is employed by the CNRS. The sketches are currently being developed by several groups of mathematicians; it is a good material for a separate article.

In Montpellier Grothendieck writes also his mathematical memoirs *Récoltes et Semailles (Harvests and Sowings)* [G1], containing marvellous pieces about his perspectives on mathematics, about "male" and "female" roots in mathematics, and hundreds of other fascinating things. The memoirs contain also a detailed account of Grothendieck's relationship with the mathematical community, as well as a very critical judgement of his former students .... But let us talk about more pleasant things. Speaking of a model mathematician, Grothendieck without hesitation names E. Galois. Of the more contemporary scientists, Grothendieck very warmly recalls J. Leray, A. Andreotti, and C. Chevalley. It is symptomatic how greatly important to Grothendieck is the human aspect of his contacts with other mathematicians. He writes in [G1]:

> *If, in "Récoltes et Semailles" I'm addressing anyone besides myself, it isn't what's called a "public". Rather I'm addressing that someone who is prepared to read me as a person, and as a solitary person.*

[18]CNRS – Centre National de la Recherche Scientifique, French institution employing scientists without formal didactic duties.

Maybe it was the loneliness experienced in all his life that made him so sensitive about it?

In 1988 Grothendieck refuses to accept a prestigious Crafoord Prize, awarded to him, jointly with Deligne, by the Royal Swedish Academy of Sciences (huge money!). Here is a quote of the most important, in my opinion, part of Grothendieck's letter to the Swedish Academy in regard to the prize (see [G2]):

I am convinced that time is the only decisive test for the fertility of new ideas or views. Fertility is measured by offspring, not by honors.

Let us add that the letter contains also his extremely critical opinion on the professional ethic of the mathematical community of the 70's and 80's ...

It is time for some summary. Here are the 12 most important topics of Grothendieck's work in mathematics, reproduced after [G1]:

1. Topological tensor products and nuclear spaces;
2. "Continuous" and "discrete" dualities (derived categories, the "six operations");
3. The Riemann-Roch-Grothendieck yoga (K-theory and its relationship to intersection theory);
4. Schemes;
5. Topos theory;
   (Toposes, as pointed out before, realize (as opposed to schemes) a "geometry without points" – see also [C1] and [C2]. Grothendieck "admired" toposes more than schemes. He valued most the topological aspects of geometry that led to the right cohomology theories.)
6. Étale cohomology and l-adic cohomology;
7. Motives, motivic Galois groups ($\otimes$-Grothendieck categories);
8. Crystals, crystalline cohomology, yoga of the De Rham coefficients, the Hodge coefficients;
9. "Topological algebra": ∞-stacks; derivations; cohomological formalism of toposes, inspiring a new conception of homotopy;
10. Mediated topology;
11. The yoga of Anabelian algebraic geometry. Galois-Teichmüller theory;
    (This point Grothendieck considered the hardest and "the deepest". Recently, important results on this subject were obtained by F. Pop.)
12. Schematic or arithmetic viewpoints on regular polyhedra and in general all regular configurations.
    (This subject was developed by Grothendieck after moving from Paris to Montpellier, in his spare time at a family vineyard.)

The work of numerous mathematicians who carried on 1.–12. has made up a significant chunk of the late XX century mathematics. Many of the Grothendieck's ideas are being actively developed nowadays and will certainly have a significant impact on the mathematics of the XXI century.

Let us name the most important continuators of Grothendieck's work (among them, a few Fields medalists):

1. P. Deligne: complete proof of the Weil conjectures in 1973 (to a large extent based on techniques of SGA);
2. G. Faltings: proof of the Mordell conjecture in 1983;
3. A. Wiles: proof of Fermat's Last Theorem in 1994;
   (it is hard to imagine 2. and 3. without EGA)
4. V. Drinfeld, L. Lafforgue: proof of the Langlands reciprocity for general linear groups over function fields;
5. V. Voevodsky: theory of motives and proof of Milnor's conjecture.

The last point is related to the following Grothendieck's "dream": there should exist an "Abelianization" of the category of algebraic varieties – a category of *motives* together with the *motivic cohomology*, from which one could read the Picard variety, the Chow groups, etc. A. Suslin and V. Voevodsky constructed motivic cohomology satisfying the postulates of Grothendieck.

In August 1991 Grothendieck suddenly abandons his house and, without a word, leaves to an unknown location somewhere in the Pyrenees. He devotes himself to philosophical meditations (free choice, determinism, and the existence of... the devil in the world); earlier, he wrote an interesting text *La clef des songes* describing his argument for the existence of God based on a dream analysis, and writes texts on physics. He wishes *no* contacts with the outside world.

We come to the end. Here is a handful of reflections.

The following words of Grothendieck, from [G1], describe what interested him most in mathematics:

That is to say that, if there is one thing in mathematics which (no doubt this has always been so) fascinates me more than anything else, it is neither "number", nor "magnitude" but above all "form". And, among the thousand and one faces that form chooses in presenting itself to our attention, the one that has fascinated me more than any other, and continues to fascinate me, is the structure buried within mathematical objects.

It is truly amazing that resulting from this reflection of Grothendieck on the "form" and "structure" are theories that provide tools (of unparalleled precision) for calculating *specific* numerical quantities and finding *explicit* algebraic relations. An example of such a tool in algebraic geometry is the Grothendieck-Riemann-Roch theorem. Another, less known example, is the language of *Grothendieck's λ-rings* [102], that allows one to treat *symmetric functions* as operators on polynomials. This in turn provides a uniform approach to numerous classical polynomials (e.g., symmetric, orthogonal) and formulas (e.g., interpolation formulas or those of the representation theory of general linear groups and symmetric groups). The polynomials and formulas are often related to the famous names such as: E. Bézout, A. Cauchy, A. Cayley, P. Chebyshev, L. Euler, C.F. Gauss, C.G. Jacobi, J. Lagrange, E. Laguerre, A.-M. Legendre, I. Newton, I. Schur, T.J. Stieltjes,

J. Stirling, J.J. Sylvester, J.M. Hoene-Wroński, and others. What's more, the language of λ-rings allows one to establish useful algebro-combinatorial generalizations of the results of these classics, see [L]. The work of Grothendieck shows that there is no essential dichotomy between the quantitative and qualitative aspects of mathematics.

Undoubtedly, Grothendieck's point of view explained above helped him to accomplish the enormous work towards the *unification* of important subjects in geometry, topology, arithmetic, and complex analysis. It also relates to Grothendieck's fondness for studying mathematical problems in their full generality.

Grothendieck's work style is well described in the following tale of his, from [G1]. Suppose one wants to prove a conjecture. There are two extreme methods to do this. First: by force. As with opening a nut: one cracks the shell with a nutcracker and gets to the fruit inside. But there is also another way. One can put a nut into a softening liquid and wait patiently until it suffices to gently press the shell and it opens all by itself. Anyone who read Grothendieck's works would have no doubt that it was the latter approach he used when working on mathematics. Cartier [C1] gives a yet more suggestive characterization of this method: it is the Joshua way of conquering Jericho. One wants to get to Jericho guarded by tall walls. If one compasses the city sufficiently many times, thus weakening their construction (by resonance), then eventually it will suffice to blow with the trumpets and shout with a great shout and ... the walls of Jericho shall fall down flat!

Let us share the following piece of advice, especially with young mathematicians. Grothendieck highly valued writing down his mathematical considerations. He regarded the process of writing and editing of mathematical papers itself an *integral* part of the research work, see [He].

Finally, let us listen to Dieudonné, a faithful witness of Grothendieck's work, and a mathematician of an immense encyclopedic knowledge. He wrote (see [D]) on the occasion of Grothendieck's 60'th birthday (that is, some 15 years ago):

There are few examples in mathematics of a theory that monumental and fruitful, done by a single man in such a short time.

He is accompanied by the editors of *The Grothendieck Festschrift* [C-R] (where [D] was published), who say in the introduction:

It is difficult to grasp fully the magnitude of Alexander Grothendieck's contribution to and influence on twentieth century mathematics. He has changed the very way we think about many branches in mathematics. Many of his ideas, revolutionary when introduced, now seem so natural as to have been inevitable. Indeed, there is a whole new generation of mathematicians for whom these ideas are part of the mathematical landscape, a generation who cannot imagine that Grothendieck's ideas were ever absent.

During the preparation of this article I asked a couple of my French friends whether Grothendieck was still alive. Their answers could be summarized as follows: "Unfortunately, the only news we will have about Grothendieck will be the notice of his death. Since we still haven't got any, he must be alive." On March 28, 2004 Grothendieck turned 76.

The bibliography of Grothendieck's work is huge and obviously stretches beyond the scope of this modest exposition. We cite only those bibliographical items to which we refer directly in the text. One can find there more detailed references to papers of Grothendieck and other authors writing about him and his work. We heartily recommend visiting the website of the Grothendieck Circle:

`http://www.grothendieck-circle.org/`

containing much interesting mathematical and biographical material about Grothendieck and his parents.

Acknowledgements. My warm thanks go to: Marcin Chałupnik, Paweł Domański, and Adrian Langer for their critical reading of previous versions of the manuscript, to Janusz Adamus for translating the text into English, and to Michel Brion and Jerzy Trzeciak for their comments that helped me to ameliorate the exposition.

Photographs used in this article are courtesy of Marie-Claude Vergne (the picture of IHES) and the website of the Grothendieck Circle (the pictures of A. Grothendieck).

References

[number] = the publication of Grothendieck with this number from his bibliography in: *The Grothendieck Festschrift*, P. Cartier et al. (eds.), vol. **1**, pp. xiii–xx, Progress in Mathematics **86**, Birkhäuser, Boston, 1990. See also: `http://www.math.columbia.edu/~lipyan/GrothBiblio.pdf`

[B] S. Banach, *Théorie des opérations linéaires*, Monografie Matematyczne, vol. **1**, Warszawa, 1932.

[BS] A. Borel, J.-P. Serre, *Le théorème de Riemann-Roch (d'après Grothendieck)*, Bull. Soc. Math. France **86** (1958), 97–136.

[BD] I. Bucur, A. Deleanu, *Introduction to the Theory of Categories and Functors*, Wiley and Sons, London, 1968.

[C-R] P. Cartier, L. Illusie, N.M. Katz, G. Laumon, Y. Manin, K.A. Ribet (eds.), *The Grothendieck Festschrift*, Progress in Mathematics **86**, Birkhäuser, Boston, 1990.

[C1] P. Cartier, *Grothendieck et les motifs*, Preprint IHES/M/00/75.

[C2] P. Cartier, *A mad day's work: From Grothendieck to Connes and Kontsevich. The evolution of concepts of space and symmetry*, Bull. Amer. Math. Soc. **38** (2001), 389–408.

[CS] P. Colmez, J.-P. Serre (eds.), *Correspondance Grothendieck-Serre*, Documents Mathématiques **2**, Soc. Math. de France, Paris, 2001.

[De] P. Deligne, *La Conjecture de Weil, I*, Publ. Math. IHES **43** (1974), 273–307.

[D] J. Dieudonné, *De l'analyse fonctionnelle aux fondements de la géométrie algébrique*, [in:] *The Grothendieck Festschrift*, P. Cartier et al. (eds.), vol. **1**, 1–14, Progress in Mathematics **86**, Birkhäuser, Boston, 1990.

[Du] E. Dumas, *Une entrevue avec Jean Giraud, à propos d'Alexandre Grothendieck*, Le journal de maths **1** no. 1 (1994), 63–65.

[G1] A. Grothendieck, *Récoltes et Semailles; Réflexions et témoignages sur un passé de mathématicien*, Preprint, Université des Sciences et Techniques du Languedoc (Montpellier) et CNRS, 1985.

[G2] A. Grothendieck, *Les dérives de la «science officielle»*, Le Monde, Paris, 4.05.1988 (see also: Math. Intelligencer **11** no. 1 (1989), 34–35).

[H] R. Hartshorne, *Algebraic Geometry*, Graduate Texts in Math. **52**, Springer, New York, 1977.

[He] A. Herreman, *Découvrir et transmettre*, Preprint IHES/M/00/75.

[L] A. Lascoux, *Symmetric functions and combinatorial operators on polynomials*, CBMS Reg. Conf. Ser. in Math. **99**, Amer. Math. Soc., Providence, 2003.

[Ma] R.D. Mauldin (ed.), *The Scottish Book. Mathematics from the Scottish Café*, Birkhäuser, Boston, 1981.

[M] J.S. Milne, *Étale Cohomology*, Princeton University Press, Princeton, 1980.

[P] A. Pełczyński, *Letter to the author*, 20.03.2004.

[PB] P. Pérez Carreras, J. Bonet, *Barrelled Locally Convex Spaces*, North-Holland, Amsterdam, 1987.

[S1] J.-P. Serre, *Faisceaux algébriques cohérents*, Ann. of Math. **61** (1955), 197–278.

[S2] J.-P. Serre, *Rapport au comité Fields sur les travaux de A. Grothendieck*, K-theory **3** (1989), 199–204.

[W] A. Weil, *Number of solutions of equations over finite fields*, Bull. Amer. Math. Soc. **55** (1949), 497–508.

Piotr Pragacz
Institute of Mathematics
Polish Academy of Sciences
Śniadeckich 8
PL-00-956 Warszawa, Poland
e-mail: P.Pragacz@impan.gov.pl

Trends in Mathematics:
Topics in Cohomological Studies of Algebraic Varieties, 1–32

Characteristic Classes of Singular Varieties

Paolo Aluffi

Preface

These five lectures aim to explain an algebro-geometric approach to the study of different notions of Chern classes for singular varieties, with emphasis on results leading to concrete computations.

My main goal in the lectures was not to summarize the history or to give a complete, detailed treatment of the subject; five lectures would not suffice for this purpose, and I doubt I would be able to accomplish it in any amount of time anyway. My goal was simply to provide enough information so that interested listeners could start working out examples on their own. As these notes are little more than a transcript of my lectures, they are bound to suffer from the same limitations. In particular, I am certainly not quoting here all the sources that should be quoted; I offer my apologies to any author that may feel his or her contribution has been neglected.

The lectures were given in the mini-school with the same title organized by Professors Pragacz and Weber at the Banach Center. Jörg Schürmann gave a parallel cycle of lectures at the same mini-school, on the same topic but from a rather different viewpoint. I believe everybody involved found the counterpoint provided by the accostment of the two approaches very refreshing. I warmly thank Piotr Pragacz and Andrzej Weber for giving us the opportunity to present this beautiful subject.

1. Lecture I

1.1. Cardinality of finite sets vs. Euler characteristic vs. Chern-Schwartz-MacPherson classes

Let $\mathcal{F}in$ denote the category of finite sets. I want to consider a functor $\mathcal{C}$ from $\mathcal{F}in$ to abelian groups, defined as follows: for S a finite set, $\mathcal{C}(S)$ denotes the group of functions $S \to \mathbb{Z}$.

Note: we could see $\mathcal{C}(S)$ as the group of linear combinations $\sum_V m_V \mathbb{1}_V$, where V runs over the subsets of S, $m_V \in \mathbb{Z}$, and $\mathbb{1}_V$ is the constant 1 on V and 0 in the complement of V. We could even select V to be the singletons $\{s\}$, with $s \in S$, if we wanted.

How do we make $\mathcal{C}$ into a functor? For $f : S \to T$ a map of finite sets, we have to decide what $\mathcal{C}(f)$ does; and for this it is enough to decide what function $T \to \mathbb{Z}$

$$\mathcal{C}(f)(\mathbb{1}_V)$$

should be, for every subset $V \subset S$; and for this, we have to decide the value of

$$\mathcal{C}(f)(\mathbb{1}_V)(t)$$

for $t \in T$. Here is the definition:

$$\mathcal{C}(f)(\mathbb{1}_V)(t) = \#(f^{-1}(t) \cap V)$$

where $\#$ denotes 'number of elements'. Exercise: this makes $\mathcal{C}$ into a functor.

This trivial observation is the source of equally trivial, but rather interesting properties of the counting function. Note that $\mathcal{C}(\{p\}) = \mathbb{Z}$, and for the constant map $\kappa : S \to \{p\}$,

$$\mathcal{C}(\kappa)(\mathbb{1}_S) = \#S.$$

So if S_1, S_2 are two subsets of S and $S = S_1 \cup S_2$, thinking about the covariance for

$$S_1 \amalg S_2 \to S = S_1 \cup S_2 \to \{p\}$$

tells us that

$$\#(S_1 \cup S_2) = \#S_1 + \#S_2 - \#(S_1 \cap S_2);$$

and, more generally, the 'inclusion-exclusion' counting principle follows.

A much more remarkable observation is that the *topological Euler characteristic* satisfies the same properties. If S admits a structure of CW complex, define $\chi(S)$ to be the number of vertices, minus the number of edges, plus the number of faces, ... Then whenever S_1, S_2, S all admit such a structure one verifies immediately that

$$\chi(S_1 \cup S_2) = \chi(S_1) + \chi(S_2) - \chi(S_1 \cap S_2);$$

and, more generally, an inclusion-exclusion principle for χ holds. So we could think of the Euler characteristic as a 'counting' function.

The main character in these lectures will be 'the next step' in this philosophy: the *Chern-Schwartz-MacPherson* class of a variety V, $c_{\mathrm{SM}}(V)$, will be an even fancier analog of 'counting', in the sense that it will satisfy the same 'inclusion-exclusion' principle. In fact, the Euler characteristic will be part of the information carried by the CSM class: for V a compact complex algebraic variety, $\chi(V)$ will be the degree $\int c_{\mathrm{SM}}(V)$ of $c_{\mathrm{SM}}(V)$, that is, the degree of the zero-dimensional part of $c_{\mathrm{SM}}(V)$. The class $c_{\mathrm{SM}}(V)$ will live in a homology theory for V.

My emphasis will be: how do we *concretely* compute such classes?

But maybe the first question should be: what does 'computing' mean?

1.2. Computer demonstration

Objection: we have not defined $c_{\mathrm{SM}}(V)$ yet, so it is unfair to ask how to compute it. This is correct. However, I have defined the topological Euler characteristic, so it is fair to ask: can we compute it now? Isn't the definition itself, as 'vertices minus sides plus faces...', already a 'computation'?

That depends. How would one use this in practice to compute the Euler characteristic of the subscheme of $\mathbb{P}^3$ given by the ideal

$$(x^2 + y^2 + z^2, xy - zw)\,?$$

The point is that what it means to 'compute' something strictly depends on what information one starts from. Of course if I start from a description of V from which a triangulation is obtained easily, then the Euler characteristic can be computed just as easily. As an algebraic geometer, however, I may have to be able to start off from the raw information of a scheme; for example, from a defining homogeneous ideal in projective space. And then? how do I 'compute' a CW-complex realization of the support of a scheme starting from its ideal? In this sense, $\chi(V) = \#\text{vertices} - \#\text{edges} + \cdots$ is not a 'computation': if I already knew so much about V as to be able to count vertices, edges, etc. then I would not gain much insight about V by applying this formula.

By contrast, *here* is what a computation is:

```
themis{aluffi}1: Macaulay2
Macaulay 2, version 0.9
--Copyright 1993-2001, D. R. Grayson and M. E. Stillman
--Singular-Factory 1.3b, copyright 1993-2001, G.-M. Greuel, et al.
--Singular-Libfac 0.3.2, copyright 1996-2001, M. Messollen

i1 : load "CSM.m2"
--loaded CSM.m2

i2 : QQ[x,y,z,w];

i3 : time CSM ideal(x^2+y^2+z^2,x*y-z*w)
                                        3     2
Chern-Schwartz-MacPherson class : H + 4H
-- used 49.73 seconds
```

this tells me that the Chern-Schwartz-MacPherson class of that scheme is $4H^2 + H^3 - 4[\mathbb{P}^1] + [\mathbb{P}^0]$ (once it is pushed forward into projective space); hence its Euler characteristic is 1.

I will have accomplished my goal in these lectures if I manage to explain how this computation is performed. As a preview of the philosophy behind the whole approach, we will 'divide and conquer': split the information in the Chern-Schwartz-MacPherson class into the sum of an 'easy' term (this will be what I will call the 'Chern-Fulton' class), and an 'interesting' one (usually called 'Milnor class') accounting specifically for the singularities of the scheme.

'Computing' the Milnor class will be the most substantial part of the work. To give an idea of how difficult it may be, here is a rather loose question:

Is there a natural scheme structure on the singularities of a given variety V, which determines the Milnor class of V?

To my knowledge, this is completely open! But it is understood rather well for hypersurfaces of nonsingular varieties, so that will be my focus in most of my lectures.

1.3. Chern-Schwartz-MacPherson classes: definition

Back to our 'counting' analogy. For V a variety, let $\mathcal{C}(V)$ denote the abelian group of finite linear combinations $\sum m_W 1\!\!1_W$, where W are (closed) subvarieties of V, $m_W \in \mathbb{Z}$, and $1\!\!1_W$ denotes the function that is 1 along W, and 0 in the complement of W. Elements of $\mathcal{C}(V)$ are called 'constructible functions' on V. How to make $\mathcal{C}$ into a functor?

For $f : V_1 \to V_2$ a *proper* function (so that the image of a closed subvariety is a closed subvariety), it suffices to define $\mathcal{C}(f)(1\!\!1_W)$ for a subvariety W of V_1; so we have to prescribe $\mathcal{C}(f)(1\!\!1_W)(p)$ for $p \in V_2$. We set

$$\mathcal{C}(f)(1\!\!1_W)(p) = \chi(f^{-1}(p) \cap W),$$

in complete analogy with the counting case. Exercise: this makes $\mathcal{C}$ into a functor.

Now observe that there are *other* functors from the category of varieties, proper maps to abelian groups. I will denote by $\mathcal{A}$ the *Chow group* functor. As a quick reminder, $\mathcal{A}(V)$ can be obtained from $\mathcal{C}(V)$ by setting to zero $\sum m_W 1\!\!1_W$ if there is a subvariety U of V and a rational function φ on U such that the divisor of φ equals $\sum m_W[W]$. The equivalence corresponding to the subgroup generated by these constructible functions is called 'rational equivalence'; $\mathcal{A}(V)$ is the abelian group of 'cycles modulo rational equivalence'. It is a functor for proper maps under a seemingly less interesting prescription: for $f : V_1 \to V_2$ proper, simply set $\mathcal{A}(f)([1\!\!1_W]) = d[1\!\!1_{f(W)}]$, where d is the degree of $f|_W$. The Chow group $\mathcal{A}(V)$ should be thought of as a 'homology'; indeed, there is a natural transformation $\mathcal{A} \to H_*$; in fact, $\mathcal{A}(V) = H_*(V)$ in many interesting case, e.g., $V = \mathbb{P}^n$.

By construction there is a map $\mathcal{C}(V) \to \mathcal{A}(V)$; but the *functors* $\mathcal{C}$ and $\mathcal{A}$ do not have so much to do with each other – this is easily seen *not* to be a natural transformation.

As a side remark, note however that even this naive recipe does define a natural transformation on the associated graded functors $G\mathcal{C} \rightsquigarrow G\mathcal{A}$ (where the G is taken with respect to the evident filtration by dimension); the objection is that this does not lift to a natural transformation $\mathcal{C} \rightsquigarrow \mathcal{A}$ in the most obvious way.

Does it lift at all? If it does, does it lift in some particularly interesting way? Let us assume that there is a lift, that is, a homomorphism $c_* : \mathcal{C}(V) \to \mathcal{A}(V)$ for all V, satisfying covariance. What can we say *a priori* about it?

Just as we did when we were playing with finite sets, consider the constant map $\kappa : V \to \{p\}$. The covariance diagram would then say that

$$\int c_*(\mathbb{1}_V) = \chi(V).$$

So whatever $c_*(\mathbb{1}_V)$ is, its degree must be the Euler characteristic of V. This should sound reminiscent of something... if for example V is nonsingular, what class canonically defined on V has the property that its degree is $\chi(V)$?

Answer: $c(TV) \cap [V]$ – in words, the 'total homology Chern class of the tangent bundle of V'. By one of the many descriptions of Chern classes, $\int c(TV) \cap [V]$ measures the number of zeros of a tangent vector field on V, counted with multiplicities; this is $\chi(V)$, by the Poincaré-Hopf theorem.

So we could make an educated guess: maybe a natural transformation c_* does exist, with the further amazing property that $c_*(\mathbb{1}_V)$ equals $c(TV) \cap [V]$ whenever V is nonsingular.

This was conjectured by Pierre Deligne and Alexandre Grothendieck, and an explicit construction of c_* was given by Robert MacPherson ([Mac74]).

Grothendieck discusses this conjecture and the context surrounding it in the note (87_1), p. 361 and ff., of part II in [ReS], one of many mathematical commentaries of great interest in this reference.

Definition. $c_{\mathrm{SM}}(V) := c_*(\mathbb{1}_V)$ is the 'Chern-Schwartz-MacPherson class' of V.

1.4. Other classes: quick tour

Thus, Chern-Schwartz-MacPherson classes are 'characteristic classes for singular varieties', in the sense that they are defined for all varieties and agree with the ordinary characteristic classes for nonsingular ones.

The definition of Chern-Schwartz-MacPherson classes I just gave is only partly useful for computations, and does not hint at the subtleties of MacPherson's construction. Neither does it say anything about the subtleties of the alternative construction given by Marie-Hélène Schwartz, which in fact predated MacPherson's contribution (but to my knowledge does not address the functorial set-up); see [BS81] and [Bra00].

MacPherson's construction is obtained by taking linear combinations of another notion of 'characteristic class' for singular varieties, also introduced by MacPherson, and usually named 'Chern-Mather class'.

To define these, assume the given variety V is embedded in a nonsingular variety M. We can map every *smooth* $v \in V$ to its tangent space T_vV, seen as a subspace of T_vM; this gives a rational map $V \dashrightarrow Grass(\dim V, TM)$. The closure $\widetilde{V}$ of the image of this map is called the *Nash blow-up* of V. It comes equipped with a projection ν to V, and with a tautological bundle $\mathcal{T}$ inherited from the Grassmann bundle. Since $\mathcal{T}$ 'agrees with' TV where the latter is defined, it seems a very sensible idea to define a characteristic class by

$$c_{\mathrm{Ma}}(V) = \nu_*(c(\mathcal{T}) \cap [\widetilde{V}]).$$

(Exercise: this is independent of the ambient variety M.) This is the Chern-Mather class. It is clear that if V is nonsingular to begin with, again $c_{\mathrm{Ma}}(V) = c(TV) \cap [V]$. Still, $c_{\mathrm{Ma}}(V) \neq c_{\mathrm{SM}}(V)$ in general. MacPherson's natural transformation can be defined in terms of Chern-Mather classes; we will come back to this later.

There are other sensible ways to define 'characteristic classes' for singular varieties – in fact, for arbitrary schemes. A seemingly very distant approach leads to classes known as *(Chern-)Fulton* and *(Chern-)Fulton-Johnson* classes. Both of these are defined as $c(TM) \cap S(V, M)$, where again we are embedding V in a nonsingular ambient M, and $S(V, M)$ is a class capturing information about the embedding.

- For Fulton classes, $S(V, M)$ is the *Segre class* of V in M. These will be very important in what follows, so I will talk about them separately.
- For Fulton-Johnson classes, $S(V, M)$ is the Segre class of the *conormal sheaf* of V in M.

If $\mathcal{I}$ is the ideal of V in M, the conormal sheaf of V in M is the coherent sheaf $\mathcal{I}/\mathcal{I}^2$; that is, $\mathcal{I}$ restricted to V (I mean: tensored by $\mathcal{O}_V = \mathcal{O}_M/\mathcal{I}$). To think about the Segre class of a coherent sheaf $\mathcal{F}$ on V, consider the corresponding 'linear fiber space' $\mathrm{Proj}(\mathrm{Sym}\mathcal{F}) \xrightarrow{p} V$; this comes with an invertible sheaf $\mathcal{O}(1)$, and we can set

$$s(\mathcal{F}) = p_* c(\mathcal{O}(-1))^{-1} \cap [\mathrm{Proj}(\mathrm{Sym}\mathcal{F})].$$

So Fulton-Johnson classes capture, up to restricting to V and standard intersection-theoretic maneuvers, the Symmetric algebra of $\mathcal{I}$.

What do Fulton classes capture? *Exactly the same kind of information,* but for the *Rees* algebra rather than the *Symmetric* algebra. This is not a minor difference in general, but it is completely invisible if $\mathcal{I}$ is (locally) a complete intersection.

Important example. If V is a hypersurface in a nonsingular variety M (or more generally a local complete intersection) then both Fulton and Fulton-Johnson classes yield $c(TM)c(N_M V)^{-1} \cap [V]$. If V is nonsingular, this is automatically $c(TV) \cap [V]$.

The main point here is that, at least when V is a local complete intersection, these classes are 'easier' than the functorial Chern-Schwartz-MacPherson classes.

So you may hit upon the idea of trying to relate these notions.

1.5. Preliminaries: Segre classes

Segre classes will show up often (they already have), and it will be important to have a certain technical mastery of them. If V is a proper closed subscheme of M, the *Segre class* of V in M is the element of $\mathcal{A}V$ (the Chow group of V) characterized by the following two requirements:

- if $V \subset M$ is a regular embedding, then $s(V, M) = c(N_V M)^{-1} \cap [V]$;

- if $\pi : M' \to M$ is a proper birational map, and $p : \pi^{-1}(V) \to V$ is the restriction of π, then $p_* s(\pi^{-1}V, M') = s(V, M)$.

These are enough to define $s(V, M)$ in any circumstance, since by the second item we can replace M with the blow-up of M along V, and V by the exceptional divisor in the blow-up; this is regularly embedded, so the first item clinches the class.

In fact this argument reduces the computation of a Segre class $s(V, M)$ to the computation when V is a *hypersurface* of M, that is, it is locally given by one equation. The shorthand for the Segre class is then

$$s(V, M) = \frac{[V]}{1 + V},$$

by which one means the sum $(1 - V + V^2 - V^3 + \cdots) \cap [V] = V - V^2 + V^3 - \cdots$.

Often one can use this in reverse.

Example. Consider the second Veronese embedding of $\mathbb{P}^2$ in $\mathbb{P}^5$. The hyperplane H in $\mathbb{P}^5$ restricts to twice the hyperplane h in $\mathbb{P}^2$. Standard sequences lead to the following computation of the Segre class of $v_2(\mathbb{P}^2)$ in $\mathbb{P}^5$:

$$s(v_2(\mathbb{P}^2), \mathbb{P}^5) = \frac{(1+h)^3}{(1+2h)^6} \cap [\mathbb{P}^2] = [\mathbb{P}^2] - 9[\mathbb{P}^1] + 51[\mathbb{P}^0].$$

Now let $\widetilde{\mathbb{P}}^5$ be the blow-up of $\mathbb{P}^5$ along the Veronese surface, and let E be the exceptional divisor. The birational invariance of Segre classes then says that $s(E, \widetilde{\mathbb{P}}^5)$ must push-forward to $s(v_2(\mathbb{P}^2), \mathbb{P}^5)$; that is,

$$E - E^2 + E^3 - E^4 + E^5 \mapsto [\mathbb{P}^2] - 9[\mathbb{P}^1] + 51[\mathbb{P}^0].$$

This is enough information to determine the Chow *ring* of the blow-up. Interpreting $\mathbb{P}^5$ as the $\mathbb{P}^5$ of conics, then a standard game in enumerative geometry would compute the number of *conics tangent to five conics in general position* as

$$(6H - 2E)^5 = 7776 - 2880 \cdot 4 + 480 \cdot 2 \cdot 9 - 32 \cdot 51 = 3264.$$

(Exercise: make sense out of this!) In fact, Segre classes provide a systematic framework for enumerative geometry computation; but this is of relatively little utility, as Segre classes are in general extremely hard to compute.

Why? Because blow-ups are hard to compute. If $\mathcal{I}$ is the ideal of V in M, 'computing' the blow-up of M along V amounts to realizing

$$\text{Proj}(\text{Rees}\mathcal{I}) = \text{Proj}(\mathcal{O} \oplus \mathcal{I} \oplus \mathcal{I}^2 \oplus \mathcal{I}^3 \oplus \cdots):$$

this is not easy.

One way to do this in practice is to see V as the vanishing of a section of a vector bundle on M: $s : M \to E$; then it is not hard to show that the blow-up of M along V is the closure of the image of M in $\mathbb{P}(E)$ by the induced rational section $\overline{s} : M \dashrightarrow \mathbb{P}(E)$. Then the line bundle of the exceptional divisor is the restriction of the tautological $\mathcal{O}(-1)$ from $\mathbb{P}(E)$. We will use this remark later on.

For applications of Segre classes to enumerative geometry, see [Ful84], Chapter 9 (and elsewhere).

2. Lecture II

2.1. (Chern-)Fulton classes vs. (Chern-)Fulton-Johnson classes

I would like to look at a simple example to explore the difference between the Chern-Fulton and Chern-Fulton-Johnson business.

Consider the planar triple point S with ideal $I = (x^2, xy, y^2)$ in the affine plane $\mathbb{A}^2$. For such an object we can be most explicit.

First of all, I is dominated by $k[x,y]^{\oplus 3}$:

$$k[x,y]\langle s,t,u\rangle \to I \to 0,$$

sending $s \mapsto x^2$, $t \mapsto xy$, $u \mapsto y^2$. Tensoring by $k[x,y]/I$:

$$(k[x,y]/I)\langle s,t,u\rangle \to I/I^2 \to 0,$$

with the same prescriptions for the map. The kernel is computed to be $(xt - ys, xu - yt)$. Thus $\text{Proj}(\text{Sym}I/I^2)$ is defined by the ideal $(xt - ys, xu - yt)$ in the product $S \times \mathbb{P}^2$. Now if I believe Macaulay2, taking primary decomposition here gives me $(x,y) \cap$ an embedded component. That is, $\text{Proj}(\text{Sym}I/I^2)$ is the *reduced* $\mathbb{P}^2$ inside the (nonreduced) $S \times \mathbb{P}^2$. The Fulton-Johnson class is then 1 times the class of the point supporting S.

What about Chern-Fulton? To obtain the Segre class of S in $\mathbb{A}^2$, blow-up $\mathbb{A}^2$ along S. We can fit the blow-up in $\mathbb{A}^2 \times \mathbb{P}^2$, and a bit of patience gives its ideal:

$$(xt - ys, xu - yt, su - t^2).$$

Note the extra generator! That is the difference between the Rees algebra of I, computed here, and the Symmetric algebra, which is what comes up in the Fulton-Johnson computation. The exceptional divisor of the blow-up is the intersection of the blow-up with $S \times \mathbb{P}^2$ (the same $S \times \mathbb{P}^2$ as before!), and taking the primary decomposition again shows it is $S \times$a conic in $\mathbb{P}^2$.

It is completely different: it was a $\mathbb{P}^2$ before, it is a curve now. Its degree is $2 \times 2 = 4$, so the Chern-Fulton class is 4 times the class of the point.

Remark. Taking away the extra generator xy leaves the multiple point (x^2, y^2), a complete intersection. It is a fact that this does not change the Chern-Fulton class (because xy is integral over this ideal); on the other hand, Chern-Fulton and Chern-Fulton-Johnson agree for complete intersections, so they must both give 4 times the class of the point in this case. Exercise: verify this explicitly.

One moral to be learned from such examples is that classes such as Chern-Fulton or Chern-Fulton-Johnson are extremely sensitive to the *scheme structure.* This is important for me, since I am aiming to develop a tool that can do computations starting from an arbitrary ideal. On the other hand, no scheme considerations have entered the discussion of Chern-Schwartz-MacPherson or Chern-Mather classes.

To cover such cases, I will simply declare that the Chern-Schwartz-MacPherson class of a possibly non-reduced scheme is the class of its *support.* This turns out to be computationally convenient, but it is not arising from any 'functoriality' considerations.

2.2. Segre classes again: inclusion-exclusion

In view of the considerations that took us here, we may now want to ask: are classes such as Chern-Fulton or Chern-Fulton-Johnson classes also a fancy form of 'counting'? That is, do they satisfy 'inclusion-exclusion'?

Since the difference between these classes and Segre classes is a common factor ($c(TM)\cap$), the question for Fulton classes is equivalent to: do Segre classes satisfy inclusion-exclusion?

Example. Take V = the union of two distinct lines L_1, L_2 in the projective plane. Then $s(V,\mathbb{P}^2) = \frac{1}{(1+2H)}\cap[V] = [L_1]+[L_2]-4[pt]$, with hopefully evident notations. On the other hand, $s(L_i,\mathbb{P}^2) = [L_i]-[pt]$. Thus

$$s(L_1,\mathbb{P}^2)+s(L_2,\mathbb{P}^2)-s(V,\mathbb{P}^2) = 2[pt] \neq s(L_1\cap L_2,\mathbb{P}^2).$$

In other words, inclusion-exclusion fails miserably for Segre classes, on the very first example one may try.

Is there a way out? Actually yes, [Alu03c]; it is completely trivial, and I didn't notice it for many years. The remark is that variations on the definition of Segre classes do tend to satisfy inclusion-exclusion.

This follows immediately from 8th grade algebra. The simplest case, which is also the only one I need in what follows, goes like this:

$$\begin{aligned}&\frac{R_1+E}{1+R_1+E}+\frac{R_2+E}{1+R_2+E}-\frac{R_1+R_2+E}{1+R_1+R_2+E}-\frac{E}{1+E}\\ &=\frac{R_1R_2(2+R_1+R_2+2E)}{(1+R_1+E)(1+R_2+E)(1+R_1+R_2+E)(1+E)}\end{aligned}$$

How can this possibly say something useful? Assume X_1, X_2 are effective Cartier divisors in an ambient scheme M, and $Y = X_1\cap X_2$ (scheme-theoretically, of course). Blow-up M along Y, and let E be the exceptional divisor. Then, by our definition of Segre classes, $s(Y,M) = p_*\frac{[E]}{1+E}$, where p is the blow-up map. On the other hand, p^*X_i will consist of E and of a proper transform R_i ('residual'). Note that $R_1R_2 = 0$, since the proper transforms do not meet (this is why I am stressing that the intersection must be 'scheme-theoretic'). Further, by the projection formula, $p_*\frac{[R_i+E]}{1+R_i+E} = \frac{[X_i]}{1+X_i} = s(X_i,M)$. That is, 8th grade algebra says

$$s(Y,M) = s(X_1,M)+s(X_2,M)-p_*\frac{[R_1]+[R_2]+[E]}{1+R_1+R_2+E}:$$

something very close to inclusion-exclusion. The funny term

$$p_*\frac{[R_1]+[R_2]+[E]}{1+R_1+R_2+E}$$

is where one would expect 'the class of the union'; this would instead equal

$$p_* \frac{[R_1] + [R_2] + 2[E]}{1 + R_1 + R_2 + 2E}.$$

My point of view is that inclusion-exclusion *does* works for Segre classes once one takes care to correct them adequately for (something like) multiple contributions of subsets. The case I just illustrated (which generalizes nicely to arbitrarily many X_i of any kind...) is one way to correct the classes. We will run into another one later on.

2.3. Remark on notations

Before massaging the formula we just obtained into something yielding any readable information, I must introduce two simple notational devices.

These concern rational equivalence classes in an ambient scheme M, which I will now assume to be pure-dimensional. The notations are best written by indexing the classes by *co*dimension: so I will write a class $A \in \mathcal{A}N$ as $A = \sum_{i \geq 0} a^i$, denoting by a^i the piece of A of codimension i. Then I will let

$$A^\vee = \sum_{i \geq 0} (-1)^i a^i,$$

and I will sloppily call this the 'dual' of A. Note that the notation hides the ambient space, which may sometime lead to confusion. The rationale for the notation is simple: if E is a vector bundle on M, or more generally a class in the K-theory of vector bundles, then with this notation one simply has

$$(c(E) \cap A)^\vee = c(E^\vee) \cap A^\vee.$$

The second piece of notation is similar, but a bit more interesting. Let $\mathcal{L}$ be a line bundle on M. I will let

$$A \otimes \mathcal{L} = \sum_{i \geq 0} \frac{a^i}{c(\mathcal{L})^i}.$$

This also hides the ambient M; when necessary, I subscript the tensor: $\otimes_M$. But in these lectures all tensors will be in the ambient variety.

The rationale for the second notation is similar to the first: if E is a class in K-theory *and of rank* 0, then one can check that

$$(c(E) \cap A) \otimes \mathcal{L} = c(E \otimes \mathcal{L}) \cap (A \otimes \mathcal{L}).$$

Watch out for the 'rank 0' part!

Of course the notation $A \otimes \mathcal{L}$ suggests an 'action', and this is easy to verify: if $\mathcal{M}$ is another line bundle, one checks that

$$(A \otimes \mathcal{L}) \otimes \mathcal{M} = A \otimes (\mathcal{L} \otimes \mathcal{M}).$$

Also, $(A \otimes \mathcal{L})^\vee = A^\vee \otimes \mathcal{L}^\vee$.

All these observations are simple algebra of summations and binomial coefficients; they are useful insofar as they compress complicated formulas into simpler ones by avoiding $\sum$'s and elementary combinatorics.

Here is an example of such manipulations, which I will need in a moment.

$$\frac{-E}{1+X-E} = \left(\frac{E}{1+E-X}\right)^{\vee} = (E \otimes \mathcal{O}(E-X))^{\vee} = ((E \otimes \mathcal{O}(E)) \otimes \mathcal{O}(-X))^{\vee}$$
$$= \left(\frac{E}{1+E} \otimes \mathcal{O}(-X)\right)^{\vee} = \left(\frac{E}{1+E}\right)^{\vee} \otimes \mathcal{O}(X).$$

2.4. Key example

Let's go back to the 'inclusion-exclusion' formula we obtained a moment ago:

$$s(Y, M) = s(X_1, M) + s(X_2, M) - p_* \frac{[R_1 + R_2 + E]}{1 + R_1 + R_2 + E}.$$

Here we had $Y = X_1 \cap X_2$, where X_1, X_2 are hypersurfaces; and E is the exceptional divisor of the blow-up along Y, R_i the proper transform of X_i. Assuming that the ambient M is nonsingular, and capping through by $c(TM)$, gives

$$c_{\mathrm{F}}(Y) = c_{\mathrm{F}}(X_1) + c_{\mathrm{F}}(X_2) - c(TM) \cap p_* \frac{[R_1 + R_2 + E]}{1 + R_1 + R_2 + E},$$

where c_{F} denotes Chern-Fulton class.

Now assume X_1, X_2, and Y are nonsingular. Then $c_{\mathrm{F}} = c_{\mathrm{SM}}$, since both equal the classes of the tangent bundle. That is:

$$c_{\mathrm{SM}}(Y) = c_{\mathrm{SM}}(X_1) + c_{\mathrm{SM}}(X_2) - c(TM) \cap p_* \frac{[R_1 + R_2 + E]}{1 + R_1 + R_2 + E}$$

in this extremely special case. On the other hand, c_{SM} satisfies inclusion-exclusion on the nose:

$$c_{\mathrm{SM}}(Y) = c_{\mathrm{SM}}(X_1) + c_{\mathrm{SM}}(X_2) - c_{\mathrm{SM}}(X_1 \cup X_2).$$

The conclusion is that if X_1, X_2 are transversal nonsingular hypersurfaces in a nonsingular ambient variety M, and $X = X_1 \cup X_2$, then

$$c_{\mathrm{SM}}(X) = c(TM) \cap p_* \frac{[R_1 + R_2 + E]}{1 + R_1 + R_2 + E}.$$

Let us work on the funny piece $\frac{[R_1+R_2+E]}{1+R_1+R_2+E}$ in this formula. First, $R_i + E$ stands for (the pull-back of) X_i; so we can rewrite this as $\frac{[R_1+R_2+E]}{1+R_1+R_2+E} = \frac{[X-E]}{1+X-E}$. Next, another bit of 8th grade algebra:

$$\frac{[X-E]}{1+X-E} = \frac{X}{1+X} + \frac{1}{1+X} \cdot \frac{-E}{1+X-E}$$

Using the example from the previous section:

$$\frac{-E}{1+X-E} = \left(\frac{E}{1+E}\right)^{\vee} \otimes \mathcal{O}(X)$$

Put everything together and use the projection formula:

$$c_{\mathrm{SM}}(X) = c(TM) \cap p_* \left(\frac{X}{1+X} + \frac{1}{1+X} \cdot \left(\left(\frac{E}{1+E} \right)^\vee \otimes \mathcal{O}(X) \right) \right)$$

$$= c(TM) \cap \left(s(X, M) + \frac{1}{1+X} \cdot (s(Y, M)^\vee \otimes \mathcal{O}(X)) \right).$$

Now we are back in M: everything relating to the blow-up has been absorbed into terms in the original ambient space. We may also note that $c(TM) \cap s(X, M) = c_{\mathrm{F}}(X)$, so that what we are really saying is that

$$c_{\mathrm{SM}}(X) = c_{\mathrm{F}}(X) + c(TM) \cap \left(\frac{1}{c(\mathcal{O}(X))} \cdot (s(Y, M)^\vee \otimes \mathcal{O}(X)) \right).$$

We have proved that this holds if X is the union of two nonsingular hypersurfaces in a nonsingular variety M, meeting transversally along Y.

2.5. Main theorem

That is a reasonably pretty formula, but how do we interpret it in more 'intrinsic' terms? 'What is' $Y = X_1 \cap X_2$, in terms of $X = X_1 \cup X_2$, when X_1 and X_2 are nonsingular and transversal?

Working locally, suppose (F_i) is the ideal of X_i. Then X has ideal $(F_1 F_2)$. The 'singularity subscheme' of a hypersurface with ideal (F) is the scheme defined by the ideal (F, dF), where dF is shorthand for the partial derivatives of F. It is clear that this scheme is supported on the singular *locus* of X; the specified ideal gives it a scheme structure (it is easy to see that this structure patches on affine overlaps).

For our X, this ideal would be

$$(F_1 F_2, F_1 dF_2 + F_2 dF_1).$$

It is clear that this is supported on $Y = X_1 \cap X_2$, since X_1 and X_2 are nonsingular. But in fact we are asking X_1 and X_2 to be *transversal:* thus dF_1 and dF_2 are independent at every point of Y. Hence the ideal of the singularity subscheme is $(F_1 F_2, F_1, F_2) = (F_1, F_2)$: that is, precisely the ideal of $X_1 \cap X_2 = Y$. In other words, *Y is the singularity subscheme of X* in this case.

Therefore, we can rephrase the formula we have obtained above: if X is a (very special) hypersurface in a nonsingular ambient variety M, and Y is the singularity subscheme of X, then

$$c_{\mathrm{SM}}(X) = c_{\mathrm{F}}(X) + c(TM) \cap \left(\frac{1}{c(\mathcal{O}(X)} \cdot (s(Y, M)^\vee \otimes \mathcal{O}(X)) \right).$$

Theorem 2.1. *This formula holds for* every *hypersurface in a nonsingular variety.*

This is the main result of the lectures: everything else I can say is simply a variation or restatement or application of this theorem.

Many proofs are known of this statement, and I will review some of them in the next lecture. The first proof, going back to 1994 ([Alu94]), proved this formula over $\mathbb{C}$ and in a weak, 'numerical' sense. The formula is in fact true in the Chow group of X, and over any algebraically closed field of characteristic 0 (the theory of CSM classes extends to this context, by work of Gary Kennedy, [Ken90]).

The 1994 proof is instructive in the sense that it clarifies what kind of information the difference $c_{\mathrm{SM}}(X) - c_{\mathrm{F}}(X)$ carries. Read the term of dimension 0:

$$\int c_{\mathrm{SM}}(X) - c_{\mathrm{F}}(X) = \chi(X) - \chi(X_g),$$

where $\int c_{\mathrm{SM}}(X) = \chi(X)$ as we have seen, and X_g stands for a general, nonsingular hypersurface in the same rational equivalence class as X (should there be one). It is well known (see for example [Ful84], p. 245–246) that if the singularities of X are isolated then this difference is (up to sign) the sum of the Milnor numbers of X. Adam Parusiński ([Par88]) defines an invariant of (not necessarily isolated) hypersurface singularities by taking it to be this difference in general. On the other hand, I had obtained a formula for Parusiński's number in terms of the Segre class of the singularity subscheme of X ([Alu95]). A bit of work using the funny $\otimes$ notations shows then that the Theorem holds for the dimension 0 component of the class. The numerical form can be obtained by reasoning in terms of general hyperplane sections.

Because of this history, it makes sense to call the difference $c_{\mathrm{SM}}(X) - c_{\mathrm{F}}(X)$ the 'Milnor class' of X (up to sign, depending on the author). I do not think Milnor has been informed...

3. Lecture III

3.1. Main theorem: Proof I

I would like to review some of the approaches developed in order to prove the formula for the Milnor class of a hypersurface X in a nonsingular variety M:

$$c_{\mathrm{SM}}(X) - c_{\mathrm{F}}(X) = c(TM) \cap \left(c(\mathcal{O}(X))^{-1} \cap (s(Y,M)^\vee \otimes \mathcal{O}(X))\right).$$

The first one ([Alu99a]) is rather technical, but has the advantage of working over an arbitrary algebraically closed field of characteristic 0, and for arbitrary hypersurfaces (allowing multiplicities for the components).

The idea is the following. It would of course be enough to show that the class defined by

$$c_?(X) = c_{\mathrm{F}}(X) + c(TM) \cap \left(c(\mathcal{O}(X))^{-1} \cap (s(Y,M)^\vee \otimes \mathcal{O}(X))\right)$$

is covariant in the same sense as the CSM class. This would be very nice, but I have never quite managed to do it directly.

Using resolution of singularity, however, it is enough to prove a weak form of covariance, and this can be done. Specifically:

- if X is a divisor with normal crossings and (possibly multiple) nonsingular components, then $c_?(X) = c_{\mathrm{SM}}(X)$;

- if $\pi : \widetilde{M} \to M$ is a blow-up along a nonsingular subvariety of the singular locus of X, then $\pi_* c_?(\pi^{-1}X) = c_?(X) + \pi_* c_{\mathrm{SM}}(\pi^{-1}X) - c_{\mathrm{SM}}(X)$.

Indeed, by resolution of singularities one can reduce to the normal crossing case after a number of blow-ups; then $c_?$ and c_{SM} agree there, by the first item; and as they map down each step of the resolution, $c_?$ and c_{SM} change in the same way, so they must agree for X. Note that this forces us to work with nonreduced objects!

The proofs of the two listed properties are rather technical. The first one in the reduced case boils down to the following: if $X = X_1 \cup \cdots \cup X_r$ is a reduced divisor with normal crossings and nonsingular components, with singularity subscheme Y, then

$$s(Y, M) = \left(\left(1 - \frac{c(\mathcal{L}^\vee)}{c(\mathcal{L}_1^\vee) \cdots c(\mathcal{L}_r^\vee)}\right) \cap [M]\right) \otimes \mathcal{L}$$

where $\mathcal{L} = \mathcal{O}(X)$ and $\mathcal{L}_i = \mathcal{O}(X_i)$. Proof: induction, and properties of $\otimes$.

The other item is more interesting. If $\widetilde{M}$ is obtained by blowing up M along a nonsingular Z of codimension d, one is reduced to showing that

$$\pi_* c_?(\pi^{-1}(X)) = c_?(X) + (d-1)c(TZ) \cap [Z].$$

This should be much easier than it is! After a number of manipulations, one is led to transferring the question to within the bundle $\mathbb{P}(\pi^* \mathcal{P}^1_M \mathcal{L} \oplus \mathcal{P}^1_{\widetilde{M}} \mathcal{L})$, where $\mathcal{P}^1$ denotes 'principal parts'. I will return to these later on, so I won't say much about them here. Suffice it to say that there are two classes in this bundle, related to the blow-up of M along the singularity subscheme Y of X, and to the blow-up of $\widetilde{M}$ along the singularity subscheme of $\pi^{-1}(X)$. The statement is translated in a suitable triviality of the *difference* between these two classes.

Once one phrases the problem in this manner, one sees right away what is the most natural tool to attack it: it's the graph construction – maybe not surprisingly, since this was the key tool in the original paper by MacPherson. Here it must be applied to the graph of the 'differential' map $\pi^* \mathcal{P}^1_M \mathcal{L} \to \mathcal{P}^1_{\widetilde{M}} \mathcal{L}$. Lots and lots of technical details later, the needed relation is proved.

3.2. Characteristic cycles

The proof I just surveyed was obtained in 1995-6, but only found the glory of the printed page in 1999: the reason is simply that referees did not like it at all. They pointed out that a different viewpoint would probably yield a much shorter and more insightful proof.

That turned out to be correct. The key is to rephrase the whole question in terms of *characteristic cycles,* a translation of the constructible function framework that goes back to Claude Sabbah and yields a very powerful approach. The subject is discussed at length in Lecture 3 of Jörg Schürmann's lecture cycle.

Sabbah summarizes the situation very well, in the following quote from [Sab85]: *la théorie des classes de Chern de* [Mac74] *se ramène à une théorie de Chow sur T^*M, qui ne fait intervenir que des classes fondamentales.* The functor of constructible functions is replaced with a functor of *Lagrangian cycles* of

T^*M (or its projectivization $\mathbb{P}(T^*M)$); then the key operations on constructible functions become more geometrically transparent, and this affords a general clarification of the theory. I can recommend [Ken90] for a good treatment of c_{SM} classes from this point of view.

I will summarize the situation here. Let M be a nonsingular variety. If $V \subset M$ is nonsingular, then we have a sequence

$$0 \to T^*_V M \to T^*M|_V \to T^*V \to 0$$

where $T^*_V M$ denotes the *conormal bundle* (or 'space') of V; we view this as a subvariety of the total space of T^*M. If V is singular, do this on its nonsingular part, then close it up in T^*M to obtain its conormal space $T^*_V M$. Linear combinations of cycles $[T^*_V M]$ (these are the *classes fondamentales* in Sabbah's quote) form an abelian group $L(M)$ (L stands for 'Lagrangian').

Now go back to the *Nash blow-up* $\nu : \widetilde{V} \to V$, with tautological bundle $\mathcal{T}$. For each $p \in V$ we can define a number as follows:

$$\mathrm{Eu}_V(p) = \int c(\mathcal{T}) \cap s(\nu^{-1}(p), \widetilde{V}).$$

This is the 'local Euler obstruction', a constructible function originally defined (in a different way) by MacPherson in his paper. As it happens, these functions span $\mathcal{C}(M)$, so we may use them to define a homomorphism Ch from $\mathcal{C}(V)$ to $\mathcal{L}(V)$: require that $\mathrm{Ch}(\mathrm{Eu}_V) = (-1)^{\dim V}[T^*_V M]$. The cycle $\mathrm{Ch}(\varphi)$ corresponding to a constructible function φ is called its *characteristic cycle.* In particular, every subvariety V of M has a characteristic cycle $\mathrm{Ch}(1\!\!1_V)$: this is a certain combination of $T^*_V M$ and of conormal spaces to subvarieties of V, according to the singularities of V.

Now all the ingredients are there. The original definition given by MacPherson for the natural transformation c_* is a combination of Chern-Mather classes, with coefficients determined by local Euler obstructions. A relatively straightforward computation shows that Chern-Mather classes can be computed in terms of conormal spaces. The homomorphism $\mathcal{C} \rightsquigarrow L$ is concocted so as to be compatible with this set-up. All in all, we get an explicit expression for c_*:

$$c_*(\varphi) = (-1)^{\dim M - 1} c(TM) \cap \pi_* \left(c(\mathcal{O}(1))^{-1} \cap [\mathbb{P}\mathrm{Ch}(\varphi)] \right),$$

where π is the projection $\mathbb{P}(T^*M) \to M$ ([PP01], p.67).

The operation on the right-hand side may look somewhat unnatural, but is on the contrary the most direct way to deal with classes in a projective bundle; we may come to this later. I call the whole operation (maybe up to some sign) *casting the shadow* of $\mathrm{Ch}(\varphi)$. Thus, the Chern-Schwartz-MacPherson class of V is nothing but the shadow of its characteristic cycle.

3.3. Main theorem: proof II

In [PP01], Adam Parusiński and Piotr Pragacz give a proof (in fact, two proofs) of a theorem which implies the formula for the difference $c_{\mathrm{SM}}(X) - c_{\mathrm{F}}(X)$ for X a hypersurface of a nonsingular variety M, at least in the reduced case.

The proof is a clever interplay of certain constructible functions, which are very close to the geometry of the hypersurface, and characteristic cycles in the sense of the previous section. I will have to pass in silence the story concerning the interesting constructible functions, but I want to expose the point of contact with [Alu99a].

The formula in [Alu99a] (writing $\mathcal{L} = \mathcal{O}(X)$):

$$c_{\mathrm{SM}}(X) = c_{\mathrm{F}}(X) + c(TM) \cap \left(c(\mathcal{L})^{-1} \cap (s(Y, M)^{\vee} \otimes \mathcal{L})\right)$$

writes the 'hard' part of c_{SM} in terms of the Segre class of the singularity subscheme Y of X. As we have seen previously, an equivalent formulation is in terms of the exceptional divisor E of the blow-up $\widetilde{M}$ of M along Y; in fact, tracing the different terms yields a rather pretty formula (Theorem I.3 in [Alu99a] works this out):

$$c_{\mathrm{SM}}(X) = c(TM) \cap \pi_* \left(\frac{[X' - E]}{1 + X' - E}\right),$$

where π is the blow-up map, and I am writing X' for the inverse image $\pi^{-1}(X)$.

Now I mentioned in one of the first sections that there is an efficient way to realize a blow-up, when the center of the blow-up is the zero-scheme of a section of a vector bundle. This is our situation: Y can be viewed as the zero-scheme of a section of the bundle $\mathcal{P}^1_M\mathcal{L}$ which made its appearance above. This bundle fits a nice exact sequence:

$$0 \to \Omega^1_M \otimes \mathcal{L} \to \mathcal{P}^1_M\mathcal{L} \to \mathcal{L} \to 0;$$

the section is obtained by combining the differential of a defining equation F for X (in the Ω^1 part) with F itself (in the cokernel). The ideal of this section is generated by F and dF, hence it gives Y.

By the general blow-up story, then, $\widetilde{M}$ lives in $\mathbb{P}(\mathcal{P}^1_M\mathcal{L})$. *Further,* the part over X lives in the subbundle $\mathbb{P}(\Omega^1_M \otimes \mathcal{L})$, since the cokernel part of the section is 0 along X. *Further still,* $\mathbb{P}(\Omega^1_M \otimes \mathcal{L}) = \mathbb{P}(T^*M)$ since tensoring by a line-bundle just changes the meaning of $\mathcal{O}(1)$.

Summarizing, we see that the cycle $[X' - E]$ lives naturally in $\mathbb{P}(T^*M)$, which is the home of the characteristic cycle of X. Once one unravels notations, the operation getting $c_{\mathrm{SM}}(X)$ from $[X' - E]$ turns out to match precisely the one getting c_* from the characteristic cycle.

In the end, and modulo another bit of 8th grade algebra, one sees that *the main formula is equivalent to the statement that $[X' - E]$ be the characteristic cycle of X:* if X is a (reduced) hypersurface in a nonsingular variety M, then $\mathrm{Ch}(1\!\!1_X) = [X' - E]$.

This framework offers a different way to prove the main formula. We have to verify that $\mathrm{Ch}^{-1}([X' - E])$ is the constant function 1 on X. Unravelling local Euler obstructions and using minimal general knowledge about Segre classes reduces the statement to the equality

$$\int \frac{s(\pi^{-1}(p), X' - E)}{1 + X' - E} = 1 \qquad \forall p \in X$$

This can be checked directly, by a multiplicity computation. The details are in [Alu00], a spin-off of [PP01].

3.4. Differential forms with logarithmic poles

Yet another viewpoint on the main formula comes from aiming to understand the class of the *complement* of X in M rather than the class of X itself: after all,

$$c_*(1\!\!1_{M\setminus X}) = c_*(1\!\!1_M) - c_*(1\!\!1_X) = (c(TM)\cap[M]) - c_{\rm SM}(X),$$

so the information can be used to recover $c_{\rm SM}(X)$. One price to be paid here is that one gets $c_{\rm SM}(X)$ as a class in the ambient variety M rather than on X itself; this is common to other approaches, and there does not seem to be a way around it.

The observation is now that it is in fact *straightforward* to compute $c_*(1\!\!1_{M\setminus X})$: it turns out that this can be realized in a way somewhat similar to the computation of Chern-Mather classes. That is, there is a blow-up $\widetilde{M}$ of M and a vector bundle on $\widetilde{M}$ whose (honest) Chern classes push forward to $c_*(1\!\!1_{M\setminus X})$. Further, this can be done even if X is not a hypersurface!

For this, blow-up X and resolve singularities, so as to obtain a variety $\pi:\widetilde{M}\to M$ such that $X'=\pi^{-1}(X)$ is a divisor with normal crossing and nonsingular components. Denote by X'' the support of X'. Then there is an interesting *locally free* sheaf on $\widetilde{M}$, denoted $\Omega^1_{\widetilde{M}}(\log X'')$: over an open set U where X'' has ideal $(x_1\cdot\dots\cdot x_r)$ (where $x_1,\dots,x_n$ are local parameters), sections of $\Omega^1_{\widetilde{M}}(\log X'')$ can be written

$$\alpha_1\frac{dx_1}{x_1}+\cdots+\alpha_r\frac{dx_r}{x_r}+\alpha_{r+1}dx_{r+1}+\cdots+\alpha_n dx_n\;:$$

that is, they consist of differential forms 'with logarithmic poles' along the components of X''.

Claim. $c_*(1\!\!1_{M\setminus X})=\pi_*(c(\Omega^1_{\widetilde{M}}(\log X'')^\vee)\cap[\widetilde{M}])$.

This is surprisingly easy, actually: put $\Omega^1_{\widetilde{M}}(\log X'')$ in a sequence, then observe that the resulting class satisfies enough 'inclusion-exclusion' as to be forced to agree with the CSM class. Details are in [Alu99b]. The same remark was made by Mark Goresky and William Pardon (a small lemma in [GP02]).

This is another case in which we have 'computed' something, but we are essentially no wiser than before. A certain amount of detective work should be performed before we can extract the information packaged in the Claim. I will summarize in the next section what is involved in this work in the case of hypersurfaces. To my knowledge, no one has tried the same for higher codimension varieties, although the Claim given above works just as well.

I should add that it is not clear that the Claim may not be computationally useful as is: I am told that embedded resolution of singularities has in fact been implemented, and it should not be hard to compute the Chern class of an explicit vector bundle and push-forward. This is another tempting project, waiting for a willing soul to pursue it.

3.5. Main theorem: proof III

How can we use differentials with logarithmic poles to effectively compute $c_{\text{SM}}(X)$ for a hypersurface? 'Effectively' means: without resolving singularities.

The question is how to identify a bundle *on* M, and some operation on this bundle that will yield the same classes as the projection of the $\Omega(\log)$ sheaf as seen above.

The operation that works is again something reminiscent of Chern-Mather classes of a variety V. Recall that these are defined as the projection of the classes of a tautological bundle $\mathcal{T}$ from the Nash blow-up $\widetilde{V} \xrightarrow{\nu} V$. This bundle agrees with ν^*TV when the latter is defined. In fact, the *co*tangent *sheaf* Ω^1_V is always defined, and by the same token must agree with $\mathcal{T}^\vee$ wherever ν is an isomorphism. We can go one step further: there is in fact an onto morphism

$$\nu^*\Omega_V \to \mathcal{T}^\vee \to 0,$$

whose kernel is torsion on $\widetilde{V}$. Thus, we can think (up to duals) of the Chern-Mather operation as something done on the sheaf of differential forms of V to 'make it locally free modulo torsion'; and the Chern-Mather classes are obtained by mod-ing out the torsion and taking ordinary Chern classes of what is left.

That can be done for every coherent sheaf $\mathcal{F}$ on V. The resulting class is called the *Chern-Mather class* $c_{\text{Ma}}(\mathcal{F})$, and has been studied by Marie-Hélène Schwartz a while back and Michał Kwieciński more recently ([Sch82] and [Kwi94]).

So the plan is to show that the push-forward of the classes of the $\Omega(\log)$ sheaf actually computes the Chern-Mather class of a sheaf that can be effectively described in M. This works out, but is as usual rather technical. The sheaf on M is very natural, given the data of a hypersurface X: I have already pointed out that X determines a section

$$0 \to \mathcal{O}_M \to \mathcal{P}^1_M\mathcal{L}$$

where $\mathcal{L} = \mathcal{O}(X)$; tensoring by $\mathcal{L}^\vee$ we have an injection

$$0 \to \mathcal{L}^\vee \to \mathcal{L}^\vee \otimes \mathcal{P}^1_M\mathcal{L};$$

define $\overline{\Omega}_X$ to be the sheaf (on M!) obtained as the cokernel of this map.

Claim. $c_{\text{Ma}}(\overline{\Omega}_X) = \pi_*(c(\Omega^1_{\widetilde{M}}(\log X'')) \cap [\widetilde{M}])$.

The obvious approach to proving this would be to construct a surjection from $\pi^*\overline{\Omega}_X$ to the $\Omega(\log)$ sheaf. I was not able to do this, but I was able to construct a locally free sheaf onto which $\pi^*\overline{\Omega}_X$ maps: after blowing up further if necessary in order to assume that $Y' := \pi^{-1}(Y)$ is a Cartier divisor, one can show that there is a surjection

$$\pi^*\overline{\Omega}_X \to \frac{\mathcal{L}^\vee \otimes \pi^*\mathcal{P}^1_M\mathcal{L}}{\mathcal{L}^\vee \otimes \mathcal{O}(Y')} \to 0,$$

so one can aim to showing that

$$c\left(\frac{\mathcal{L}^\vee \otimes \pi^*\mathcal{P}^1_M\mathcal{L}}{\mathcal{L}^\vee \otimes \mathcal{O}(Y')}\right) = c(\Omega^1_{\widetilde{M}}(\log X'')).$$

This does work out – the key to it, after some standard set-up work, is again the graph construction.

At the same time, the classes on the left-hand side can be computed 'formally'; when one does this, the main theorem comes to light. So this gives another proof of our main statement for hypersurfaces.

Can the same trick be used in higher codimension? I do not know!

4. Lecture IV

4.1. The higher codimension puzzle

Once more, here is the main theorem of these lectures: if X is a hypersurface in a nonsingular variety M, then

$$c_{\mathrm{SM}}(X) = c_{\mathrm{F}}(X) + c(TM) \cap \left(c(\mathcal{O}(X))^{-1} \cap (s(Y, M)^\vee \otimes \mathcal{O}(X))\right).$$

The next natural question is: what about higher codimension? can one remove the hypothesis that X be a hypersurface?

In a very weak sense the answer is: yes, cf. the approach via differential forms with logarithmic poles: the main formula there works for arbitrary X. But this doesn't teach us much, since we have almost no access to the resolution necessary in order to apply that formula. One would want to have a statement that remains within M, just as the main formula given above.

The first natural step to take is to go from hypersurfaces to complete intersections, or local complete intersections if one is brave. For the alternative viewpoints on Milnor classes mentioned in the previous lecture this has been carried out, first by Jean-Paul Brasselet et al., [BLSS02]; and now it is part of Jörg Schürmann powerful theory. Schürmann's work generalizes the approach by Parusiński and Pragacz in the sense that it concentrates on the 'interesting constructible functions' mentioned in the previous lecture; this is one of the main topics of Lecture 5 in Schürmann's cycle. One way to rephrase that approach is to view the Milnor class as arising from a constructible function corresponding to vanishing cycles; as Schürmann explains, this can be done for arbitrary complete intersections.

From this perspective, our 'Main Theorem' shows that *for hypersurfaces* the vanishing cycle information is essentially captured by the Segre class of the singularity subscheme. This statement cries out for a generalization, but one is still lacking.

Thus, a puzzle remains for higher codimensions, even in the cases covered by these approaches. For this, I must go back to a question I posed in one of the first sections:

Is there a natural scheme structure on the singularities of a given variety V, which determines the Milnor class of V?

The Milnor class is, up to sign, simply the difference $c_{\mathrm{SM}}(X) - c_{\mathrm{F}}(X)$ (recall that for local complete intersection Fulton and Fulton-Johnson classes agree, so this convention is compatible with the one used by other authors in that context).

The main formula answers this puzzle for hypersurfaces: the Milnor class of a hypersurface X is determined by the *singularity subscheme* of X, that is, the subscheme of X defined by the partials of a defining equation.

It is very tempting to guess that a similar statement should hold for more general varieties. Candidates for immediate generalizations of the 'singularity subscheme' are not hard to find: for example, we could take the base scheme of the rational map from V to its Nash blow-up. Still, I do not know of any formula that will start from this scheme, perform an intersection-theoretic operation analogous to taking the Segre class and manipulating the result, and would thereby yield the Milnor class of V.

Some comments on what may be behind such an operation will have to wait until the last lecture. In this lecture I will leave the main philosophical question aside, and use a brute-force approach to obtain the information for higher codimension without doing any new work at all.

4.2. Segre classes: Inclusion-exclusion again

Brute force means: go back to inclusion-exclusion considerations. I have already pointed out that a variation on the definition of Segre class provides a notion that satisfies a certain kind of inclusion-exclusion.

There happens to be another variation on the theme of a Segre class that satisfies inclusion-exclusion. We can in fact force inclusion-exclusion down the throat of Segre classes, by recasting the main formula (for hypersurfaces) as part of the definition.

Explicitly, define the *SM-Segre class* of a hypersurface X in a nonsingular variety M, with singularity subscheme Y, to be

$$s^{\circ}(X, M) := s(X, M) + c(\mathcal{O}(X))^{-1} \cap (s(Y, M)^{\vee} \otimes \mathcal{O}(X));$$

and for a proper subscheme Z of M, say

$$Z = X_1 \cap \cdots \cap X_r,$$

with X_i hypersurfaces, set

$$s^{\circ}(Z, M) := \sum_{s=1}^{r} (-1)^{s-1} \sum_{i_1 < \cdots < i_s} s^{\circ}(X_{i_1} \cup \cdots \cup X_{i_s}, M).$$

Here $X_{i_1} \cup \cdots \cup X_{i_s}$ is any hypersurfaces supported on the union.

There are a few points to be made here. The first is a formal exercise: check that $s^{\circ}(Z, M)$ satisfies inclusion-exclusion (so $s^{\circ}(Z_1 \cap Z_2, M) = s^{\circ}(Z_1, M) + s^{\circ}(Z_2, M) - s^{\circ}(Z_1 \cup Z_2, M)$, for a suitable $Z_1 \cup Z_2$). The second point is more striking: this definition should look *extremely unlikely,* due to the amazing range of choices involved in it. What if I change the X_i to another collection of hypersurfaces cutting out Z? (for example, what if I throw in a few inessential generators of the ideal of Z?) what if at each stage in the $\sum$ I choose some other hypersurface supported on $X_{i_1} \cup \cdots \cup X_{i_s}$?

What magic property of Segre classes ensures that this definition does not depend on these choices?

I do not know! It must be something powerful indeed, but I have no clue as to its true nature.

So how do I know that the definition of $s^\circ(Z,M)$ really does not depend on the choices leading to it? Simply because

$$c_{\mathrm{SM}}(Z) = c(TM) \cap s^\circ(Z,M).$$

Exercise: realize that this is completely obvious, modulo the main result. (The impatient reader may look up [Alu03b], Theorem 3.1.)

One last, equally obvious, point: if I can compute $s^\circ(X,M)$ for hypersurfaces, then I can compute it for everything. There is nothing to this, but it has an important implication: I have almost bypassed the higher codimension puzzle. The SM-Segre class $s^\circ(Z,M)$ stands to $c_{\mathrm{SM}}(Z)$ in essentially the same way as the ordinary $s(Z,M)$ stands to $c_{\mathrm{F}}(Z)$.

The qualifiers 'almost...', 'essentially...' are there for an important reason: as defined, the class $s^\circ(Z,M)$ lives *in* M, not *in* Z. After the fact I know that there is a class on Z which agrees with $s^\circ(Z,M)$ after push-forward to M, but in terms of computations I won't be able to squeeze the class down to its proper place. In fact, this seems to be another puzzle about the definition that must go back to some mysterious and powerful property of Segre classes.

In any case, these considerations force me to shift my aim a little, and focus on computing the *push-forward of* $c_{\mathrm{SM}}(Z)$ *to the ambient variety* M. This seems to be the most natural question at this point.

4.3. Putting together the algorithm

At this point the algorithm behind the `Macaulay2` code I showed in action in the first lecture should have few secrets left. *Suppose you can compute Segre classes* in some ambient space M; then formal manipulations (8th grade algebra again) will allow you to compute $s^\circ(Z,M)$. For this, you will need to

- obtain a set of hypersurfaces cutting out Z (for example, a homogeneous ideal for Z if $M = \mathbb{P}^n$);
- for each subset S of this set, compute the singularity subscheme Y_S of the union X_S of the hypersurfaces in S;
- compute the ordinary Segre class of Y, and use it to obtain $s^\circ(X_S,M)$;
- put all the information together and obtain $s^\circ(Z,M)$.

Obtaining $c_{\mathrm{SM}}(Z)$ from $s^\circ(Z,M)$ is trivial, as pointed out in the previous section.

In short: *if you are able to compute ordinary Segre classes, then you can compute Chern-Schwartz-MacPherson classes.* This means that after all this work we are back to the very beginning: how to compute Segre classes? as I mentioned in the first lecture, this is in general a very hard problem.

Since the actual implementation of the algorithm works in $M = \mathbb{P}^n$, it is worth rewriting the main formula for hypersurface in this case. So let X be a hypersurface in $\mathbb{P}^n$; then $\mathcal{O}(X) = \mathcal{O}(d+1)$, where $d = \deg X - 1$ (the shift makes some other formulas a little more pleasant). If $F \in k[z_0, \ldots, z_n]$ is a homogeneous polynomial defining X, then the singularity subscheme Y of X is defined by taking the ideal $(\frac{\partial F}{\partial z_0}, \ldots, \frac{\partial F}{\partial z_n})$ (by Euler's formula, this contains F). So we may see Y as a zero of a section

$$\mathbb{P}^n \to \mathcal{O}(d)^{n+1} :$$

again our trick to realize a blow-up tells us what to do – the blow-up of $\mathbb{P}^n$ along Y is the closure of the graph of the induced rational map

$$\mathbb{P}^n \dashrightarrow \mathbb{P}^n .$$

Let $\Gamma \subset \mathbb{P}^n \times \mathbb{P}^n$ be this closure. Denote by H, K respectively the pull-backs of the hyperplane class from the first, resp. second factor. As Γ is a class of dimension n, it is determined by the $(n+1)$ coefficients in

$$[\Gamma] = g_0 K^n + g_1 H K^{n-1} + \cdots + g_n H^n .$$

Claim. The push-forward of $c_{\mathrm{SM}}(X)$ to $\mathbb{P}^n$ is

$$(1+H)^{n+1} - \sum_{d=0}^{n} g_d \, (-H)^d (1+H)^{n-d}$$

as an element of $\mathbb{Z}[H]/(H^{n+1}) = \mathcal{A}\mathbb{P}^n$.

This is of course no more and no less than the main theorem again. The switch from $s(Y, M)$ to the information about Γ is something that has to do with Segre classes, so I will talk about it separately. Everything else is 8th grade algebra, and is explained more in detail in [Alu03a].

4.4. Segre classes in projective space: concrete computations

We are back to Segre classes, and how to compute them in practice. We have already seen where the main information will be, but it only takes a moment to repeat the set-up more in general.

So let Z be a proper subscheme of $\mathbb{P}^n$, with homogeneous ideal $(f_0, \ldots, f_r)$. We may assume all generators have the same degree: indeed, otherwise we can multiply a generator of lower degree by powers of all homogeneous coordinates and bring its degree up.

This realizes Z as the zero-scheme of a section of $\mathcal{O}(d)^{r+1}$, for some d. Once more, the blow-up of $\mathbb{P}^n$ along Z will simply be the closure Γ of the graph of the rational map

$$\mathbb{P}^n \dashrightarrow \mathbb{P}(\mathcal{O}(d)^{r+1}) = \mathbb{P}^n \times \mathbb{P}^r \to \mathbb{P}^r$$

defined by the f_i's.

Incidentally, it is not conceptually hard (but computationally *very* demanding) to compute equations for Γ. The key step is performed by elimination theory, a slow but well understood process. It is easy to instruct Macaulay2 to do this.

As before, the class $[\Gamma]$ is determined by the $(n+1)$ numbers g_i for which

$$[\Gamma] = g_0 K^r + g_1 K^{r-1} H + g_2 K^{r-2} H^2 + \cdots :$$

these can be viewed as obtained by intersecting Γ with general copies of the hyperplane K, then pushing forward to $\mathbb{P}^n$ and computing degrees. Again, a tool like Macaulay2 can perform these operations.

How do we reconstruct the Segre class $s(Z, \mathbb{P}^n)$ (or rather its push-forward to $\mathbb{P}^n$) from these numbers? Assemble them together into a class $G = g_0 + g_1 H + g_2^2 H^2 + \cdots$ in $\mathcal{A}\mathbb{P}^n$.

Claim. The push-forward of $s(Z, \mathbb{P}^n)$ is $1 - c(\mathcal{O}(dH))^{-1} \cap (G \otimes \mathcal{O}(dH))$.

To see this, write $s(Z, \mathbb{P}^n)$ as the push forward of $\frac{[E]}{1+E}$ from Γ, where E is the exceptional divisor in the blow-up. The class of E is the hyperplane class in $\mathbb{P}^n \times \mathbb{P}^r = \mathbb{P}(\mathcal{O}(dH)^{r+1})$: chasing classes around shows that we need the push-forward of

$$\frac{[dH - K]}{1 + dH - K}.$$

And this game should now look familiar. Omitting evident notations (using the projection formula, etc.):

$$\begin{aligned}\frac{dH - K}{1 + dH - K} &= 1 - \frac{1}{1 + dH - K} = 1 - \frac{1}{1 + dH} \cdot \frac{1 + dH}{1 + dH - K} \\ &= 1 - c(\mathcal{O}(dH))^{-1} \left(\frac{1}{1 - K} \otimes \mathcal{O}(dH) \right) \\ &= 1 - c(\mathcal{O}(dH))^{-1} \cdot (G \otimes \mathcal{O}(dH))\end{aligned}$$

since $\frac{1}{1-K} = (1 + K + K^2 + \cdots)$ does precisely what it should (intersect Γ with higher and higher powers of K to get the components of G).

It is not hard to instruct Macaulay2 to perform these operations (see [Alu03a]) although computation time takes off exponentially, making any serious application impractical at the moment.

4.5. 'Almost unrelated' example: enumerative geometry

The most immediate application of an algorithm computing Segre classes would be enumerative geometry. This is not immediately related to characteristic classes of singular variety (but see the next lecture), but it is quite a bit of fun. So here is a simple example which is actually within reach of the Macaulay2 routine. I am taking this almost verbatim from [Alu03a].

The enumerative question has to do with configurations of point on $\mathbb{P}^1$, and is a baby version of a much harder question in $\mathbb{P}^2$. The harder question is: how many translations of a given curve in $\mathbb{P}^2$ contain 8 general points?

Why 8? because 'translation' means 'in the same PGL(3) orbit', and PGL(3) has dimension 8; so it is natural to expect that the orbit should have dimension 8, and that 8 conditions should determine a finite number.

Why is this interesting? For example, for a general plane quartic the answer is the degree of the rational map from a general $\mathbb{P}^6$ inside the $\mathbb{P}^{14}$ of quartics to

$\mathcal{M}_3$. More generally, the question should be seen as an 'isotrivial' variation on the Gromov-Witten invariants theme.

It is a little harder to defend why the analogous question should be interesting in dimension 1; but it is in fact of some interest in representation theory, and it is a good warm-up case for the harder question. In $\mathbb{P}^1$, the question becomes: how many PGL(2) translates of a given configuration of points (allowing and counting multiplicities) contain 3 points in general position?

The question translates easily into an intersection question in the $\mathbb{P}^3$ of 2×2 matrices: if $(F(s:t))$ is the ideal defining the 'configuration of points', then the condition that a translate contains the point p is written $F(\varphi(p)) = 0$, where φ varies in $\mathbb{P}^3$. This is a certain surface, and we are after 'honest' intersections of three general such surfaces.

The matter of 'honesty' has to do with the fact that if φ is a matrix whose image is a zero of F, then $F(\varphi(p)) = 0$ no matter what p is. That is, the intersection of any number of surfaces of the type we are considering contains a certain 'base scheme' S, of which it is not hard to compute the ideal in any given case. The intersections we want to compute are those that are *not* contained in S.

In fact, we know what the 'total intersection number' of three surfaces should be: if d is the degree of F, then the surfaces have degree d, so by Bézout's theorem three of them meet at d^3 points, counting multiplicities. The question is, what is the contribution of S to this number?

Fulton-MacPherson's intersection theory gives a neat answer: the contribution is

$$\int (1 + dH)^3 \cap s(S, \mathbb{P}^3).$$

Thus if we can compute the Segre class of S in $\mathbb{P}^3$ then we can answer our enumerative question. I tried this for the specific $F = s(s+3t)^2(s+5t)(s+16t)$, letting the characteristic of the ground field vary. For example, this collapses to two points over $\mathbb{F}_2$, it's three points with multiplicities $3, 1, 1$ over $\mathbb{F}_3$, etc. My Macaulay2 routine can compute these Segre classes in (very) small characteristic, and the result matches precisely the general formula I had obtained with Carel Faber a long time ago ([AF93]).

I must add that this is really an extremely small example. We would need computers millions of times larger and faster than the ones we have, in order to solve serious enumerative questions this way.

5. Lecture V

5.1. Multiplicity of discriminants

I would now like to talk about a few items that are somewhat loosely related with characteristic classes of singular varieties. They are all rather concrete questions, so they should serve as 'after-the-fact motivation'.

I will begin with a very concrete computation. Consider the space of plane curves of degree d: a projective space $\mathbb{P}^N$, with $N = \frac{d(d+3)}{2}$. In this projective space there is a *discriminant* hypersurface D, consisting of all *singular* plane curves. If X is a singular curve, then I can think of it as a point of D. What is the multiplicity of the hypersurface D at the point X?

It has been known for a long time that if X has no multiple components then the multiplicity is simply the sum of the Milnor numbers of X. So this has the potential of having to do with Milnor classes, and characteristic classes of singular variety...

It is in fact very easy to give a formula for this multiplicity. The hypersurface D can be realized as the projection from $\mathbb{P}^2 \times \mathbb{P}^N$ of a locus D' of codimension 3; and in fact equations for this locus are nothing but the 3 partial derivatives of the generic homogeneous polynomial of degree d in the homogeneous coordinates $\mathbb{P}^2$. Denoting by H, K respectively the hyperplane classes in $\mathbb{P}^2$ and $\mathbb{P}^N$, each derivative is a polynomial of class $(d-1)H + K$, so the normal bundle of D' has class $(1 + (d-1)H + K)^3$.

Now we want the multiplicity $m_X D$ of X in D. Segre classes compute multiplicities: $s(X, D) = m_X D[X]$. Segre classes are birational invariants, so $s(X, D) = \pi_* s(\pi^{-1} X, D')$, where $\pi : D' \to D$ is the projection. Capping with the classes of the ambient gives the intrinsic Fulton class, hence: $c(TD') \cap s(\pi^{-1}X, D') = c_{\mathrm{F}}(\pi^{-1}X) = c(T\mathbb{P}^2 \times \mathbb{P}^N) \cap s(\pi^{-1}X, \mathbb{P}^2 \times \mathbb{P}^N)$ or in other words

$$s(\pi^{-1}X, D') = (1 + (d-1)H + K)^3 \cap s(\pi^{-1}X, \mathbb{P}^2 \times \mathbb{P}^N).$$

There is more: $\pi^{-1}X$ is manifestly the *singularity subscheme* Y of X, and it sits in a fiber $\mathbb{P}^2$ so its Segre class does not depend on the $\mathbb{P}^N$ factor (another consequence of the fact that Fulton classes are intrinsic), and further K does nothing to it. Thus

$$s(\pi^{-1}X, D') = (1 + (d-1)H)^3 \cap s(Y, \mathbb{P}^2).$$

This should start looking familiar. Of course $(1 + (d-1)H)^3$ is nothing but the Chern class of the trivial extension of $T^*\mathbb{P}^2$ tensored by $\mathcal{L} = \mathcal{O}(d)$:

$$s(\pi^{-1}X, D') = c(\mathcal{L})c(T^*M \otimes \mathcal{L}) \cap s(Y, M)$$

where I have written $M = \mathbb{P}^2$. The multiplicity of the discriminant is the degree of this class.

Claim. This holds in general, for a hypersurface X in a nonsingular variety M, with singularity subscheme Y.

The proof is not hard – just trace the computation in the general case; details can be found in [AC93]. What does this have to do with Milnor classes? The point I want to make is that $s(\pi^{-1}X, D') = s(Y, D')$ 'is' the Milnor class! More precisely:

Claim. $c_{\mathrm{SM}}(X) - c_{\mathrm{F}}(X) = \left(\dfrac{s(Y, D')}{c(\mathcal{L})^{\dim M}} \right)^{\vee} \otimes \mathcal{L}.$

Proof: Exercise. Use the above formula for $s(\pi^{-1}X, D')$ and the properties of $^{\vee}$ and $\otimes$ to reduce to the main formula.

'Undoing this', it is very easy to provide a formula for the multiplicity of a discriminant in terms of c_{SM} and c_{F}. It seems that discriminants naturally embody a lot of information about characteristic classes of what they parametrize.

5.2. Constraints on singular loci of hypersurfaces

One class that has surfaced in the considerations in the previous section:

$$c(T^*M \otimes \mathcal{L}) \cap s(Y, M)$$

turns out to pack a good amount of information. After the fact this is easy to believe, since it *is* the Milnor class up to 8th grade algebra and notation juggling. But when I run into it at first I was not aware of the connection with characteristic classes, and I studied it independently of such considerations, cf. [Alu95], calling it the 'μ-class of Y with respect to $\mathcal{L}$', $\mu_{\mathcal{L}}(Y)$. It is in fact possible to figure out the exact dependence of this class with respect to $\mathcal{L}$, but this is not important here.

It should come to no surprise that the Milnor class can be written in terms of the μ class; the precise formula is

$$c_{\mathrm{SM}}(X) - c_{\mathrm{F}}(X) = c(\mathcal{L})^{\dim X} \cap (\mu_{\mathcal{L}}(Y)^{\vee} \otimes \mathcal{L})$$

(exercise). That also means that almost everything I say here could be directly transcribed in terms of Milnor classes.

The subtlest property of the μ-class is that its notation makes sense: that is, that the ambient variety M does not matter. In other words, if you manage to realize Y as the singularity subscheme of a hypersurface with bundle $\mathcal{L}$ (restricted to Y) in an*other* nonsingular ambient M, and compute the μ-class there, you will get the same class. Note that this isn't so clear, even after the connection with Milnor classes has been established: who knows what has happened to X in the process?

But it is true. One very special case of this fact leads to very nice applications: what if Y (the singularity subscheme of X) is itself *nonsingular*? Then I can pathologically consider it as the singularity subscheme of the 'hypersurface' with equation $0 = 0$ on Y itself, and the main property of μ-classes then tells me

$$c(T^*M \otimes \mathcal{L}) \cap s(Y, M) = \mu_{\mathcal{L}}(Y) = c(T^*Y \otimes \mathcal{L}) \cap [Y]$$

(since $s(Y, Y) = [Y]$).

One particularly pretty consequence of this formula is that it poses constraints on 'what' nonsingular Y can be singularity subschemes of hypersurfaces in a given M.

Example. The twisted cubic (with the reduced structure) cannot be the singularity subscheme of a hypersurface in $M = \mathbb{P}^n$.

Indeed, the twisted cubic is abstractly $\mathbb{P}^1$, embedded as a curve of degree 3 in $\mathbb{P}^n$. If it were the singularity subscheme of a hypersurface of degree d, then we would have:

$$c(T^*\mathbb{P}^n \otimes \mathcal{O}(X)) \cap s(\mathbb{P}^1, \mathbb{P}^n) = \mu_{\mathcal{O}(3d)}(\mathbb{P}^1) = c(T^*\mathbb{P}^1 \otimes \mathcal{O}(3d)) \cap [\mathbb{P}^1]$$

that is, calling h the hyperplane on $\mathbb{P}^1$:

$$\frac{(1+3(d-1)h)^{n+1}}{1+3dh}\frac{(1+h)^2}{(1+3h)^{n+1}}\cap[\mathbb{P}^1]=\frac{(1+(3d-1)h)^2}{1+3dh}\cap[\mathbb{P}^1]$$

and finally (take degree 1 terms):

$$(3d-6)n-4=3d-2$$

which is nonsense (read modulo 3).

Many such examples can be found. The moral is that it is *very unlikely* for a singularity subscheme to be nonsingular. A 'general' example (whatever this may mean) must have embedded components somewhere. The topologist's viewpoint on this should be that the 'general' singular hypersurface must have a 'complicated' Whitney stratification. I don't know if there are explicit results in this direction.

5.3. Enumerative geometry again

I brought up enumerative geometry already a couple of times, in ways that were only tangentially related to characteristic classes. But characteristic classes can actually be used to attack some enumerative problems, at least in comparatively simple instances.

Many enumerative problems can be interpreted as asking what is the degree of a specific locus in a discriminant. For example, the number of nonsingular plane cubics tangent to 9 lines (that is: my Ph.D. thesis) can be computed as the degree of the closure of the locus of plane *sextic* curves with exactly 9 cusps. Even the celebrated numbers of rational plane curves of a given degree containing the right number of general points is manifestly the degree of a geometrically significant locus in a discriminant.

One difficulty is that these problems are completely unrelated: what connection can there be between the degree of one locus and the degree of another locus, even within the same hypersurface in a variety? (The discriminant hypersurface in this case.) Well, there is a way out of this bind: what if we asked for the c_{SM} class instead?

So I would like to pose a variation on the enumerative problem. For every 'geometrically significant' locus X in a discriminant (for example: for every closure of a PGL(3) orbit in the discriminant of plane curves of a given degree), aim to computing $c_{\mathrm{SM}}(X)$, as a class in the ambient projective space.

This is at least as difficult as computing degrees, since the very first term in $c_{\mathrm{SM}}(X)$ will carry the degree information. *On the other hand,* the information of c_{SM} classes *is* structured: because every class corresponding to a known constructible function on the discriminant will give a relation between these c_{SM} classes.

The problem is now that there are not that many choice of 'geometrically significant' constructible functions on a discriminant D. Two possibilities are $1\!\!1_D$ and Eu_D, so this is a natural place to start: what can one say about $c_{\mathrm{SM}}(D)$ and $c_{\mathrm{Ma}}(D)$, and can one use this knowledge to get any enumerative information?

If one can play this game at all, it is because the Nash blow-up of a discriminant is easily accessible. We saw this already in one previous section, where I computed multiplicities: the variety D' I used there was nothing but the Nash blow-up. The tautological bundle can also be computed with ease, and this gives access to the function Eu_D:

Claim. If $X \in D$, then $\mathrm{Eu}_D(X) = \int \mu_{\mathcal{O}(X)}(Y)$.

Phrased otherwise, this says that the local Euler obstruction of a discriminant at a hypersurface X is nothing but Parusiński's Milnor number of X. (This holds provided D is a hypersurface.) Now studying the situation in low codimension says that $\mathrm{Eu}_D = 1\!\!1_D + 1\!\!1_{\overline{C}} + 1\!\!1_{\overline{G}} + \cdots$, where C is the locus of 'cuspidal' hypersurfaces, and G is the locus of 'binodal' hypersurfaces. One could go further down the list of singularities, although this gets combinatorially messy.

But this is enough for some applications. It says that

$$c_{\mathrm{SM}}(D) = c_{\mathrm{Ma}}(D) - [\overline{C}] - [\overline{G}] + \text{higher codimension terms.}$$

Playing the same game with other classes gives more such relations. When enough relations are known, one can start solving the relations, getting information about the loci C, G, etc. This way one can recover classical formulas for cuspidal and binodal loci, and do a little more – see [Alu98] for more examples.

It would be fun to carry this out for more loci, or to actually compute c_{SM} of all orbits within a discriminant, for instance for low degree plane curves. I have done it for degree 3, and it is a very pleasant computation.

5.4. What now? Shadows of blow-up algebras...

What keeps me glued to the main theorem is the high codimension puzzle: even if other approaches are successful in dealing with more general cases than hypersurfaces, I simply cannot believe that the formula that has come up over and over in these lectures does not admit a straightforward generalization.

As I see it, the factors that prevent a generalization of the formula are its dependence on the ambient space and on the line bundle: these are elements that make sense only if X is a hypersurface. A satisfactory formula should not depend on this information. There must be a way to reformulate the main statement, independent of these factors.

The most convincing thought I have had recently in this direction is the following. First, I need a name for the operation involved in the Lagrangian formulation (back from the third lecture). If $\mathbb{P}E \xrightarrow{\epsilon} S$ is a projective bundle of rank e over a scheme S, then there is a structure theorem for its Chow group: every class $C \in A_r\mathbb{P}E$ can be written uniquely as $C = \sum_{j=0}^{e} c_1(\mathcal{O}(1))^j \cap \epsilon^*(C_{r-e+j}))$ where $\mathcal{O}(1)$ is the tautological line bundle and $C_{r-e+j} \in A_{r-e+j}S$ ([Ful84], §3.3). So every class in $\mathbb{P}E$ has a counterpart in S, although dimensions 'get spread out', introducing lots of fuzz. The class C can be reconstructed from $C_{r-e} + \cdots + C_r$ if the dimension of C is known, but not otherwise. I call this counterpart in $\mathcal{A}S$ the 'shadow' of C.

Now I claim that we have encountered this operation already. The Lagrangian setup expressed $c_*(\varphi)$ in terms of a certain cycle $\mathrm{Ch}(\varphi)$ associated with the constructible function φ, in a projective bundle (Ch was defined in a vector bundle but was conical, so its information is equivalent to the corresponding cycle in the projectivization, and now I will put it there). Actually one has to introduce a dualization in the process, changing every other sign (something like $^\vee$ in my favorite notations). So I will call $\check{c}_*(\varphi)$ this 'dualized' class.

Claim. $\check{c}_*(\varphi)$ is the shadow of the characteristic cycle $\mathrm{Ch}(\varphi)$.

This remark has the only beneficial effect of providing me with an element of language that does not immediately seem to depend on the ambient variety. If I manage to realize the characteristic cycle of X inside any bundle, I will know what to do in order to extract the information from it without invoking '$c(TM) \cap \cdots$'.

But by far the most important item is the realization of the characteristic cycle. As we have seen, the formula is equivalent to the statement that the characteristic cycle of X can be obtained by pulling X back via the blow-up of the ambient M along Y, and then taking away one copy of the exceptional divisor. Every word in this sentence requires M to be there.

I have found a way around this, but (to my total amazement) it seems to rely on a strange condition on X. I have called this the $\times$-condition, and because nodal curves are the prototypical example of hypersurfaces that satisfies it. In its gory details, the condition relies on the system of differential equations satisfied by a defining equation for the hypersurface; more geometrically (but probably not equivalently), the $\times$-condition is satisfied if the singularity subscheme is *linearly embedded* in the ambient space. Many examples of this are known (see [Alu02]) Exercise: find a hypersurface that does *not* satisfy the $\times$-condition.

What's good about surfaces satisfying the $\times$-condition?

Claim. Up to signs and duals, $c_{\mathrm{SM}}(X)$ is the shadow of $\mathbb{P}(\mathrm{Sym}_X(Y))$.

The algebra $\mathrm{Sym}_X(Y)$ could not care less about X being a hypersurface. So this goes one step in the right direction. Note: $\mathbb{P}(\mathrm{Rees}_X(Y))$ gives $c_{\mathrm{Ma}}(X)$, a further indication that I may be on the right track.

5.5. ... and maybe a new Chern class

That looks promising, although there are still a couple of problems with it: (1) I do not understand the '$\times$-condition'; (2) the whole operation is still not really independent of the ambient variety.

Comments: (1) if a hypersurface X does not satisfy the $\times$-condition, then I can construct an algebra whose shadow is $c_{\mathrm{SM}}(X)$, but this algebra is not as simple-minded as $\mathrm{Sym}_X Y$ then; it 'interpolates' between $\mathrm{Rees}_X(Y)$ and $\mathrm{Sym}_X(Y)$, with which it agrees when Y is linearly embedded in the ambient variety. Details are in [Alu02]. I am hoping commutative algebraists will find some use for this new 'blow-up algebra'.

(2) There is a subtle point about the operation: the shadow does depend on the bundle in whose projectivization $\mathbb{P}(\mathrm{Sym}_X Y)$ is immersed, and there are more than one choice for this. Knowing which one to choose seems to require knowing the line bundle of X, hence, it still relies on X being a hypersurface.

I can be a little more precise. The ideal of Y in X is generated by the components of a section of $\mathcal{P}^1_M\mathcal{L}|_X$; this realizes $\mathrm{Sym}_Y X$ as a quotient of $\mathrm{Sym}(\mathcal{P}^1_M\mathcal{L}|_X)^\vee$, hence places the proj in $\mathbb{P}(\mathcal{P}^1_M\mathcal{L}|_X)$. We have already seen that it can be juggled back into $\mathbb{P}(T^*M)$, and *this* is where I want to cast the shadow. But to go from one to the other I have to tensor by $\mathcal{L}^\vee$: how to do that if I don't have a $\mathcal{L}$ around?

Here is a way out. If X is *any* subscheme of a variety M, with ideal $\mathcal{I}$, there is a surjection $\Omega^1_M|_X \to \Omega^1_X \to 0$; taking $\mathrm{Hom}(-,\mathcal{O}_X)$, we get a morphism

$$\mathrm{Hom}(\Omega^1_X,\mathcal{O}_X) \to \mathrm{Hom}(\Omega^1_M|_X,\mathcal{O}_X).$$

What could be more natural than giving the following:

Definition. Denote by $\mathcal{N}_X M$ the cokernel of this morphism.

What good does this do for us?

Claim. If X is a hypersurface, then $\mathrm{Proj}(\mathrm{Sym}_X Y) = \mathrm{Proj}(\mathrm{Sym}_X(\mathcal{N}_X M))$.

This very simple observation solves the line bundle puzzle: because $\mathcal{N}_X M$ is defined for *every* X, and it already lives in the right place (no $\mathcal{L}$ involved!). Putting everything together, the main statement becomes

Theorem 5.1. *If X is a hypersurface satisfying the $\times$-condition, then $c_{\mathrm{SM}}(X)$ is (essentially) the shadow of* $\mathrm{Proj}(\mathrm{Sym}_X(\mathcal{N}_X M))$.

The point here is of course that the shadow of $\mathrm{Proj}(\mathrm{Sym}_X(\mathcal{N}_X M))$ is defined *even if X is not a hypersurface!* Might this be the statement I have been looking for? Not quite: for example, because of the qualifier '... satisfying the $\times$-condition'. I feel that any advance on a statement of this sort has to wait until the $\times$-condition is better understood.

But this construction raises a tantalizing point. The 'shadow' in the last statement can be seen as an operation performed on Ω^1_X: dominate it with the locally free $\Omega^1_M|_X$, then take kernel, then $\mathrm{Hom}(-,\mathcal{O}_X)$, etc. This operation can be performed on *every* coherent sheaf $\mathcal{F}$, and the upshot of the above considerations is that if X is a hypersurface satisfying the $\times$-condition, then doing it to Ω^1_X essentially yields $c_{\mathrm{SM}}(X)$.

Exercise: if $\mathcal{F}$ is locally free, this operation gives (essentially) the total 'homology' Chern class of $\mathcal{F}$. So this is a new 'Chern class' for coherent sheaves, on possibly singular schemes, which is closely related to Chern-Schwartz-MacPherson classes in an interesting class of examples.

I will then end with what is now the obvious next question: *what is the exact relation between the new Chern class and the functorial Chern-Schwartz-MacPherson class?*

References

[AC93] Paolo Aluffi and Fernando Cukierman. Multiplicities of discriminants. *Manuscripta Math.*, 78(3):245–258, 1993.

[AF93] Paolo Aluffi and Carel Faber. Linear orbits of d-tuples of points in $\mathbf{P}^1$. *J. Reine Angew. Math.*, 445:205–220, 1993.

[Alu94] Paolo Aluffi. MacPherson's and Fulton's Chern classes of hypersurfaces. *Internat. Math. Res. Notices*, (11):455–465, 1994.

[Alu95] Paolo Aluffi. Singular schemes of hypersurfaces. *Duke Math. J.*, 80(2):325–351, 1995.

[Alu98] Paolo Aluffi. Characteristic classes of discriminants and enumerative geometry. *Comm. Algebra*, 26(10):3165–3193, 1998.

[Alu99a] Paolo Aluffi. Chern classes for singular hypersurfaces. *Trans. Amer. Math. Soc.*, 351(10):3989–4026, 1999.

[Alu99b] Paolo Aluffi. Differential forms with logarithmic poles and Chern-Schwartz-MacPherson classes of singular varieties. *C. R. Acad. Sci. Paris Sér. I Math.*, 329(7):619–624, 1999.

[Alu00] Paolo Aluffi. Weighted Chern-Mather classes and Milnor classes of hypersurfaces. In *Singularities – Sapporo 1998*, pages 1–20. Kinokuniya, Tokyo, 2000.

[Alu02] Paolo Aluffi. Shadows of blow-up algebras. Tohoku Math. J. (2) **56** (2004), no. 4, 593–619.

[Alu03a] Paolo Aluffi. Computing characteristic classes of projective schemes. *J. Symbolic Comput.*, 35(1):3–19, 2003.

[Alu03b] Paolo Aluffi. Inclusion-exclusion and Segre classes. *Comm. Algebra*, 31(8):3619–3630, 2003. Special issue in honor of Steven L. Kleiman.

[Alu03c] Paolo Aluffi. Inclusion-exclusion and Segre classes. II. In *Topics in algebraic and noncommutative geometry (Luminy/Annapolis, MD, 2001)*, volume 324 of *Contemp. Math.*, pages 51–61. Amer. Math. Soc., Providence, RI, 2003.

[BLSS02] J.-P. Brasselet, D. Lehmann, J. Seade, and T. Suwa. Milnor classes of local complete intersections. *Trans. Amer. Math. Soc.*, 354(4):1351–1371 (electronic), 2002.

[Bra00] Jean-Paul Brasselet. From Chern classes to Milnor classes – a history of characteristic classes for singular varieties. In *Singularities – Sapporo 1998*, pages 31–52. Kinokuniya, Tokyo, 2000.

[BS81] J.-P. Brasselet and M.-H. Schwartz. Sur les classes de Chern d'un ensemble analytique complexe. In *The Euler-Poincaré characteristic (French)*, pages 93–147. Soc. Math. France, Paris, 1981.

[Ful84] William Fulton. *Intersection theory.* Springer-Verlag, Berlin, 1984.

[GP02] Mark Goresky and William Pardon. Chern classes of automorphic vector bundles. *Invent. Math.*, 147:561–612, 2002.

[ReS] A. Grothendieck,*Récoltes et Semailles; Réflexions et Témoignages sur un passé de mathématicien*, Reprint, Université des Sciences et Techniques du Languedoc (Montpellier) et CNRS, 1985.

[Ken90] Gary Kennedy. MacPherson's Chern classes of singular algebraic varieties. *Comm. Algebra*, 18(9):2821–2839, 1990.

[Kwi94] Michał Kwieciński. *Sur le transformé de Nash et la construction du graphe de MacPherson.* PhD thesis, Université de Provence (Aix-Marseille I), 1994.

[Mac74] R.D. MacPherson. Chern classes for singular algebraic varieties. *Ann. of Math. (2)*, 100:423–432, 1974.

[Par88] Adam Parusiński. A generalization of the Milnor number. *Math. Ann.*, 281(2):247–254, 1988.

[PP01] Adam Parusiński and Piotr Pragacz. Characteristic classes of hypersurfaces and characteristic cycles. *J. Algebraic Geom.*, 10(1):63–79, 2001.

[Sab85] C. Sabbah. Quelques remarques sur la géométrie des espaces conormaux. *Astérisque*, (130):161–192, 1985. Differential systems and singularities (Luminy, 1983).

[Sch82] Marie-Hélène Schwartz. Classes et caractères de Chern-Mather des espaces linéaires. *C. R. Acad. Sci. Paris Sér. I Math.*, 295(5):399–402, 1982.

Paolo Aluffi
Max-Planck-Institut für Mathematik
Bonn, Germany

and

Florida State University
Tallahassee
Florida, USA

Trends in Mathematics:
Topics in Cohomological Studies of Algebraic Varieties, 33–85

Lectures on the Geometry of Flag Varieties

Michel Brion

Introduction

In these notes, we present some fundamental results concerning flag varieties and their Schubert varieties. By a flag variety, we mean a complex projective algebraic variety X, homogeneous under a complex linear algebraic group. The orbits of a Borel subgroup form a stratification of X into Schubert cells. These are isomorphic to affine spaces; their closures in X are the Schubert varieties, generally singular.

The classes of the Schubert varieties form an additive basis of the cohomology ring $H^*(X)$, and one easily shows that the structure constants of $H^*(X)$ in this basis are all non-negative. Our main goal is to prove a related, but more hidden, statement in the Grothendieck ring $K(X)$ of coherent sheaves on X. This ring admits an additive basis formed of structure sheaves of Schubert varieties, and the corresponding structure constants turn out to have alternating signs.

These structure constants admit combinatorial expressions in the case of Grassmannians: those of $H^*(X)$ (the Littlewood-Richardson coefficients) have been known for many years, whereas those of $K(X)$ were only recently determined by Buch [10]. This displayed their alternation of signs, and Buch conjectured that this property extends to all the flag varieties. In this setting, the structure constants of the cohomology ring (a fortiori, those of the Grothendieck ring) are yet combinatorially elusive, and Buch's conjecture was proved in [6] by purely algebro-geometric methods.

Here we have endeavored to give a self-contained exposition of this proof. The main ingredients are geometric properties of Schubert varieties (e.g., their normality), and vanishing theorems for cohomology of line bundles on these varieties (these are deduced from the Kawamata-Viehweg theorem, a powerful generalization of the Kodaira vanishing theorem in complex geometry). Of importance are also the intersections of Schubert varieties with opposite Schubert varieties. These "Richardson varieties" are systematically used in these notes to provide geometric explanations for many formulae in the cohomology or Grothendieck ring of flag varieties.

The prerequisites are familiarity with algebraic geometry (for example, the contents of the first three chapters of Hartshorne's book [30]) and with some algebraic topology (e.g., the book [26] by Greenberg and Harper). But no knowledge of algebraic groups is required. In fact, we have presented all the notations and results in the case of the general linear group, so that they may be extended readily to arbitrary connected, reductive algebraic groups by readers familiar with their structure theory.

Thereby, we do not allow ourselves to use the rich algebraic and combinatorial tools which make Grassmannians and varieties of complete flags so special among all the flag varieties. For these developments of Schubert calculus and its generalizations, the reader may consult the seminal article [43], the books [21], [23], [49], and the notes of Buch [11] and Tamvakis [68] in this volume. On the other hand, the notes of Duan in this volume [18] provide an introduction to the differential topology of flag varieties, regarded as homogeneous spaces under compact Lie groups, with applications to Schubert calculus.

The present text is organized as follows. The first section discusses Schubert cells and varieties, their classes in the cohomology ring, and the Picard group of flag varieties. In the second section, we obtain restrictions on the singularities of Schubert varieties, and also vanishing theorems for the higher cohomology groups of line bundles on these varieties. The third section is devoted to a degeneration of the diagonal of a flag variety into unions of products of Schubert varieties, with applications to the Grothendieck group. In the fourth section, we obtain several "positivity" results in this group, including a solution of Buch's conjecture. Each section begins with a brief overview of its contents, and ends with bibliographical notes and open problems.

These notes grew out of courses at the Institut Fourier (Grenoble) in the spring of 2003, and at the mini-school "Schubert Varieties" (Banach Center, Warsaw) in May 2003. I am grateful to the organizers of this school, Piotr Pragacz and Andrzej Weber, for their invitation and encouragements. I also thank the auditors of both courses, especially Dima Timashev, for their attention and comments.

Conventions. Throughout these notes, we consider algebraic varieties over the field $\mathbb{C}$ of complex numbers. We follow the notation and terminology of [30]; in particular, varieties are assumed to be irreducible. Unless otherwise stated, subvarieties are assumed to be closed.

1. Grassmannians and flag varieties

We begin this section by reviewing the definitions and fundamental properties of Schubert varieties in Grassmannians and varieties of complete flags. Then we introduce the Schubert classes in the cohomology ring of flag varieties, and we study their multiplicative properties. Finally, we describe the Picard group of flag varieties, first in terms of Schubert divisors, and then in terms of homogeneous line bundles; we also sketch the relation of the latter to representation theory.

1.1. Grassmannians

The *Grassmannian* $\mathrm{Gr}(d,n)$ is the set of d-dimensional linear subspaces of $\mathbb{C}^n$. Given such a subspace E and a basis $(v_1,\ldots,v_d)$ of E, the exterior product $v_1 \wedge \cdots \wedge v_d \in \bigwedge^d \mathbb{C}^n$ only depends on E up to a non-zero scalar multiple. In other words, the point

$$\iota(E) := [v_1 \wedge \cdots \wedge v_d]$$

of the projective space $\mathbb{P}(\bigwedge^d \mathbb{C}^n)$ only depends on E. Further, $\iota(E)$ uniquely determines E, so that the map ι identifies $\mathrm{Gr}(d,n)$ with the image in $\mathbb{P}(\bigwedge^d \mathbb{C}^n)$ of the cone of decomposable d-vectors in $\bigwedge^d \mathbb{C}^n$. It follows that $\mathrm{Gr}(d,n)$ is a subvariety of the projective space $\mathbb{P}(\bigwedge^d \mathbb{C}^n)$; the map

$$\iota : \mathrm{Gr}(d,n) \to \mathbb{P}(\bigwedge^d \mathbb{C}^n)$$

is the *Plücker embedding*.

The general linear group

$$G := GL_n(\mathbb{C})$$

acts on the variety

$$X := \mathrm{Gr}(d,n)$$

via its natural action on $\mathbb{C}^n$. Clearly, X is a unique G-orbit, and the Plücker embedding is equivariant with respect to the action of G on $\mathbb{P}(\bigwedge^d \mathbb{C}^n)$ arising from its linear action on $\bigwedge^d \mathbb{C}^n$. Let $(e_1,\ldots,e_n)$ denote the standard basis of $\mathbb{C}^n$, then the isotropy group of the subspace $\langle e_1,\ldots,e_d\rangle$ is

$$P := \left\{ \begin{pmatrix} a_{1,1} & \cdots & a_{1,d} & a_{1,d+1} & \cdots & a_{1,n} \\ \vdots & \ddots & \vdots & \vdots & \ddots & \vdots \\ a_{d,1} & \cdots & a_{d,d} & a_{d,d+1} & \cdots & a_{d,n} \\ 0 & \cdots & 0 & a_{d+1,d+1} & \cdots & a_{d+1,n} \\ \vdots & \ddots & \vdots & \vdots & \ddots & \vdots \\ 0 & \cdots & 0 & a_{n,d+1} & \cdots & a_{n,n} \end{pmatrix} \right\}$$

(this is a *maximal parabolic subgroup* of G). Thus, X is the homogeneous space G/P. As a consequence, the algebraic variety X is nonsingular, of dimension $\dim(G) - \dim(P) = d(n-d)$.

For any multi-index $I := (i_1,\ldots,i_d)$, where $1 \le i_1 < \cdots < i_d \le n$, we denote by E_I the corresponding *coordinate subspace* of $\mathbb{C}^n$, i.e., $E_I = \langle e_{i_1},\ldots,e_{i_d}\rangle \in X$. In particular, $E_{1,2,\ldots,d}$ is the *standard coordinate subspace* $\langle e_1,\cdots,e_d\rangle$. We may now state the following result, whose proof is straightforward.

1.1.1. Proposition.

(i) *The* E_I *are precisely the* T*-fixed points in* X*, where*

$$T := \left\{ \begin{pmatrix} a_{1,1} & 0 & \dots & 0 \\ 0 & a_{2,2} & \dots & 0 \\ \vdots & \vdots & \ddots & \vdots \\ 0 & 0 & \dots & a_{n,n} \end{pmatrix} \right\} \subseteq GL_n(\mathbb{C})$$

is the subgroup of diagonal matrices (this is a maximal torus *of* G*).*

(ii) X *is the disjoint union of the orbits* BE_I*, where*

$$B := \left\{ \begin{pmatrix} a_{1,1} & a_{1,2} & \dots & a_{1,n} \\ 0 & a_{2,2} & \dots & a_{2,n} \\ \vdots & \vdots & \ddots & \vdots \\ 0 & 0 & \dots & a_{n,n} \end{pmatrix} \right\} \subseteq GL_n(\mathbb{C})$$

is the subgroup of upper triangular matrices (this is a Borel subgroup *of* G*).*

1.1.2. Definition. The *Schubert cells* in the Grassmannian are the orbits $C_I := BE_I$, i.e., the B-orbits in X. The closure in X of the Schubert cell C_I (for the Zariski topology) is called the *Schubert variety* $X_I := \overline{C_I}$.

Note that B is the semi-direct product of T with the normal subgroup

$$U := \left\{ \begin{pmatrix} 1 & a_{1,2} & \dots & a_{1,n} \\ 0 & 1 & \dots & a_{2,n} \\ \vdots & \vdots & \ddots & \vdots \\ 0 & 0 & \dots & 1 \end{pmatrix} \right\}$$

(this is a *maximal unipotent subgroup* of G). Thus, we also have $C_I = UE_I$: the Schubert cells are just the U-orbits in X.

Also, the isotropy group U_{E_I} is the subgroup of U where $a_{ij} = 0$ whenever $i \notin I$ and $j \in I$. Let U^I be the "complementary" subset of U, defined by $a_{ij} = 0$ if $i \in I$ or $j \notin I$. Then one checks that U^I is a subgroup of U, and the map $U^I \to X$, $g \mapsto gE_I$ is a locally closed embedding with image C_I. It follows that C_I is a locally closed subvariety of X, isomorphic to the affine space $\mathbb{C}^{|I|}$, where $|I| := \sum_{j=1}^{d}(i_j - j)$. Thus, its closure X_I is a projective variety of dimension $|I|$.

Next we present a geometric characterization of Schubert cells and varieties (see, e.g., [21] 9.4).

1.1.3. Proposition.

(i) C_I *is the set of* d*-dimensional subspaces* $E \subset \mathbb{C}^n$ *such that*

$$\dim(E \cap \langle e_1, \dots, e_j \rangle) = \#\{k \mid 1 \le k \le d,\ i_k < j\}, \quad \textit{for} \quad j = 1, \dots, n.$$

(ii) X_I *is the set of* d*-dimensional subspaces* $E \subset \mathbb{C}^n$ *such that*

$$\dim(E \cap \langle e_1, \dots, e_j \rangle) \ge \#\{k \mid 1 \le k \le d,\ i_k < j\}, \quad \textit{for} \quad j = 1, \dots, n.$$

Thus, we have

$$X_I = \bigcup_{J \le I} C_J,$$

where $J \le I$ *if and only if* $j_k \le i_k$ *for all* k.

1.1.4. Examples. 1) For $d = 1$, the Grassmannian is just the projective space $\mathbb{P}^{n-1}$, and the Schubert varieties form a flag of linear subspaces $X_1 \subset \cdots \subset X_n$, where $X_j \cong \mathbb{P}^{j-1}$.

2) For $d = 2$ and $n = 4$ one gets the following poset of Schubert varieties:

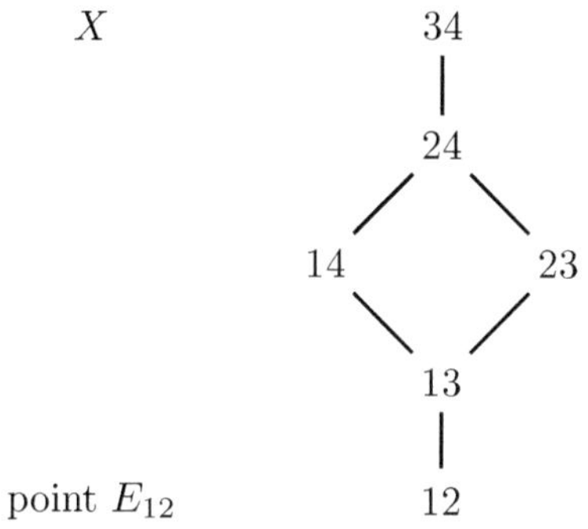

Further, the Schubert variety X_{24} is singular.

Indeed, one checks that $X \subset \mathbb{P}(\bigwedge^2 \mathbb{C}^4) = \mathbb{P}^5$ is defined by one quadratic equation (the Plücker relation). Further, X_{24} is the intersection of X with its tangent space at the point E_{12}. Thus, X_{24} is a quadratic cone with vertex E_{12}, its unique singular point.

3) For arbitrary d and n, the Schubert variety $X_{1,2,\ldots,d}$ is just the point $E_{1,2,\ldots,d}$, whereas $X_{n-d+1,n-d+2,\ldots,n}$ is the whole Grassmannian. On the other hand, $X_{n-d,n-d+2,\ldots,n}$ consists of those d-dimensional subspaces E that meet $\langle e_1, \ldots, e_{n-d}\rangle$: it is the intersection of X with the hyperplane of $\mathbb{P}(\bigwedge^d \mathbb{C}^n)$ where the coordinate on $e_{n-d+1} \wedge \cdots \wedge e_n$ vanishes.

Since X is the disjoint union of the open Schubert cell $C_{n-d+1,n-d+2,\ldots,n} \cong \mathbb{C}^{d(n-d)}$ with the irreducible divisor $D := X_{n-d,n-d+2,\ldots,n}$, any divisor in X is linearly equivalent to a unique integer multiple of D. Equivalently, any line bundle on X is isomorphic to a unique tensor power of the line bundle $L := \mathcal{O}_X(D)$, the pull-back of $\mathcal{O}(1)$ via the Plücker embedding. Thus, *the Picard group* $\mathrm{Pic}(X)$ *is freely generated by the class of the very ample line bundle* L.

We may re-index Schubert varieties in two ways:

1. By partitions: with any multi-index $I = (i_1, \ldots, i_d)$ we associate the partition $\lambda = (\lambda_1,\ldots,\lambda_d)$, where $\lambda_j := i_j - j$ for $j = 1,\ldots,d$. We then write X_λ instead of X_I.

This yields a bijection between the set of multi-indices $I = (i_1, \dots, i_d)$ such that $1 \le i_1 < \cdots < i_d \le n$, and the set of tuples of integers $\lambda = (\lambda_1, \dots, \lambda_d)$ satisfying $0 \le \lambda_1 \le \cdots \le \lambda_d \le n - d$. This is the set of *partitions with* $\le d$ *parts of size* $\le n - d$.

The *area* of the partition λ is the number $|\lambda| := \sum_{j=1}^{d} \lambda_j = |I|$. With this indexing, the dimension of X_λ is the area of λ; further, $X_\mu \subseteq X_\lambda$ if and only if $\mu \le \lambda$, that is, $\mu_j \le \lambda_j$ for all j.

Alternatively, one may associate with any multi-index $I = (i_1, \dots, i_d)$ the *dual partition* $(n - i_d, n - 1 - i_{d-1}, \dots, n - d + 1 - i_1)$. This is still a partition with $\le d$ parts of size $\le n - d$, but now its area is the codimension of the corresponding Schubert variety. This indexing is used in the notes of Buch [11] and Tamvakis [68].

2. By permutations: with a multi-index $I = (i_1, \dots, i_d)$ we associate the permutation w of the set $\{1, 2, \dots, n\}$, defined as follows: $w(k) = i_k$ for $k = 1, \dots, d$, whereas $w(d + k)$ is the kth element of the ordered set $\{1, \dots, n\} \setminus I$ for $k = 1, \dots, n - d$. This sets up a bijection between the multi-indices and the permutations w such that $w(1) < w(2) < \cdots < w(d)$ and $w(d + 1) < \cdots < w(n)$. These permutations form a system of representatives of the coset space $S_n/(S_d \times S_{n-d})$, where S_n denotes the permutation group of the set $\{1, 2, \dots, n\}$, and $S_d \times S_{n-d}$ is its subgroup stabilizing the subset $\{1, 2, \dots, d\}$ (and $\{d+1, d+2, \dots, n\}$). Thus, we may parametrize the T-fixed points of X, and hence the Schubert varieties, by the map $S_n/(S_d \times S_{n-d}) \to X$, $w(S_d \times S_{n-d}) \mapsto E_{w(1),\dots,w(d)}$. This parametrization will be generalized to all flag varieties in the next subsection.

1.2. Flag varieties

Given a sequence $(d_1, \dots, d_m)$ of positive integers with sum n, a *flag of type* $(d_1, \dots, d_m)$ *in* $\mathbb{C}^n$ is an increasing sequence of linear subspaces

$$0 = V_0 \subset V_1 \subset V_2 \subset \cdots \subset V_m = \mathbb{C}^n$$

such that $\dim(V_j/V_{j-1}) = d_j$ for $j = 1, \dots, m$. The *coordinate flags* are those consisting of coordinate subspaces.

Let $X(d_1, \dots, d_m)$ denote the set of flags of type $(d_1, \dots, d_m)$. For example, $X(d, n - d)$ is just the Grassmannian $\mathrm{Gr}(d, n)$. More generally, $X(d_1, \dots, d_m)$ is a subvariety of the product of the Grassmannians $\mathrm{Gr}(d_i, n)$, called the *partial flag variety* of type $(d_1, \dots, d_m)$.

The group $G = \mathrm{GL}_n(\mathbb{C})$ acts transitively on $X(d_1, \dots, d_m)$. Let $P = P(d_1, \dots, d_m)$ be the isotropy group of the *standard flag* (consisting of the standard coordinate subspaces). Then $P(d_1, \dots, d_m)$ consists of the block upper triangular invertible matrices with diagonal blocks of sizes $d_1, \dots, d_m$. In particular, $P(d_1, \dots, d_m)$ contains B; in fact, all subgroups of G containing B occur in this way. (These subgroups are the *standard parabolic subgroups of* G). Since $X \cong G/P$, it follows that X is nonsingular of dimension $\sum_{1 \le i < j \le m} d_i d_j$.

In particular, we have the variety $X := X(1, \dots, 1)$ of *complete flags*, also called the *full flag variety*; it is the homogeneous space G/B, of dimension

$n(n-1)/2$. By sending any complete flag to the corresponding partial flag of a given type $(d_1,\ldots,d_m)$, we obtain a morphism

$$f : X = G/B \to G/P(d_1,\ldots,d_m) = X(d_1,\ldots,d_m).$$

Clearly, f is G-equivariant with fiber P/B at the base point B/B (the standard complete flag). Thus, f is a fibration with fiber being the product of varieties of complete flags in $\mathbb{C}^{d_1}$, ..., $\mathbb{C}^{d_m}$. This allows us to reduce many questions regarding flag varieties to the case of the variety of complete flags; see Example 1.2.3 below for details on this reduction. Therefore, we will mostly concentrate on the full flag variety.

We now introduce Schubert cells and varieties in G/B. Observe that the complete coordinate flags correspond to the permutations of the set $\{1,\ldots,n\}$, by assigning to the flag

$$0 \subset \langle e_{i_1}\rangle \subset \cdots \subset \langle e_{i_1}, e_{i_2}, \ldots, e_{i_k}\rangle \subset \cdots$$

the permutation w such that $w(k) = i_k$ for all k. We regard the permutation group S_n as a subgroup of $\mathrm{GL}_n(\mathbb{C})$ via its natural action on the standard basis $(e_1,\ldots,e_n)$. Then the (complete) coordinate flags are exactly the $F_w := wF$, where F denotes the standard complete flag. Further, S_n identifies to the quotient $W := N_G(T)/T$, where $N_G(T)$ denotes the normalizer of T in G. (In other words, S_n is the *Weyl group of G with respect to T*).

We may now formulate an analogue of Proposition 1.1.1 (see, e.g., [21] 10.2 for a proof).

1.2.1. Proposition.

(i) *The fixed points of T in X are the coordinate flags F_w, $w \in W$.*

(ii) *X is the disjoint union of the orbits $C_w := BF_w = UF_w$, where $w \in W$.*

(iii) *Let $X_w := \overline{C_w}$ (closure in the Zariski topology of X), then*

$$X_w = \bigcup_{v\in W,\ v\leq w} C_v,$$

where $v \leq w$ if and only if we have

$$(v(1),\ldots,v(d))_{\text{r.t.i.v.}} \leq (w(1),\ldots,w(d))_{\text{r.t.i.v.}} \quad \textit{for } d = 1,\ldots,n-1$$

(here r.t.i.v. stands for "reordered to increasing values").

1.2.2. Definition. $C_w := BF_w$ is a *Schubert cell*, and $X_w := \overline{C_w}$ is the corresponding *Schubert variety.* The partial ordering $\leq$ on W is the *Bruhat order.*

By the preceding proposition, we have $X_v \subseteq X_w$ if and only if this holds for the images of X_v and X_w in $\mathrm{Gr}(d,n)$, where $d = 1,\ldots,n-1$. Together with Proposition 1.1.3, this yields a geometric characterization of the Bruhat order on Schubert varieties. Also, note that the T-fixed points in X_w are the coordinate flags F_v, where $v \in W$ and $v \leq w$.

We now describe the Schubert cells UF_w. Note that the isotropy group

$$U_{F_w} = U \cap wUw^{-1} =: U_w$$

is defined by $a_{i,j} = 0$ whenever $i < j$ and $w^{-1}(i) < w^{-1}(j)$. Let U^w be the "complementary" subset of U, defined by $a_{ij} = 0$ whenever $i < j$ and $w^{-1}(i) > w^{-1}(j)$. Then $U^w = U \cap wU^- w^{-1}$ is a subgroup, and one checks that the product map $U^w \times U_w \to U$ is an isomorphism of varieties. Hence the map $U^w \to C_w$, $g \mapsto gF_w$ is an isomorphism as well. It follows that each C_w is an affine space of dimension

$$\begin{aligned}\#\{(i,j) \mid 1 \le i < j \le n,\ w^{-1}(i) > w^{-1}(j)\} \\ = \#\{(i,j) \mid 1 \le i < j \le n,\ w(i) > w(j)\}.\end{aligned}$$

The latter set consists of the *inversions* of the permutation w; its cardinality is the *length* of w, denoted by $\ell(w)$. Thus, $C_w \cong \mathbb{C}^{\ell(w)}$.

More generally, we may define Schubert cells and varieties in any partial flag variety $X(d_1,\ldots,d_m) = G/P$, where $P = P(d_1,\ldots,P_m)$; these are parametrized by the coset space $S_n/(S_{d_1} \times \cdots \times S_{d_m}) =: W/W_P$.

Specifically, each right coset mod W_P contains a unique permutation w such that we have $w(1) < \cdots < w(d_1)$, $w(d_1+1) < \cdots < w(d_1+d_2)$, ..., $w(d_1+\cdots+d_{m-1}+1) < \cdots < w(d_1+\cdots+d_m) = w(n)$. Equivalently, $w \le wv$ for all $v \in W_P$. This defines the set W^P of *minimal representatives* of W/W_P.

Now the Schubert cells in G/P are the orbits $C_{wP} := BwP/P = UwP/P \subset G/P$ $(w \in W^P)$, and the Schubert varieties X_{wP} are their closures. One checks that the map $f : G/B \to G/P$ restricts to an isomorphism $C_w = BwB/B \cong BwP/P = C_{wP}$, and hence to a birational morphism $X_w \to X_{wP}$ for any $w \in W^P$.

1.2.3. Examples. 1) The Bruhat order on S_2 is just

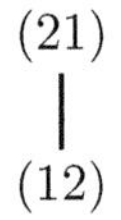

The picture of the Bruhat order on S_3 is

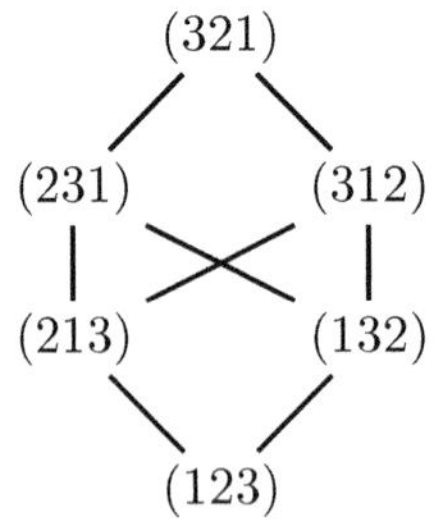

2) Let $w_o := (n, n-1, \ldots, 1)$, the order-reversing permutation. Then $X = X_{w_o}$, i.e., w_o is the unique maximal element of the Bruhat order on W. Note that $w_o^2 = \mathrm{id}$, and $\ell(w_o w) = \ell(w_o) - \ell(w)$ for any $w \in W$.

3) The permutations of length 1 are exactly the *elementary transpositions* $s_1, \ldots, s_{n-1}$, where each s_i exchanges the indices i and $i+1$ and fixes all other indices. The corresponding Schubert varieties are the *Schubert curves* $X_{s_1}, \ldots, X_{s_{n-1}}$. In fact, X_{s_i} may be identified with the set of i-dimensional subspaces $E \subset \mathbb{C}^n$ such that

$$\langle e_1, \ldots, e_{i-1} \rangle \subset E \subset \langle e_1, \ldots, e_{i+1} \rangle.$$

Thus, X_{s_i} is the projectivization of the quotient space $\langle e_1, \ldots, e_{i+1} \rangle / \langle e_1, \ldots, e_{i-1} \rangle$, so that $X_{s_i} \cong \mathbb{P}^1$.

4) Likewise, the Schubert varieties of codimension 1 are $X_{w_o s_1}, \ldots, X_{w_o s_{n-1}}$, also called the *Schubert divisors.*

5) Apart from the Grassmannians, the simplest partial flag variety is the *incidence variety* $I = I_n$ consisting of the pairs (V_1, V_{n-1}), where $V_1 \subset \mathbb{C}^n$ is a line, and $V_{n-1} \subset \mathbb{C}^n$ is a hyperplane containing V_1. Denote by $\mathbb{P}^{n-1} = \mathbb{P}(\mathbb{C}^n)$ (resp. $\check{\mathbb{P}}^{n-1} = \mathbb{P}((\mathbb{C}^n)^*)$) the projective space of lines (resp. hyperplanes) in $\mathbb{C}^n$, then $I \subset \mathbb{P}^{n-1} \times \check{\mathbb{P}}^{n-1}$ is defined by the bi-homogeneous equation

$$x_1 y_1 + \cdots + x_n y_n = 0,$$

where $x_1, \ldots, x_n$ are the standard coordinates on $\mathbb{C}^n$, and $y_1, \ldots, y_n$ are the dual coordinates on $(\mathbb{C}^n)^*$.

One checks that the Schubert varieties in I are the

$$I_{i,j} := \{(V_1, V_{n-1}) \in I \quad | \quad V_1 \subseteq E_{1,\ldots,i} \quad \text{and} \quad E_{1,\ldots,j-1} \subseteq V_{n-1}\},$$

where $1 \le i, j \le n$ and $i \ne j$. Thus, $I_{i,j} \subseteq I$ is defined by the equations

$$x_{i+1} = \cdots = x_n = y_1 = \cdots = y_{j-1} = 0.$$

It follows that $I_{i,j}$ *is singular for* $1 < j < i < n$ *with singular locus* $I_{j-1,i+1}$, *and is nonsingular otherwise.*

6) For any partial flag variety G/P and any $w \in W^P$, the pull-back of the Schubert variety X_{wP} under $f : G/B \to G/P$ is easily seen to be the Schubert variety $X_{ww_{0,P}}$, where $w_{0,P}$ denotes the maximal element of W_P. Specifically, if $P = P(d_1, \ldots, d_m)$ so that $W_P = S_{d_1} \times \cdots \times S_{d_m}$, then $w_{0,P} = (w_{0,d_1}, \ldots, w_{0,d_m})$ with obvious notation. The products $ww_{0,P}$, where $w \in W^P$, are the *maximal representatives* of the cosets modulo W_P. Thus, f restricts to a locally trivial fibration $X_{ww_{0,P}} \to X_{wP}$ with fiber P/B.

In particular, the preceding example yields many singular Schubert varieties in the variety of complete flags, by pull-back from the incidence variety.

1.2.4. Definition. The *opposite Schubert cell (resp. variety)* associated with $w \in W$ is $C^w := w_o C_{w_o w}$ (resp. $X^w := w_o X_{w_o w}$).

Observe that $C^w = B^- F_w$, where

$$B^- := \left\{ \begin{pmatrix} a_{1,1} & 0 & \dots & 0 \\ a_{2,1} & a_{2,2} & \dots & 0 \\ \vdots & \vdots & \ddots & \vdots \\ a_{n,1} & a_{n,2} & \dots & a_{n,n} \end{pmatrix} \right\} = w_o B w_o$$

(this is the *opposite Borel subgroup* to B containing the maximal torus T). Also, X^w has codimension $\ell(w)$ in X.

For example, $C^{\mathrm{id}} \cong U^-$ via the map $U^- \to X, g \mapsto gF$, where $U^- := w_o U w_o$. Further, this map is an open immersion. Since $X = G/B$, this is equivalent to the fact that the product map $U^- \times B \to G$ is an open immersion (which, of course, may be checked directly). It follows that the quotient $q : G \to G/B$, $g \mapsto gB$, is a trivial fibration over C^{id}; thus, by G-equivariance, q is locally trivial for the Zariski topology. This also holds for any partial flag variety G/P with the same proof. Likewise, the map $f : G/B \to G/P$ is a locally trivial fibration with fiber P/B.

1.3. Schubert classes

This subsection is devoted to the cohomology ring of the full flag variety. We begin by recalling some basic facts on the homology and cohomology of algebraic varieties, referring for details to [21] Appendix B or [23] Appendix A. We will consider (co)homology groups with integer coefficients.

Let X be a projective nonsingular algebraic variety of dimension n. Then X (viewed as a compact differentiable manifold of dimension $2n$) admits a canonical orientation, hence a canonical generator of the homology group $H_{2n}(X)$: the *fundamental class* $[X]$. By Poincaré duality, the map $H^j(X) \to H_{2n-j}(X)$, $\alpha \mapsto \alpha \cap [X]$ is an isomorphism for all j.

Likewise, any nonsingular subvariety $Y \subseteq X$ of dimension p has a fundamental class in $H_{2p}(Y)$. Using Poincaré duality, the image of this class in $H_{2p}(X)$ yields the *fundamental class* $[Y] \in H^{2c}(X)$, where $c = n - p$ is the codimension of Y. In particular, we obtain the fundamental class of a point $[x]$, which is independent of x and generates the group $H^{2n}(X)$. More generally, one defines the fundamental class $[Y] \in H^{2c}(X)$ for any (possibly singular) subvariety Y of codimension c.

Given α, β in the cohomology ring $H^*(X)$, let $\langle \alpha, \beta \rangle$ denote the coefficient of the class $[x]$ in the cup product $\alpha \cup \beta$. Then $\langle , \rangle$ is a bilinear form on $H^*(X)$ called the *Poincaré duality pairing.* It is non-degenerate over the rationals, resp. over the integers in the case where the group $H^*(X)$ is torsion-free.

For any two subvarieties Y, Z of X, each irreducible component C of $Y \cap Z$ satisfies $\dim(C) \geq \dim(Y) + \dim(Z)$, i.e., $\operatorname{codim}(C) \leq \operatorname{codim}(Y) + \operatorname{codim}(Z)$. We say that *$Y$ and Z meet properly in X*, if $\operatorname{codim}(C) = \operatorname{codim}(Y) + \operatorname{codim}(Z)$ for each C. Then we have in $H^*(X)$:

$$[Y] \cup [Z] = \sum_C m_C \, [C],$$

where the sum is over all irreducible components of $Y \cap Z$, and m_C is the *intersection multiplicity of* Y *and* Z *along* C, a positive integer. Further, $m_C = 1$ if and only if Y *and* Z *meet transversally along* C, i.e., there exists a point $x \in C$ such that: x is a nonsingular point of Y and Z, and the tangent spaces at x satisfy $T_xY + T_xZ = T_xX$. Then x is a nonsingular point of C, and $T_xC = T_xY \cap T_xZ$.

In particular, if Y and Z are subvarieties such that $\dim(Y) + \dim(Z) = \dim(X)$, then Y meets Z properly if and only if their intersection is finite. In this case, we have $\langle [Y], [Z] \rangle = \sum_{x \in Y \cap Z} m_x$, where m_x denotes the intersection multiplicity of Y and Z at x. In the case of transversal intersection, this simplifies to $\langle [Y], [Z] \rangle = \#(Y \cap Z)$.

Returning to the case where X is a flag variety, we have the cohomology classes of the Schubert subvarieties, called the *Schubert classes.* Since X is the disjoint union of the Schubert cells, *the Schubert classes form an additive basis of* $H^*(X)$; in particular, this group is torsion-free.

To study the cup product of Schubert classes, we will need a version of Kleiman's transversality theorem, see [35] or [30] Theorem III.10.8.

1.3.1. Lemma. *Let* Y, Z *be subvarieties of a flag variety* X *and let* $Y_0 \subseteq Y$ *(resp.* $Z_0 \subseteq Z$*) be nonempty open subsets consisting of nonsingular points. Then there exists a nonempty open subset* Ω *of* G *such that: for any* $g \in \Omega$, Y *meets* gZ *properly, and* $Y_0 \cap gZ_0$ *is nonsingular and dense in* $Y \cap gZ$. *Thus,* $[Y] \cup [Z] = [Y \cap gZ]$ *for all* $g \in \Omega$.

In particular, if $\dim(Y) + \dim(Z) = \dim(X)$, *then* Y *meets* gZ *transversally* for general $g \in G$, *that is, for all* g *in a nonempty open subset* Ω *of* G. *Thus,* $Y \cap gZ$ *is finite and* $\langle [Y], [Z] \rangle = \#(Y \cap gZ)$, *for general* $g \in G$.

Proof. Consider the map $m : G \times Z \to X$, $(g, z) \mapsto gz$. This is a surjective morphism, equivariant for the action of G on $G \times Z$ by left multiplication on the first factor. Since $X = G/P$, it follows that m is a locally trivial fibration for the Zariski topology. Thus, its scheme-theoretic fibers are varieties of dimension $\dim(G) + \dim(Z) - \dim(X)$.

Next consider the fibered product $V := (G \times Z) \times_X Y$ and the pull-back $\mu : V \to Y$ of m. Then μ is also a locally trivial fibration with fibers being varieties. It follows that the scheme V is a variety of dimension $\dim(G) + \dim(Z) - \dim(X) + \dim(Y)$.

Let $\pi : V \to G$ be the composition of the projections $(G \times Z) \times_X Y \to G \times Z \to G$. Then the fiber of π at any $g \in G$ may be identified with the scheme-theoretic intersection $Y \cap gZ$. Further, there exists a nonempty open subset Ω of G such that the fibers of π at points of Ω are either empty or equidimensional of dimension $\dim(Y) + \dim(Z) - \dim(X)$, i.e., of codimension $\mathrm{codim}(Y) + \mathrm{codim}(Z)$. This shows that Y meets gZ properly for any $g \in \Omega$.

Likewise, the restriction $m_0 : G \times Z_0 \to X$ is a locally trivial fibration with nonsingular fibers, so that the fibered product $V_0 := (G \times Z_0) \times_X Y_0$ is a nonempty open subset of V consisting of nonsingular points. By generic smoothness, it follows that $Y_0 \cap gZ_0$ is nonsingular and dense in $Y \cap gZ$, for all g in a (possibly smaller) nonempty open subset of G. This implies, in turn, that all intersection multiplicities of $Y \cap Z$ are 1.

Thus, we have $[Y] \cup [gZ] = [Y \cap gZ]$ for any $g \in \Omega$. Further, $[Z] = [gZ]$ as G is connected, so that $[Y] \cup [Z] = [Y \cap gZ]$. □

As a consequence, in the full flag variety X, any Schubert variety X_w meets properly any opposite Schubert variety X^v. (Indeed, the open subset Ω meets the open subset $BB^- = BU^- \cong B \times U^-$ of G; further, X_w is B-invariant, and X^v is B^--invariant). Thus, $X_w \cap X^v$ is equidimensional of dimension $\dim(X_w) + \dim(X^v) - \dim(X) = \ell(w) - \ell(v)$. Moreover, the intersection $C_w \cap C^v$ is nonsingular and dense in $X_w \cap X^v$. In fact, we have the following more precise result which may be proved by the argument of Lemma 1.3.1; see [9] for details.

1.3.2. Proposition. *For any $v, w \in W$, the intersection $X_w \cap X^v$ is non-empty if and only if $v \leq w$; then $X_w \cap X^v$ is a variety.*

1.3.3. Definition. Given v, w in W such that $v \leq w$, the corresponding *Richardson variety* is $X_w^v := X_w \cap X^v$.

Note that X_w^v is T-invariant with fixed points being the coordinate flags $F_x = xB/B$, where $x \in W$ satisfies $v \leq x \leq w$. It follows that $X_w^v \subseteq X_{w'}^{v'}$ if and only if $v' \leq v \leq w \leq w'$. Thus, the Richardson varieties may be viewed as geometric analogues of intervals for the Bruhat order.

1.3.4. Examples. 1) As special cases of Richardson varieties, we have the Schubert varieties $X_w = X_w^{\mathrm{id}}$ and the opposite Schubert varieties $X^v = X_{w_o}^v$. Also, note that the Richardson variety X_w^w is just the T-fixed point F_w, the transversal intersection of X_w and X^w.

2) Let X_w^v be a Richardson variety of dimension 1, that is, $v \leq w$ and $\ell(v) = \ell(w) - 1$. Then X_w^v is isomorphic to the projective line, and $v = ws$ for some transposition $s = s_{ij}$ (exchanging i and j, and fixing all the other indices). More generally, any T-invariant curve $Y \subset X$ is isomorphic to $\mathbb{P}^1$ and contains exactly two T-fixed points v, w, where $v = ws$ for some transposition s. (Indeed, after multiplication by an element of W, we may assume that Y contains the standard flag F. Then $Y \cap C^{\mathrm{id}}$ is a T-invariant neighborhood of F in Y, and is also a T-invariant curve in $C^{\mathrm{id}} \cong U^-$ (where T acts by conjugation). Now any such curve is a "coordinate line" given by $a_{i,j} = 0$ for all $(i,j) \neq (i_0, j_0)$, for some (i_0, j_0) such that $1 \leq j_0 < i_0 \leq n$. The closure of this line in X has fixed points F and $s_{i_0,j_0}F$.)

Richardson varieties may be used to describe the local structure of Schubert varieties along Schubert subvarieties, as follows.

1.3.5. Proposition. *Let v, $w \in W$ such that $v \leq w$. Then $X_w \cap vC^{\mathrm{id}}$ is an open T-invariant neighborhood of the point F_v in X_w, which meets X_w^v along $X_w \cap C^v$. Further, the map*

$$(U \cap vU^- v^{-1}) \times (X_w \cap C^v) \to X_w, \quad (g, x) \mapsto gx$$

is an open immersion with image $X_w \cap vC^{\mathrm{id}}$. *(Recall that* $U \cap vU^-v^{-1}$ *is isomorphic to* $\mathbb{C}^{\ell(v)}$ *as a variety, and that the map* $U \cap vU^-v^{-1} \to X$, $g \mapsto gF_v$ *is an isomorphism onto* C_v.*)*

If, in addition, $\ell(v) = \ell(w) - 1$, *then* $X_w \cap C^v$ *is isomorphic to the affine line. As a consequence,* X_w *is nonsingular along its* Schubert divisor X_v.

Proof. Note that vC^{id} is an open T-invariant neighborhood of F_v in X, isomorphic to the variety vU^-v^{-1}. In turn, the latter is isomorphic to $(U \cap vU^-v^{-1}) \times (U^- \cap vU^-v^{-1})$ via the product map; and the map $U^- \cap vU^-v^{-1} \to X$, $g \mapsto gF_v$ is a locally closed immersion with image C^v. It follows that the map

$$(U \cap vU^-v^{-1}) \times C^v \to X, \quad (g, x) \mapsto gx$$

is an open immersion with image vF^{id}, and that $vF^{\mathrm{id}} \cap X^v = C^v$. Intersecting with the subvariety X_w (invariant under the subgroup $U \cap vU^-v^{-1}$) completes the proof of the first assertion. The second assertion follows from the preceding example. □

Richardson varieties also appear when multiplying Schubert classes. Indeed, by Proposition 1.3.2, we have in $H^*(X)$:

$$[X_w] \cup [X^v] = [X_w^v].$$

Since $\dim(X_w^v) = \ell(w) - \ell(v)$, it follows that the Poincaré duality pairing $\langle [X_w], [X^v] \rangle$ equals 1 if $w = v$, and 0 otherwise. This implies easily the following result.

1.3.6. Proposition.

(i) *The bases* $\{[X_w]\}$ *and* $\{[X^w]\} = \{[X_{w_o w}]\}$ *of* $H^*(X)$ *are dual for the Poincaré duality pairing.*

(ii) *For any subvariety* $Y \subseteq X$, *we have*

$$[Y] = \sum_{w \in W} a^w(Y)\,[X_w],$$

where $a^w(Y) = \langle [Y], [X^w] \rangle = \#(Y \cap gX^w)$ *for general* $g \in G$. *In particular, the coefficients of* $[Y]$ *in the basis of Schubert classes are non-negative.*

(iii) *Let*

$$[X_v] \cup [X_w] = \sum_{x \in W} a_{vw}^x\,[X_x] \quad \text{in} \quad H^*(X),$$

then the structure constants a_{vw}^x *are non-negative integers.*

Note finally that all these results adapt readily to any partial flag variety G/P. In fact, the map $f : G/B \to G/P$ induces a ring homomorphism $f^* : H^*(G/P) \to H^*(G/B)$ which sends any Schubert class $[X_{wP}]$ to the Schubert class $[X_{ww_{o,P}}]$, where $w \in W^P$. In particular, f^* is injective.

1.4. The Picard group

In this subsection, we study the Picard group of the full flag variety $X = G/B$. We first give a very simple presentation of this group, viewed as the group of divisors modulo linear equivalence. The Picard group and divisor class group of Schubert varieties will be described in Subsection 2.2.

1.4.1. Proposition. *The group* $\operatorname{Pic}(X)$ *is freely generated by the classes of the Schubert divisors* $X_{w_o s_i}$ *where* $i = 1, \dots, n-1$. *Any ample (resp. generated by its global sections) divisor on* X *is linearly equivalent to a positive (resp. non-negative) combination of these divisors. Further, any ample divisor is very ample.*

Proof. The open Schubert cell C_{w_o} has complement the union of the Schubert divisors. Since C_{w_o} is isomorphic to an affine space, its Picard group is trivial. Thus, the classes of $X_{w_o s_1}, \dots, X_{w_o s_{n-1}}$ generate the group $\operatorname{Pic}(X)$.

If we have a relation $\sum_{i=1}^{n-1} a_i X_{w_o s_i} = 0$ in $\operatorname{Pic}(X)$, then there exists a rational function f on X having a zero or pole of order a_i along each $X_{w_o s_i}$, and no other zero or pole. In particular, f is a regular, nowhere vanishing function on the affine space C_{w_o}. Hence f is constant, and $a_i = 0$ for all i.

Each Schubert divisor $X_{w_o s_d}$ is the pull-back under the projection $X \to \operatorname{Gr}(d,n)$ of the unique Schubert divisor in $\operatorname{Gr}(d,n)$. Since the latter divisor is a hyperplane section in the Plücker embedding, it follows that $X_{w_o s_d}$ is generated by its global sections. As a consequence, any non-negative combination of Schubert divisors is generated by its global sections. Further, the divisor $\sum_{d=1}^{n-1} X_{w_o s_d}$ is very ample, as the product map $X \to \prod_{d=1}^{n-1} \operatorname{Gr}(d,n)$ is a closed immersion. Thus, any positive combination of Schubert divisors is very ample.

Conversely, let $D = \sum_{i=1}^{n-1} a_i X_{w_o s_i}$ be a globally generated (resp. ample) divisor on X. Then for any curve Y on X, the intersection number $\langle [D], [Y] \rangle$ is non-negative (resp. positive). Now take for Y a Schubert curve X_{s_j}, then

$$\langle [D], [Y] \rangle = \langle \sum_{i=1}^{n-1} a_i [X_{w_o s_i}], [X_{s_j}] \rangle = \sum_{i=1}^{n-1} a_i \langle [X^{s_i}], [X_{s_j}] \rangle = a_j.$$

This completes the proof. □

1.4.2. Remark. We may assign to each divisor D on X, its cohomology class $[D] \in H^2(X)$. Since linearly equivalent divisors are homologically equivalent, this defines the cycle map $\operatorname{Pic}(X) \to H^2(X)$, which is an isomorphism by Proposition 1.4.1.

More generally, assigning to each subvariety of X its cohomology class yields the *cycle map* $A^*(X) \to H^{2*}(X)$, where $A^*(X)$ denotes the *Chow ring* of rational equivalence classes of algebraic cycles on X (graded by the codimension; in particular, $A^1(X) = \operatorname{Pic}(X)$). Since X has a "cellular decomposition" by Schubert cells, *the cycle map is a ring isomorphism* by [22] Example 19.1.11.

We will see in Section 4 that *the ring* $H^*(X)$ *is generated by* $H^2(X) \cong \operatorname{Pic}(X)$, *over the rationals.* (In fact, this holds over the integers for the variety of complete flags, as follows easily from its structure of iterated projective space bundle.)

Next we obtain an alternative description of $\mathrm{Pic}(X)$ in terms of homogeneous line bundles on X; these can be defined as follows. Let λ be a *character* of B, i.e., a homomorphism of algebraic groups $B \to \mathbb{C}^*$. Let B act on the product $G \times \mathbb{C}$ by $b(g,t) := (gb^{-1}, \lambda(b)t)$. This action is free, and the quotient $L_\lambda = G \times^B \mathbb{C} := (G \times \mathbb{C})/B$ maps to G/B via $(g,t)B \mapsto gB$. This makes L_λ the total space of a line bundle over G/B, the *homogeneous line bundle associated to the weight* λ.

Note that G acts on L_λ via $g(h,t)B := (gh,t)B$, and that the projection $f : L_\lambda \to G/B$ is G-equivariant; further, any $g \in G$ induces a linear map from the fiber $f^{-1}(x)$ to $f^{-1}(gx)$. In other words, L_λ is a *G-linearized line bundle* on X.

We now describe the characters of B. Note that any such character λ is uniquely determined by its restriction to T (since $B = TU$, and U is isomorphic to an affine space, so that any regular invertible function on U is constant). Further, one easily sees that the characters of the group T of diagonal invertible matrices are precisely the maps

$$\mathrm{diag}(t_1, \dots, t_n) \mapsto t_1^{\lambda_1} \cdots t_n^{\lambda_n},$$

where $\lambda_1, \dots, \lambda_n$ are integers. This identifies the multiplicative group of characters of B (also called *weights*) to the additive group $\mathbb{Z}^n$.

Next we express the *Chern classes* $c_1(L_\lambda) \in H^2(X) \cong \mathrm{Pic}(X)$ in the basis of Schubert divisors. More generally, we obtain the *Chevalley formula* which decomposes the products $c_1(L_\lambda) \cup [X_w]$ in this basis.

1.4.3. Proposition. *For any weight λ and any $w \in W$, we have*

$$c_1(L_\lambda) \cup [X_w] = \sum (\lambda_i - \lambda_j)\, [X_{ws_{ij}}],$$

where the sum is over the pairs (i,j) such that $1 \le i < j \le n$, $ws_{ij} < w$, and $\ell(ws_{ij}) = \ell(w) - 1$ (that is, $X_{ws_{ij}}$ is a Schubert divisor in X_w). In particular,

$$c_1(L_\lambda) = \sum_{i=1}^{n-1} (\lambda_i - \lambda_{i+1})\, [X_{w_o s_i}] = \sum_{i=1}^{n-1} (\lambda_i - \lambda_{i+1})\, [X^{s_i}].$$

Thus, the map $\mathbb{Z}^n \to \mathrm{Pic}(X)$, $\lambda \mapsto c_1(L_\lambda)$ is a surjective group homomorphism, and its kernel is generated by $(1, \dots, 1)$.

Proof. We may write

$$c_1(L_\lambda) \cup [X_w] = \sum_{v \in W} a_v\, [X_v],$$

where the coefficients a_v are given by

$$a_v = \langle c_1(L_\lambda) \cup [X_w], [X^v] \rangle = \langle c_1(L_\lambda), [X_w] \cup [X^v] \rangle = \langle c_1(L_\lambda), [X_w^v] \rangle.$$

Thus, a_v is the degree of the restriction of L_λ to X_w^v if $\dim(X_w^v) = 1$, and is 0 otherwise. Now $\dim(X_w^v) = 1$ if and only if: $v < w$ and $\ell(v) = \ell(w) - 1$. Then $v = ws_{ij}$ for some transposition s_{ij}, and X_w^v is isomorphic to $\mathbb{P}^1$, by Example 1.3.4.2. Further, one checks that the restriction of L_λ to X_w^v is isomorphic to the line bundle $\mathcal{O}_{\mathbb{P}^1}(\lambda_i - \lambda_j)$ of degree $\lambda_i - \lambda_j$. □

This relation between weights and line bundles motivates the following

1.4.4. Definition. We say that the weight $\lambda = (\lambda_1, \dots, \lambda_n)$ is *dominant* (resp. *regular dominant*), if $\lambda_1 \geq \cdots \geq \lambda_n$ (resp. $\lambda_1 > \cdots > \lambda_n$).

The *fundamental weights* are the weights $\chi_1, \dots, \chi_{n-1}$ such that

$$\chi_j := (1, \dots, 1 \ (j \text{ times}), 0, \dots, 0 \ (n-j \text{ times})).$$

The *determinant* is the weight $\chi_n := (1, \dots, 1)$. We put

$$\rho := \chi_1 + \cdots + \chi_{n-1} = (n-1, n-2, \dots, 1, 0).$$

By Propositions 1.4.1 and 1.4.3, the line bundle L_λ is globally generated (resp. ample) if and only if the weight λ is dominant (resp. regular dominant). Further, the dominant weights are the combinations $a_1\chi_1 + \cdots + a_{n-1}\chi_{n-1} + a_n\chi_n$, where $a_1, \dots, a_{n-1}$ are non-negative integers, and a_n is an arbitrary integer; χ_n is the restriction to T of the determinant function on G. For $1 \leq d \leq n-1$, the line bundle $L(\chi_d)$ is the pull-back of $\mathcal{O}(1)$ under the composition $X \to \mathrm{Gr}(d, n) \to \mathbb{P}(\bigwedge^d \mathbb{C}^n)$. Further, we have by Proposition 1.4.3:

$$c_1(L_{\chi_d}) \cup [X_w] = [X_{w_o s_d}] \cup [X_w] = \sum_v [X_v],$$

the sum over the $v \in W$ such that $v \leq w$, $\ell(v) = \ell(w) - 1$, and $v = ws_{ij}$ with $i < d < j$.

We now consider the spaces of global sections of homogeneous line bundles. For any weight λ, we put

$$H^0(\lambda) := H^0(X, L_\lambda).$$

This is a finite-dimensional vector space, as X is projective. Further, since the line bundle L_λ is G-linearized, the space $H^0(\lambda)$ is a *rational G-module*, i.e., G acts linearly on this space and the corresponding homomorphism $G \to \mathrm{GL}(H^0(\lambda))$ is algebraic. Further properties of this space and a refinement of Proposition 1.4.3 are given by the following:

1.4.5. Proposition. *The space $H^0(\lambda)$ is non-zero if and only if λ is dominant. Then $H^0(\lambda)$ contains a unique line of eigenvectors of the subgroup B^-, and the corresponding character of B^- is $-\lambda$. The divisor of any such eigenvector p_λ satisfies*

$$\mathrm{div}(p_\lambda) = \sum_{i=1}^{n-1} (\lambda_i - \lambda_{i+1})\, X^{s_i}.$$

More generally, for any $w \in W$, the G-module $H^0(\lambda)$ contains a unique line of eigenvectors of the subgroup wB^-w^{-1}, and the corresponding weight is $-w\lambda$. Any such eigenvector $p_{w\lambda}$ has a non-zero restriction to X_w, with divisor

$$\mathrm{div}(p_{w\lambda}|_{X_w}) = \sum (\lambda_i - \lambda_j)\, X_{ws_{ij}},$$

the sum over the pairs (i, j) such that $1 \leq i < j \leq n$ and $X_{ws_{ij}}$ is a Schubert divisor in X_w. (This makes sense as X_w is nonsingular in codimension 1, see Proposition 1.3.5.)

In particular, taking $\lambda = \rho$, the zero locus of $p_{w\rho}|_{X_w}$ is exactly the union of all the Schubert divisors in X_w.

Proof. If λ is dominant, then we know that L_λ is generated by its global sections, and hence admits a non-zero section. Conversely, if $H^0(\lambda) \neq 0$ then L_λ has a section σ which does not vanish at some point of X. Since X is homogeneous, the G-translates of σ generate L_λ. Thus, L_λ is dominant.

Now choose a dominant weight λ and put $D := \sum_{i=1}^{n-1}(\lambda_i - \lambda_{i+1})\, X^{s_i}$. By Proposition 1.4.3, we have $L_\lambda \cong \mathcal{O}_X(D)$, so that L_λ admits a section σ with divisor D. Since D is B^--invariant, σ is a B^--eigenvector; in particular, a T-eigenvector. And since D does not contain the standard flag F, it follows that $\sigma(F) \neq 0$. Further, T acts on the fiber of L_λ at F by the weight λ, so that σ has weight $-\lambda$. If σ' is another B^--eigenvector in $H^0(\lambda)$, then the quotient σ'/σ is a rational function on X, which is U^--invariant as σ and σ' are. Since the orbit U^-F is open in X, it follows that the function σ'/σ is constant, i.e., σ' is a scalar multiple of σ.

By G-equivariance, it follows that $H^0(\lambda)$ contains a unique line of eigenvectors of the subgroup wB^-w^{-1}, with weight $-w\lambda$. Let $p_{w\lambda}$ be such an eigenvector, then $p_{w\lambda}$ does not vanish at F_w, hence (by T-equivariance) it has no zero on C_w. So the zero locus of the restriction $p_{w\lambda}|_{X_w}$ has support in $X_w \setminus C_w$ and hence is B-invariant. The desired formula follows by the above argument together with Proposition 1.4.3. □

1.4.6. Remark. For any dominant weight λ, the G-module $H^0(\lambda)$ contains a unique line of eigenvectors for $B = w_o B^- w_o$, of weight $-w_o\lambda$. On the other hand, the evaluation of sections at the base point B/B yields a non-zero linear map $H^0(\lambda) \to \mathbb{C}$ which is a B-eigenvector of weight λ. In other words, the dual G-module

$$V(\lambda) := H^0(\lambda)^*$$

contains a canonical B-eigenvector of weight λ.

One can show that both G-modules $H^0(\lambda)$ and $V(\lambda)$ are *simple*, i.e., they admit no non-trivial proper submodules. Further, any simple rational G-module V is isomorphic to $V(\lambda)$ for a unique dominant weight λ, the *highest weight* of V. The T-module $V(\lambda)$ is the sum of its weight subspaces, and the corresponding weights lie in the convex hull of the orbit $W\lambda \subset \mathbb{Z}^n \subset \mathbb{R}^n$. For these results, see, e.g., [21] 8.2 and 9.3.

1.4.7. Example. For $d = 1, \ldots, n-1$, the space $\bigwedge^d \mathbb{C}^n$ has a basis consisting of the vectors

$$e_I := e_{i_1} \wedge \cdots \wedge e_{i_d},$$

where $I = (i_1, \ldots, i_d)$ and $1 \leq i_1 < \cdots < i_d \leq n$. These vectors are T-eigenvectors with pairwise distinct weights, and they form a unique orbit of W. It follows easily that the G-module $\bigwedge^d \mathbb{C}^n$ is simple with highest weight χ_d (the weight of the unique B-eigenvector $e_{1\ldots d}$). In other words, we have $V(\chi_d) = \bigwedge^d \mathbb{C}^n$, so that $H^0(\chi_d) = (\bigwedge^d \mathbb{C}^n)^*$.

Denote by $p_I \in (\bigwedge^d \mathbb{C}^n)^*$ the elements of the dual basis of the basis $\{e_I\}$ of $\bigwedge^d \mathbb{C}^n$. The p_I are homogeneous coordinates on $\mathrm{Gr}(d,n)$, the *Plücker coordinates.* From the previous remark, one readily obtains that

$$\mathrm{div}(p_I|_{X_I}) = \sum_{J,\ J<I,\ |J|=|I|-1} X_J.$$

This is a refined version of the formula

$$c_1(L) \cup [X_I] = \sum_{J,\ J<I,\ |J|=|I|-1} [X_J]$$

in $H^*(\mathrm{Gr}(d,n))$, where L denotes the pull-back of $\mathcal{O}(1)$ via the Plücker embedding. Note that $c_1(L)$ is the class of the unique Schubert divisor.

Notes. The results of this section are classical; they may be found in more detail in [21], [23] and [49], see also [68]. We refer to [66] Chapter 8 for an exposition of the theory of reductive algebraic groups with some fundamental results on their Schubert varieties. Further references are the survey [67] of Schubert varieties and their generalizations in this setting, and the book [39] regarding the general framework of Kac-Moody groups.

The irreducibility of the intersections $X_w \cap X^v$ is due to Richardson [64], whereas the intersections $C_w \cap C^v$ have been studied by Deodhar [16]. In fact, the Richardson varieties in the Grassmannians had appeared much earlier, in Hodge's geometric proof [32] of the Pieri formula which decomposes the product of an arbitrary Schubert class with the class of a "special" Schubert variety (consisting of those subspaces having a nontrivial intersection with a given standard coordinate subspace). The Richardson varieties play an important role in several recent articles, in relation to standard monomial theory; see [48], [41], [40], [9].

The decomposition of the products $c_1(L_\lambda) \cup [X_w]$ in the basis of Schubert classes is due to Monk [56] for the variety of complete flags, and to Chevalley [12] in general, see [61] for an algebraic proof. The Chevalley formula is equivalent to the decomposition into Schubert classes of the products of classes of Schubert divisors with arbitrary Schubert classes. This yields a closed formula for certain structure constants a_{vw}^x of $H^*(X)$; specifically, those where $v = w_o s_d$ for some elementary transposition s_d.

More generally, closed formulae for all the structure constants have been obtained by several mathematicians, see [37], [17], [61]. The latter paper presents a general formula and applies it to give an algebro-combinatorial proof of the Pieri formula. We refer to [31], [58], [59], [60] for generalizations of the Pieri formula to the isotropic Grassmannians which yield combinatorial (in particular, positive) expressions for certain structure constants.

However, the only known proof of the positivity of the general structure constants is geometric. In fact, an important problem in Schubert calculus is to find a combinatorial expression of these constants which makes their positivity evident.

2. Singularities of Schubert varieties

As seen in Examples 1.1.4 and 1.2.3, Schubert varieties are generally singular. In this section, we show that their singularities are rather mild. We begin by showing that they are normal. Then we introduce the Bott-Samelson desingularizations, and we establish the rationality of singularities of Schubert varieties. In particular, these are Cohen-Macaulay; we also describe their dualizing sheaf, Picard group, and divisor class group. Finally, we obtain the vanishing of all higher cohomology groups $H^j(X_w, L_\lambda)$, where λ is any dominant weight, and the surjectivity of the restriction map $H^0(\lambda) = H^0(X, L_\lambda) \to H^0(X_w, L_\lambda)$.

2.1. Normality

First we review an inductive construction of Schubert cells and varieties. Given $w \in W$ and an elementary transposition s_i, we have either $\ell(s_i w) = \ell(w) - 1$ (and then $s_i w < w$), or $\ell(s_i w) = \ell(w) + 1$ (and then $s_i w > w$). In the first case, we have $Bs_iC_w = C_w \cup C_{s_iw}$, whereas $Bs_iC_w = C_{s_iw}$ in the second case. Further, if $w \neq \mathrm{id}$ (resp. $w \neq w_o$), then there exists an index i such that the first (resp. second) case occurs. (These properties of the Bruhat decomposition are easily checked in the case of the general linear group; for arbitrary reductive groups, see, e.g., [66].)

Next let P_i be the subgroup of $G = \mathrm{GL}_n(\mathbb{C})$ generated by B and s_i. (This is a *minimal parabolic subgroup* of G.) Then P_i is the stabilizer of the partial flag consisting of all the standard coordinate subspaces, except $\langle e_1, \ldots, e_i \rangle$. Further, P_i/B is the Schubert curve $X_{s_i} \cong \mathbb{P}^1$, and $P_i = B \cup Bs_iB$ is the closure in G of Bs_iB.

The group B acts on the product $P_i \times X_w$ by $b(g, x) := (gb^{-1}, bx)$. This action is free; denote the quotient by $P_i \times^B X_w$. Then the map

$$P_i \times X_w \to P_i \times X, \quad (g, x) \mapsto (g, gx)$$

yields a map

$$\iota : P_i \times^B X_w \to P_i/B \times X, \quad (g, x)B \mapsto (gB, gx).$$

Clearly, ι is injective and its image consists of those pairs $(gB, x) \in P_i/B \times X$ such that $g^{-1}x \in X_w$; this defines a closed subset of $P_i/B \times X$. It follows that $P_i \times^B X_w$ is a projective variety equipped with a proper morphism

$$\pi : P_i \times^B X_w \to X$$

with image P_iX_w, and with a morphism

$$f : P_i \times^B X_w \to P_i/B \cong \mathbb{P}^1.$$

The action of P_i by left multiplication on itself yields an action on $P_i \times^B X_w$; the maps π and f are P_i-equivariant. Further, f is a locally trivial fibration with fiber $B \times^B X_w \cong X_w$.

In particular, P_iX_w is closed in X, and hence is the closure of Bs_iC_w. If $s_iw < w$, then $P_iX_w = X_w$. Then one checks that $P_i \times^B X_w$ identifies to $P_i/B \times X_w$, so that π becomes the second projection. On the other hand, if $s_i w > w$, then $P_i X_w = X_{s_i w}$. Then one checks that π restricts to an isomorphism

$$Bs_iB \times^B C_w \to Bs_iC_w = C_{s_i w},$$

so that π is birational onto its image $X_{s_i w}$.

We are now in a position to prove

2.1.1. Theorem. *Any Schubert variety X_w is normal.*

Proof. We argue by decreasing induction on $\dim(X_w) = \ell(w) =: \ell$. In the case where $\ell = \dim(X)$, the variety $X_w = X$ is nonsingular and hence normal. So we may assume that $\ell < \dim(X)$ and that all Schubert varieties of dimension $> \ell$ are normal. Then we may choose an elementary transposition s_i such that $s_i w > w$. We divide the argument into three steps.

Step 1. We show that the morphism $\pi : P_i \times^B X_w \to X_{s_i w}$ satisfies $R^j\pi_*\mathcal{O}_{P_i \times^B X_w} = 0$ for all $j \geq 1$.

Indeed, π factors as the closed immersion

$$\iota : P_i \times^B X_w \to P_i/B \times X_{s_i w} \cong \mathbb{P}^1 \times X_{s_i w}, \quad (g, x)B \mapsto (gB, gx)$$

followed by the projection

$$p : \mathbb{P}^1 \times X_{s_i w} \to X_{s_i w}, \quad (z, x) \mapsto x.$$

Thus, the fibers of π are closed subschemes of $\mathbb{P}^1$ and it follows that

$$R^j\pi_*\mathcal{O}_{P/B \times X_w} = 0 \quad \text{for } j > 1 = \dim \mathbb{P}^1.$$

It remains to check the vanishing of $R^1\pi_*\mathcal{O}_{P_i \times^B X_w}$. For this, we consider the following short exact sequence of sheaves:

$$0 \to \mathcal{I} \to \mathcal{O}_{\mathbb{P}^1 \times X_{s_i w}} \to \iota_*\mathcal{O}_{P_i \times^B X_w} \to 0,$$

where $\mathcal{I}$ denotes the ideal sheaf of the subvariety $P_i \times^B X_w$ of $\mathbb{P}^1 \times X_{s_i w}$. The derived long exact sequence for p yields an exact sequence

$$R^1p_*\mathcal{O}_{\mathbb{P}^1 \times X_{s_i w}} \to R^1p_*(\iota_*\mathcal{O}_{P_i \times^B X_w}) \to R^2p_*\mathcal{I}.$$

Further, $R^1p_*\mathcal{O}_{\mathbb{P}^1 \times X_{s_i w}} = 0$ as $H^1(\mathbb{P}^1, \mathcal{O}_{\mathbb{P}^1}) = 0$; $R^1p_*(\iota_*\mathcal{O}_{P_i \times^B X_w}) = R^1\pi_*\mathcal{O}_{P_i \times^B X_w}$ as ι is a closed immersion; and $R^2p_*\mathcal{I} = 0$ as all the fibers of p have dimension 1. This yields the desired vanishing.

Step 2. We now analyze the normalization map

$$\nu : \tilde{X}_w \to X_w.$$

We have an exact sequence of sheaves

$$0 \to \mathcal{O}_{X_w} \to \nu_*\mathcal{O}_{\tilde{X}_w} \to \mathcal{F} \to 0,$$

where $\mathcal{F}$ is a coherent sheaf with support the non-normal locus of X_w. Further, the action of B on X_w lifts to an action on $\tilde{X}_w$ so that ν is equivariant. Thus, both sheaves $\mathcal{O}_{X_w}$ and $\nu_*\mathcal{O}_{\tilde{X}_w}$ are B-linearized; hence $\mathcal{F}$ is B-linearized as well. (See [8] § 2 for details on linearized sheaves.)

Now any B-linearized coherent sheaf $\mathcal{G}$ on X_w yields an "induced " P_i-linearized sheaf $P_i \times^B \mathcal{G}$ on $P_i \times^B X_w$ (namely, the unique P_i-linearized sheaf which pulls back to the B-linearized sheaf $\mathcal{G}$ under the inclusion $X_w \cong B \times^B X_w \to P_i \times^B X_w$). Further, the assignment $\mathcal{G} \mapsto P_i \times^B \mathcal{G}$ is exact. Therefore, one obtains a short exact sequence of P_i-linearized sheaves on $P_i \times^B X_w$:

$$0 \to \mathcal{O}_{P_i \times^B X_w} \to (P_i \times^B \nu)_* \mathcal{O}_{P_i \times^B \tilde{X}_w} \to P_i \times^B \mathcal{F} \to 0.$$

Apply π_*, we obtain an exact sequence of sheaves on $X_{s_i w}$:

$$0 \to \pi_* \mathcal{O}_{P_i \times^B X_w} \to \pi_*(P_i \times^B \nu)_* \mathcal{O}_{P_i \times^B \tilde{X}_w} \to \pi_*(P_i \times^B \mathcal{F}) \to R^1 \pi_* \mathcal{O}_{P_i \times^B X_w}.$$

Now $\pi_* \mathcal{O}_{P_i \times^B X_w} = \mathcal{O}_{X_{s_i w}}$ by Zariski's main theorem, since $\pi : P_i \times^B X_w \to X_{s_i w}$ is a proper birational morphism, and $X_{s_i w}$ is normal by the induction assumption. Likewise, $\pi_*(P_i \times^B \nu)_* \mathcal{O}_{P_i \times^B \tilde{X}_w} = \mathcal{O}_{X_{s_i w}}$. Further, $R^1 \pi_* \mathcal{O}_{P_i \times^B X_w} = 0$ by Step 1. It follows that $\pi_*(P_i \times^B \mathcal{F}) = 0$.

Step 3. Finally, we assume that X_w is non-normal and we derive a contradiction.

Recall that the support of $\mathcal{F}$ is the non-normal locus of X_w. By assumption, this is a non-empty B-invariant closed subset of X. Thus, the irreducible components of $\operatorname{supp}(\mathcal{F})$ are certain Schubert varieties X_v. Choose such a v and let $\mathcal{F}_v$ denote the subsheaf of $\mathcal{F}$ consisting of sections killed by the ideal sheaf of X_v in X_w. Then $\operatorname{supp}(\mathcal{F}_v) = X_v$, since X_v is an irreducible component of $\operatorname{supp}(\mathcal{F})$. Further, $\pi_*(P_i \times^B \mathcal{F}_v) = 0$, since $\mathcal{F}_v$ is a subsheaf of $\mathcal{F}$.

Now choose the elementary transposition s_i such that $v < s_i v$. Then $w < s_i w$ (otherwise, $P_i X_w = X_w$, so that P_i stabilizes the non-normal locus of X_w; in particular, P_i stabilizes X_v, whence $s_i v < v$). Thus, the morphism $\pi : P_i \times^B X_v \to X_{s_i v}$ restricts to an isomorphism above $C_{s_i v}$. Since $\operatorname{supp}(P_i \times^B \mathcal{F}_v) = P_i \times^B X_v$, it follows that the support of $\pi_*(P_i \times^B \mathcal{F}_v)$ contains $C_{s_i v}$, i.e., this support is the whole $X_{s_i v}$. In particular, $\pi_*(P_i \times^B \mathcal{F}_v)$ is non-zero, which yields the desired contradiction. □

2.2. Rationality of singularities

Let $w \in W$. If $w \neq \mathrm{id}$ then there exists a simple transposition s_{i_1} such that $\ell(s_{i_1} w) = \ell(w) - 1$. Applying this to $s_{i_1} w$ and iterating this process, we obtain a decomposition

$$w = s_{i_1} s_{i_2} \cdots s_{i_\ell}, \quad \text{where} \quad \ell = \ell(w).$$

We then say that the sequence of simple transpositions

$$\underline{w} := (s_{i_1}, s_{i_2}, \ldots, s_{i_\ell})$$

is a *reduced decomposition* of w.

For such a decomposition, we have $X_w = P_{i_1} X_{s_{i_1} w} = P_{i_1} P_{i_2} \cdots P_{i_\ell}/B$. We put $v := s_{i_1} w$ and $\underline{v} := (s_{i_2}, \ldots, s_{i_\ell})$, so that $\underline{w} = (s_{i_1}, \underline{v})$ and $X_w = P_{i_1} X_v$. We define inductively the *Bott-Samelson variety* $Z_{\underline{w}}$ by

$$Z_{\underline{w}} := P_{i_1} \times^B Z_{\underline{v}}.$$

Thus, $Z_{\underline{w}}$ is equipped with an equivariant fibration to $P_{i_1}/B \cong \mathbb{P}^1$ with fiber $Z_{\underline{v}}$ at the base point. Further, $Z_{\underline{w}}$ is the quotient of the product $P_{i_1} \times \cdots \times P_{i_\ell}$ by the action of B^ℓ via

$$(b_1, b_2, \ldots, b_{\ell-1}, b_\ell)(g_1, g_2, \ldots, g_\ell) = (g_1 b_1^{-1}, b_1 g_2 b_2^{-1}, \ldots, b_{\ell-1} g_\ell b_\ell^{-1}).$$

The following statement is easily checked.

2.2.1. Proposition.

(i) *The space $Z_{\underline{w}}$ is a nonsingular projective B-variety of dimension ℓ, where B acts via $g(g_1, g_2, \ldots, g_\ell)B^\ell := (gg_1, g_2, \ldots, g_\ell)B^\ell$. For any subsequence $\underline{v}$ of $\underline{w}$, we have a closed B-equivariant immersion $Z_{\underline{v}} \to Z_{\underline{w}}$.*

(ii) *The map*

$$Z_{\underline{w}} \to (G/B)^\ell = X^\ell, \quad (g_1, g_2, \ldots, g_\ell)B^\ell \mapsto (g_1 B, g_1 g_2 B, \ldots, g_1 \cdots g_\ell B)$$

is a closed B-equivariant embedding.

(iii) *The map*

$$\varphi : Z_{\underline{w}} = Z_{s_{i_1},\ldots,s_{i_\ell}} \to Z_{s_{i_1},\ldots,s_{i_{\ell-1}}}, \quad (g_1, \ldots, g_{\ell-1}, g_\ell)B^\ell \mapsto (g_1, \ldots, g_{\ell-1})B^{\ell-1}$$

is a B-equivariant locally trivial fibration with fiber $P_{i_\ell}/B \cong \mathbb{P}^1$.

(iv) *The map*

$$\pi = \pi_{\underline{w}} : Z_{\underline{w}} \to P_{i_1} \cdots P_{i_\ell}/B = X_w, \quad (g_1, \ldots, g_\ell)B^\ell \mapsto g_1 \cdots g_\ell B,$$

is a proper B-equivariant morphism, and restricts to an isomorphism over C_w. In particular, π is birational.

An interesting combinatorial consequence of this proposition is the following description of the Bruhat order (which may also be proved directly).

2.2.2. Corollary. *Let $v, w \in W$. Then $v \leq w$ if and only if there exist a reduced decomposition $\underline{w} = (s_{i_1}, \ldots, s_{i_\ell})$, and a subsequence $\underline{v} = (s_{j_1}, \ldots, s_{j_m})$ with product v. Then there exists a reduced subsequence $\underline{v}$ with product v.*

As a consequence, $v < w$ if and only if there exists a sequence $(v_1, \ldots, v_k)$ in W such that $v = v_1 < \cdots < v_k = w$, and $\ell(v_{j+1}) = \ell(v_j) + 1$ for all j.

Proof. Since π is a proper T-equivariant morphism, any fiber at a T-fixed point contains a fixed point (by Borel's fixed point theorem, see, e.g., [66] Theorem 6.2.6). But the fixed points in X_w (resp. $Z_{\underline{w}}$) correspond to the $v \in W$ such that $v \leq w$ (resp. to the subsequences of $\underline{w}$). This proves the first assertion.

If $v = s_{j_1} \cdots s_{j_m}$, then the product $Bs_{j_1}B \cdots Bs_{j_m}B/B$ is open in X_v. By induction on m, it follows that there exists a reduced subsequence $(s_{k_1}, \ldots, s_{k_n})$ of $(s_{j_1}, \ldots, s_{j_m})$ such that $Bs_{k_1}B \cdots Bs_{k_n}B/B$ is open in X_v; then $v = s_{k_1} \cdots s_{k_n}$. This proves the second assertion.

The final assertion follows from the second one. Alternatively, one may observe that the complement $X_w \setminus C_w$ has pure codimension one in X_w, since C_w is an affine open subset of X_w. Thus, for any $v < w$ there exists $x \in W$ such that $v \leq x < w$ and $\ell(x) = \ell(w) - 1$. Now induction on $\ell(w) - \ell(v)$ completes the proof. □

2.2.3. Theorem. *The morphism* $\pi : Z_{\underline{w}} \to X_w$ *satisfies* $\pi_* \mathcal{O}_{Z_{\underline{w}}} = \mathcal{O}_{X_w}$, *and* $R^j \pi_* \mathcal{O}_{Z_{\underline{w}}} = 0$ *for all* $j \geq 1$.

Proof. We argue by induction on $\ell = \ell(w)$, the case where $\ell = 0$ being trivial. For $\ell \geq 1$, we may factor $\pi = \pi_{\underline{w}}$ as

$$P_{i_1} \times^B \pi_{\underline{v}} : P_{i_1} \times^B Z_{\underline{v}} \to P_{i_1} \times^B X_v, \quad (g, z)B \mapsto (g, \pi_{\underline{v}}(z))B$$

followed by the map

$$\pi_1 : P_{i_1} \times^B X_v \to X_w, \quad (g, x)B \mapsto gx.$$

By the induction assumption, the morphism $\pi_{\underline{v}}$ satisfies the conclusions of the theorem. It follows easily that so does the induced morphism $P_{i_1} \times^B \pi_{\underline{v}}$. But the same holds for the morphism π_1, by the first step in the proof of Theorem 2.1.1. Now the Grothendieck spectral sequence for the composition $\pi_1 \circ (P_{i_1} \times^B \pi_{\underline{v}}) = \pi_{\underline{w}}$ (see [28] Chapter II) yields the desired statements. □

Thus, π is a desingularization of the Schubert variety X_w, and the latter has rational singularities in the following sense (see [34] p. 49).

2.2.4. Definition. A *desingularization* of an algebraic variety Y consists of a nonsingular algebraic variety Z together with a proper birational morphism $\pi : Z \to Y$. We say that Y has *rational singularities*, if there exists a desingularization $\pi : Z \to Y$ satisfying $\pi_* \mathcal{O}_Z = \mathcal{O}_Y$ and $R^j \pi_* \mathcal{O}_Z = 0$ for all $j \geq 1$.

Note that the equality $\pi_* \mathcal{O}_Z = \mathcal{O}_Y$ is equivalent to the normality of Y, by Zariski's main theorem. Also, one can show that Y has rational singularities if and only if $\pi_* \mathcal{O}_Z = \mathcal{O}_Y$ and $R^j \pi_* \mathcal{O}_Z = 0$ for all $j \geq 1$, where $\pi : Z \to Y$ is *any* desingularization.

Next we recall the definition of the *canonical sheaf* ω_Y of a normal variety Y. Let $\iota : Y^{\text{reg}} \to Y$ denote the inclusion of the nonsingular locus, then $\omega_Y := \iota_* \omega_{Y^{\text{reg}}}$, where $\omega_{Y^{\text{reg}}}$ denotes the sheaf of differential forms of maximal degree on the nonsingular variety Y^{reg}. Since the sheaf $\omega_{Y^{\text{reg}}}$ is invertible and $\text{codim}(Y - Y^{\text{reg}}) \geq 2$, the canonical sheaf is the sheaf of local sections of a Weil divisor K_Y: the *canonical divisor*, defined up to linear equivalence. If, in addition, Y is Cohen-Macaulay, then ω_Y is its *dualizing sheaf*.

For any desingularization $\pi : Z \to Y$ where Y is normal, we have an injective trace map $\pi_* \omega_Z \to \omega_Y$. Further, $R^j \pi_* \omega_Z = 0$ for any $j \geq 1$, by the Grauert-Riemenschneider theorem (see [19] p. 59). We may now formulate the following characterization of rational singularities, proved, e.g., in [34] p. 50.

2.2.5. Proposition. *Let* Y *be a normal variety. Then* Y *has rational singularities if and only if:* Y *is Cohen-Macaulay and* $\pi_* \omega_Z = \omega_Y$ *for any desingularization* $\pi : Z \to Y$.

In particular, any Schubert variety X_w is Cohen-Macaulay, and its dualizing sheaf may be determined from that of a Bott-Samelson desingularization $Z_{\underline{w}}$. To describe the latter, put $Z := Z_{\underline{w}}$ and for $1 \leq j \leq \ell$, let $Z^j \subset Z$ be the Bott-Samelson subvariety associated with the subsequence $\underline{w}^j := (s_{i_1} \ldots, \widehat{s_{i_j}}, \ldots, s_{i_\ell})$ obtained by suppressing s_{i_j}.

2.2.6. Proposition.

(i) *With the preceding notation,* $Z^1, \ldots, Z^\ell$ *identify to nonsingular irreducible divisors in* Z*, which meet transversally at a unique point (the class of* B^ℓ*).*

(ii) *The complement in* Z *of the boundary*

$$\partial Z := Z^1 \cup \cdots \cup Z^\ell$$

equals $\pi^{-1}(C_w) \cong C_w$.

(iii) *The classes* $[Z^j]$, $j = 1, \ldots, \ell$, *form a basis of the Picard group of* Z.

Indeed, (i) follows readily from the construction of Z; (ii) is a consequence of Proposition 2.2.1, and (iii) is checked by the argument of Proposition 1.4.1.

Next put

$$\partial X_w := X_w \setminus C_w = \bigcup_{v \in W,\, v<w} X_v,$$

this is the *boundary* of X_w. By Corollary 2.2.2, ∂X_w is the union of all the Schubert divisors in X_w. Further, $\pi^{-1}(\partial X_w) = \partial Z$ (as sets).

We may now describe the dualizing sheaves of Bott-Samelson and Schubert varieties.

2.2.7. Proposition.

(i) $\omega_Z \cong (\pi^* L_{-\rho})(-\partial Z)$.

(ii) $\omega_{X_w} \cong L_{-\rho}|_{X_w}(-\partial X_w)$. *In particular,* $\omega_X \cong L_{-2\rho}$.

(iii) *The reduced subscheme* ∂X_w *is Cohen-Macaulay.*

Proof. (i) Consider the curves $C_j := Z_{s_j} = P_j/B$ for $j = 1, \ldots, \ell$. We may regard each C_j as a subvariety of Z, namely, the transversal intersection of the Z^k for $k \neq j$. We claim that any divisor D on Z such that $\langle [D], [C_j] \rangle = 0$ for all j is principal.

To see this, note that $\langle [Z^j], [C_j] \rangle = 1$ for all j, by Proposition 2.2.6. On the other hand, $\langle [Z^j], [C_k] \rangle = 0$ for all $j < k$. Indeed, we have a natural projection $\varphi_j : Z = Z_{s_{i_1}, \ldots, s_{i_\ell}} \to Z_{s_{i_1}, \ldots, s_{i_j}}$ such that Z^j is the pull-back of the corresponding divisor $Z^j_{s_{i_1}, \ldots, s_{i_j}}$. Moreover, φ_j maps C_k to a point whenever $k > j$. Since the $[Z^j]$ generate freely $\mathrm{Pic}(Z)$ by Proposition 2.2.6, our claim follows.

By this claim, it suffices to check the equality of the degrees of the line bundles $\omega_Z(\partial Z)$ and $\pi^* L_{-\rho}$ when restricted to each curve C_j. Now we obtain

$$\omega_Z(\partial Z)|_{C_j} \cong \omega_{C_j}(\partial C_j),$$

by the adjunction formula. Further, $C_j \cong \mathbb{P}^1$, and ∂C_j is one point, so that $\omega_{C_j}(\partial C_j) \cong \mathcal{O}_{\mathbb{P}^1}(-1)$. On the other hand, π maps C_j isomorphically to the Schubert curve X_{s_j}, and $\mathcal{L}_\rho|_{X_{s_j}} \cong \mathcal{O}_{\mathbb{P}^1}(1)$, so that $\pi^* L_{-\rho}|C_j \cong \mathcal{O}_{\mathbb{P}^1}(-1)$. This shows the desired equality.

(ii) Since X_w has rational singularities, we have $\omega_{X_w} = \pi_*\omega_Z$. Further, the projection formula yields $\omega_{X_w} \cong L_{-\rho} \otimes \pi_*\mathcal{O}_Z(-\partial Z)$, and $\pi_*\mathcal{O}_Z(-\partial Z) \cong \mathcal{O}_{X_w}(-\partial X_w)$ as $\pi^{-1}(\partial X_w) = \partial Z$.

(iii) By (ii), the ideal sheaf of ∂X_w in X_w is locally isomorphic to the dualizing sheaf ω_{X_w}. Therefore, this ideal sheaf is Cohen-Macaulay of depth $\dim(X_w)$. Now the exact sequence

$$0 \to \mathcal{O}_{X_w}(-\partial X_w) \to \mathcal{O}_{X_w} \to \mathcal{O}_{\partial X_w} \to 0$$

yields that the sheaf $\mathcal{O}_{\partial X_w}$ is Cohen-Macaulay of depth $\dim(X_w) - 1 = \dim(\partial X_w)$. □

We also determine the Picard group $\mathrm{Pic}(X_w)$ and divisor class group $\mathrm{Cl}(X_w)$ of any Schubert variety. These groups are related by an injective map $\mathrm{Pic}(X_w) \to \mathrm{Cl}(X_w)$ which may fail to be surjective (e.g., for $X_{24} \subset \mathrm{Gr}(4,2)$).

2.2.8. Proposition.

(i) *The classes of the Schubert divisors in X_w form a basis of the divisor class group $\mathrm{Cl}(X_w)$.*

(ii) *The restriction $\mathrm{Pic}(X) \to \mathrm{Pic}(X_w)$ is surjective, and its kernel consists of the classes L_λ, where the weight λ satisfies $\lambda_i = \lambda_{i+1}$ whenever $s_i \le w$. Further, each globally generated (resp. ample) line bundle on X_w extends to a globally generated (resp. ample) line bundle on X.*

(iii) *The map $\mathrm{Pic}(X_w) \to \mathrm{Cl}(X_w)$ sends the class of any L_λ to $\sum(\lambda_i - \lambda_j)\, X_{ws_{ij}}$ (the sum over the pairs (i,j) such that $1 \le i < j \le n$ and $X_{ws_{ij}}$ is a Schubert divisor in X_w).*

(iv) *A canonical divisor for X_w is $-\sum(j-i+1)\, X_{ws_{ij}}$ (the sum as above). In particular, a canonical divisor for the full flag variety X is $-2\sum_{i=1}^{n-1} X_{w_o s_i}$.*

Proof. (i) is proved by the argument of Proposition 1.4.1.

(ii) Let L be a line bundle in X_w and consider its pull-back π^*L under a Bott-Samelson desingularization $\pi : Z_{\underline{w}} \to X_w$. By the argument of Proposition 2.2.7, the class of π^*L in $\mathrm{Pic}(Z_{\underline{w}})$ is uniquely determined by its intersection numbers $\langle c_1(\pi^*L), [C_j]\rangle$. Further, the restriction $\pi : C_j \to \pi(C_j)$ is an isomorphism onto a Schubert curve, and all the Schubert curves in X_w arise in this way. Thus, $\langle c_1(\pi^*L), [C_j]\rangle$ equals either 0 or $\langle c_1(L), [X_{s_i}]\rangle$ for some i such that $X_{s_i} \subseteq X_w$, i.e., $s_i \le w$. We may find a weight λ such that $\lambda_i - \lambda_{i+1} = \langle c_1(L), [X_{s_i}]\rangle$ for all such indices i; then $\pi^*L_\lambda = \pi^*L$ in $\mathrm{Pic}(Z_{\underline{w}})$, whence $L = L_\lambda$ in $\mathrm{Pic}(X_w)$.

If, in addition, L is globally generated (resp. ample), then $\langle c_1(L), [X_{s_i}]\rangle \ge 0$ (resp. > 0) for each Schubert curve $X_{s_i} \subseteq X_w$. Thus, we may choose λ to be dominant (resp. regular dominant).

(iii) follows readily from Proposition 1.4.5, and (iv) from Proposition 2.2.7. □

2.3. Cohomology of line bundles

The aim of this subsection is to prove the following

2.3.1. Theorem. *Let λ be a dominant weight and let $w \in W$. Then the restriction map $H^0(\lambda) \to H^0(X_w, L_\lambda)$ is surjective. Further, $H^j(X_w, L_\lambda) = 0$ for any $j \geq 1$.*

Proof. We first prove the second assertion in the case where $X_w = X$ is the full flag variety. Then $\omega_X \cong L_{-2\rho}$, so that $\omega_X^{-1} \otimes L_\lambda \cong L_{\lambda+2\rho}$ is ample. Thus, the assertion follows from the *Kodaira vanishing theorem*: $H^j(X, \omega_X \otimes \mathcal{L}) = 0$ for $j \geq 1$, where $\mathcal{L}$ is any ample line bundle on any projective nonsingular variety X.

Next we prove the second assertion for arbitrary X_w. For this, we will apply a generalization of the Kodaira vanishing theorem to a Bott-Samelson desingularization of X_w. Specifically, choose a reduced decomposition $\underline{w}$ and let $\pi : Z_{\underline{w}} \to X_w$ be the corresponding morphism. Then the projection formula yields isomorphisms

$$R^i\pi_*(\pi^* L_\lambda) \cong L_\lambda \otimes R^i\pi_*\mathcal{O}_{Z_{\underline{w}}}$$

for all $i \geq 0$. Together with Theorem 2.2.3 and the Leray spectral sequence for π, this yields isomorphisms

$$H^j(Z_{\underline{w}}, \pi^* L_\lambda) \cong H^j(X, L_\lambda)$$

for all $j \geq 0$.

We now recall a version of the *Kawamata-Viehweg vanishing theorem*, see [19] § 5. Consider a nonsingular projective variety Z, a line bundle $\mathcal{L}$ on Z, and a family $(D_1, \ldots, D_\ell)$ of nonsingular divisors on Z intersecting transversally. Put $D := \sum_i \alpha_i D_i$, where $\alpha_1, \ldots, \alpha_\ell$ are positive integers. Let N be an integer such that $N > \alpha_i$ for all i, and put $\mathcal{M} := \mathcal{L}^N(-D)$. Assume that some positive tensor power of the line bundle $\mathcal{M}$ is globally generated, and that the corresponding morphism to a projective space is generically finite over its image (e.g., $\mathcal{M}$ is ample). Then $H^j(Z, \omega_Z \otimes \mathcal{L}) = 0$ for all $j \geq 1$.

We apply this result to the variety $Z{:=}Z_{\underline{w}}$, the line bundle $\mathcal{L}{:=}(\pi^* L_{\lambda+\rho})(\partial Z)$, and the divisor $D := \sum_i (N - b_i) Z^i$ where $b_1, \ldots, b_\ell$ are positive integers such that $\sum_i b_i Z^i$ is ample (these exist by Lemma 2.3.2 below). Then $\mathcal{L}^N(-D) = (\pi^* L_{N(\lambda+\rho)})(b_1 Z^1 + \cdots + b_\ell Z^\ell)$ is ample, and $\omega_Z \otimes \mathcal{L} = \pi^* L_\lambda$. This yields the second assertion.

To complete the proof, it suffices to show that the restriction map

$$H^0(X_w, L_\lambda) \to H^0(X_v, L_\lambda)$$

is surjective whenever $w = s_i v > v$ for some elementary transposition s_i. As above, this reduces to checking the surjectivity of the restriction map $H^0(Z, \pi^* L_\lambda) \to H^0(Z^1, \pi^* L_\lambda)$. For this, by the long exact sequence

$$\begin{aligned} 0 \to H^0(Z, (\pi^* L_\lambda)(-Z^1)) &\to H^0(Z, \pi^* L_\lambda) \\ \to H^0(Z^1, \pi^* L_\lambda) &\to H^1(Z, (\pi^* L_\lambda)(-Z^1)), \end{aligned}$$

it suffices in turn to show the vanishing of $H^1(Z, (\pi^* L_\lambda)(-Z^1))$.

We will deduce this again from the Kawamata-Viehweg vanishing theorem. Let $a_1, \ldots, a_\ell$ be positive integers such that the line bundle $(\pi^* L_{\lambda+a_1\rho})(a_2 Z^2 + \cdots + a_\ell Z^\ell)$ is ample (again, these exist by Lemma 2.3.2 below). Put $\mathcal{L} := (\pi^* L_{\lambda+\rho})(Z^2 + \cdots + Z^\ell)$ and $D := \sum_{i=2}^{\ell}(N - a_i)Z^i$, where $N > a_1, a_2, \ldots, a_\ell$. Then $\mathcal{L}^N(-D) = (\pi^* L_{N(\lambda+\rho)})(a_2 Z^2 + \cdots + a_\ell Z^\ell)$ is ample, and $\omega_Z \otimes \mathcal{L} = (\pi^* L_\lambda)(-Z^1)$. Thus, we obtain $H^j(Z, (\pi^* L_\lambda)(-Z^1)) = 0$ for all $j \geq 1$. □

2.3.2. Lemma. *Let $Z = Z_{\underline{w}}$ with boundary divisors $Z^1, \ldots, Z^\ell$. Then there exist positive integers $a_1, \ldots, a_\ell$ such that the line bundle $(\pi^* L_{a_1\rho})(a_2 Z^2 + \cdots + a_\ell Z^\ell)$ is ample. Further, there exist positive integers $b_1, \ldots, b_\ell$ such that the divisor $b_1 Z^1 + \cdots + b_\ell Z^\ell$ is ample.*

Proof. We prove the first assertion by induction on ℓ. If $\ell = 1$, then π embeds Z into X, so that $\pi^* L_{a_1\rho}$ is ample for any $a_1 > 0$. In the general case, the map

$$\varphi : Z \to Z^\ell = (P_{i_1} \times \cdots \times P_{i_{\ell-1}})/B^{\ell-1}, \quad (g_1, \ldots, g_\ell)B^\ell \mapsto (g_1, \ldots, g_{\ell-1})B^{\ell-1}$$

fits into a cartesian square

$$\begin{array}{ccc} Z & \xrightarrow{\varphi} & Z^\ell \\ {\scriptstyle\pi}\big\downarrow & & \big\downarrow{\scriptstyle\psi} \\ G/B & \xrightarrow{f} & G/P_{i_\ell}, \end{array}$$

where $\psi((g_1, \ldots, g_{\ell-1})B^{\ell-1}) = g_1 \cdots g_{\ell-1} P_{i_\ell}$. Further, the boundary divisors $Z^{1,\ell}$, …, $Z^{\ell-1,\ell}$ of Z^ℓ satisfy $\varphi^* Z^{i,\ell} = Z^i$. Denote by

$$\pi_\ell : Z^\ell = Z_{(s_{i_1}, \ldots, s_{i_{\ell-1}})} \to X_{s_{i_1} \cdots s_{i_{\ell-1}}} = X_{w s_{i_\ell}}$$

the natural map. By the induction assumption, there exist positive integers a_1, a_2, …, $a_{\ell-1}$ such that the line bundle $(\pi_\ell^* L_{a_1\rho})(a_2 Z^{1,\ell} + \cdots + a_{\ell-1} Z^{\ell-1,\ell})$ is very ample on Z^ℓ. Hence its pull-back

$$\varphi^*((\pi_\ell^* L_{a_1\rho})(a_2 Z^{2,\ell} + \cdots + a_{\ell-1} Z^{\ell-1,\ell})) = (\varphi^* \pi_\ell^* L_{a_1\rho})(a_2 Z^2 + \cdots + a_{\ell-1} Z^{\ell-1})$$

is a globally generated line bundle on Z. Thus, it suffices to show that the line bundle $\pi^* L_{b\rho} \otimes (\varphi^* \pi_\ell^* L_{-a_1\rho})(a_1 Z^\ell)$ is globally generated and φ-ample for $b \gg a_1$. (Indeed, if $\mathcal{M}$ is a globally generated, φ-ample line bundle on Z, and $\mathcal{N}$ is an ample line bundle on Z^ℓ, then $\mathcal{M} \otimes \varphi^* \mathcal{N}$ is ample on Z). Equivalently, it suffices to show that $\pi^* L_{c\rho} \otimes \pi^* L_\rho \otimes (\varphi^* \pi^* L_{-\rho})(Z^\ell)$ is globally generated and φ-ample for $c \gg 0$. But we have by Proposition 2.2.6:

$$\begin{aligned} \pi^* L_\rho \otimes (\varphi^* \pi_\ell^* L_{-\rho})(Z^\ell) &= \omega_Z^{-1}(-\partial Z) \otimes \varphi^*(\omega_{Z^\ell}(\partial Z^\ell))(Z^\ell) \\ &= \omega_Z^{-1} \otimes \varphi^* \omega_{Z^\ell} = \omega_\varphi^{-1} = \pi^* \omega_f^{-1}, \end{aligned}$$

where ω_φ (resp. ω_f) denotes the relative dualizing sheaf of the morphism φ (resp. f). Further, $L_{c\rho} \otimes \omega_f^{-1}$ is very ample on G/B for $c \gg 0$, as L_ρ is ample. Thus, $\pi^*(L_{c\rho} \otimes \omega_f^{-1})$ is globally generated and φ-ample. This completes the proof of the first assertion.

The second assertion follows by recalling that the restriction of L_ρ to X_w admits a section vanishing exactly on ∂X_w (Remark 1.4.6 2). Thus, $\pi^* L_\rho$ admits a section vanishing exactly on $\partial Z = Z^1 \cup \cdots \cup Z^\ell$. □

Next we consider a regular dominant weight λ and the corresponding very ample homogeneous line bundle L_λ. This defines a projective embedding

$$X \to \mathbb{P}(H^0(X, L_\lambda)^*) = \mathbb{P}(V(\lambda))$$

and, in turn, a subvariety $\tilde{X} \subseteq V(\lambda)$, invariant under the action of $G \times \mathbb{C}^*$, where $\mathbb{C}^*$ acts by scalar multiplication. We say that $\tilde{X}$ is the *affine cone over* X associated with this projective embedding. Likewise, we have the affine cones $\tilde{X}_w$ over Schubert varieties.

2.3.3. Corollary. *For any regular dominant weight λ, the affine cone over X_w in $V(\lambda)$ has rational singularities. In particular, X_w is projectively normal in its embedding into $\mathbb{P}(V(\lambda))$.*

Proof. Consider the total space Y_w of the line bundle $L_\lambda^{-1}|_{X_w}$. We have a proper morphism

$$\pi : Y_w \to \tilde{X}_w$$

which maps the zero section to the origin, and restricts to an isomorphism from the complement of the zero section to the complement of the origin. In particular, π is birational. Further, Y_w has rational singularities, since it is locally isomorphic to $X_w \times \mathbb{C}$. Thus, it suffices to show that the natural map $\mathcal{O}_{\tilde{X}_w} \to \pi_* \mathcal{O}_{Y_w}$ is surjective, and $R^j \pi_* \mathcal{O}_{Y_w} = 0$ for any $j \geq 1$. Since $\tilde{X}_w$ is affine, this amounts to: the algebra $H^0(Y_w, \mathcal{O}_{Y_w})$ is generated by the image of $H^0(\lambda)$, and $H^j(Y_w, \mathcal{O}_{Y_w}) = 0$ for $j \geq 1$. Further, since the projection $f : Y_w \to X_w$ is affine and satisfies

$$f_* \mathcal{O}_{Y_w} = \bigoplus_{n=0}^{\infty} L_\lambda^{\otimes n} = \bigoplus_{n=0}^{\infty} L_{n\lambda},$$

we obtain

$$H^j(Y_w, \mathcal{O}_{Y_w}) = \bigoplus_{n=0}^{\infty} H^j(X_w, L_{n\lambda}).$$

So $H^j(Y_w, \mathcal{O}_{Y_w}) = 0$ for $j \geq 1$, by Theorem 2.3.1. To complete the proof, it suffices to show that the algebra $\bigoplus_{n=0}^{\infty} H^0(X_w, L_{n\lambda})$ is generated by the image of $H^0(\lambda)$. Using the surjectivity of the restriction maps $H^0(L_{n\lambda}) \to H^0(X_w, L_{n\lambda})$ (Theorem 2.3.1 again), it is enough to consider the case where $X_w = X$. Now the multiplication map

$$H^0(\lambda)^{\otimes n} \to H^0(n\lambda), \quad \sigma_1 \otimes \cdots \otimes \sigma_n \mapsto \sigma_1 \cdots \sigma_n$$

is a non-zero morphism of G-modules. Since $H^0(n\lambda)$ is simple, this morphism is surjective, which completes the proof. □

Notes. In their full generality, the results of this section were obtained by many mathematicians during the mid-eighties. Their most elegant proofs use reduction to positive characteristics and the techniques of Frobenius splitting, see [55], [62], [63].

Here we have presented alternative proofs: for normality and rationality of singularities, we rely on an argument of Seshadri [65] simplified in [8], which is also valid in arbitrary characteristics. For cohomology of line bundles, our approach (based on the Kawamata-Viehweg vanishing theorem) is a variant of that of Kumar; see [39].

The construction of the Bott-Samelson varieties is due to ... Bott and Samelson [4] in the framework of compact Lie groups, and to Hansen [29] and Demazure [14] in our algebro-geometric setting. The original construction of Bott and Samelson is also presented in [18] with applications to the multiplication of Schubert classes.

The line bundles on Bott-Samelson varieties have been studied by Lauritzen and Thomsen in [45]; in particular, they determined the globally generated (resp. ample) line bundles. On the other hand, the description of the Picard group and divisor class group of Schubert varieties is due to Mathieu in [53]; it extends readily to any Schubert variety Y in any flag variety $X = G/P$. One may also show that the boundary of Y is Cohen-Macaulay, see [8] Lemma 4. But a simple formula for the dualizing sheaf of Y is only known in the case where X is the full flag variety.

An important open question is the explicit determination of the singular locus of a Schubert variety, and of the corresponding generic singularities (i.e., the singularities along each irreducible component of the singular locus). The book [1] by Billey and Lakshmibai is a survey of this question, which was recently solved (independently and simultaneously) by several mathematicians in the case of the general linear group; see [2], [13], [33], [50], [51]. The generic singularities of Richardson varieties are also worth investigating.

3. The diagonal of a flag variety

Let $X = G/B$ be the full flag variety and denote by $\operatorname{diag}(X)$ the diagonal in $X \times X$. In this section, we construct a degeneration of $\operatorname{diag}(X)$ in $X \times X$ to the union of all the products $X_w \times X^w$, where the X_w (resp. X^w) are the Schubert (resp. opposite Schubert) varieties.

Specifically, we construct a subvariety $\mathcal{X} \subseteq X \times X \times \mathbb{P}^1$ such that the fiber of the projection $\pi : \mathcal{X} \to \mathbb{P}^1$ at any $t \neq 0$ is isomorphic to $\operatorname{diag}(X)$, and we show that the fiber at 0 (resp. ∞) is the union of all the $X_w \times X^w$ (resp. $X^w \times X_w$). For this, we use the normality of $\mathcal{X}$ which is deduced from a general normality criterion for varieties with group actions, obtained in turn by adapting the argument for the normality of Schubert varieties.

Then we turn to applications to the Grothendieck ring $K(X)$. After a brief presentation of the definition and main properties of Grothendieck rings, we obtain two additive bases of $K(X)$ which are dual for the bilinear pairing given by the Euler characteristic of the product. Further applications will be given in Section 4.

3.1. A degeneration of the diagonal

We begin by determining the cohomology class of $\mathrm{diag}(X)$ in $X \times X$, where X is the full flag variety.

3.1.1. Lemma. *We have* $[\mathrm{diag}(X)] = \sum_{w \in W} [X_w \times X^w]$ *in* $H^*(X \times X)$.

Proof. By the results in Subsection 1.3 and the Künneth isomorphism, a basis for the abelian group $H^*(X \times X)$ consists of the classes $[X_w \times X^v]$, where $v, w \in W$. Further, the dual basis (with respect to the Poincaré duality pairing) consists of the $[X^w \times X_v]$. Thus, we may write

$$[\mathrm{diag}(X)] = \sum_{v,w \in W} a_{wv}\, [X_w \times X^v],$$

where the coefficients a_{wv} are given by

$$a_{wv} = \langle [\mathrm{diag}(X)], [X^w \times X_v] \rangle.$$

Further, since X^w meets X_v properly along X_v^w with intersection multiplicity 1, it follows that $\mathrm{diag}(X)$ meets $X^w \times X_v$ properly along $\mathrm{diag}(X_v^w)$ in $X \times X$ with intersection multiplicity 1. This yields

$$[\mathrm{diag}(X)] \cup [X^w \times X_v] = [\mathrm{diag}(X_w^v)].$$

And since $\dim(X_v^w) = 0$ if and only if $v = w$, we see that a_{wv} equals 1 if $v = w$, and 0 otherwise. □

This formula suggests the existence of a degeneration of $\mathrm{diag}(X)$ to $\bigcup_{w \in W} X_w \times X^w$. We now construct such a degeneration. The idea is to move $\mathrm{diag}(X)$ in $X \times X$ by a general one-parameter subgroup of the torus T acting on $X \times X$ via its action on the second copy, and to take limits.

Specifically, let

$$\lambda : \mathbb{C}^* \to T, \quad t \mapsto (t^{a_1}, \ldots, t^{a_n})$$

where $a_1, \ldots, a_n$ are integers satisfying $a_1 > \cdots > a_n$. Define $\mathcal{X}$ to be the closure in $X \times X \times \mathbb{P}^1$ of the subset

$$\{(x, \lambda(t)x, t) \mid x \in X,\ t \in \mathbb{C}^*\} \subseteq X \times X \times \mathbb{C}^*.$$

Then $\mathcal{X}$ is a projective variety, and the fibers of the projection $\pi : \mathcal{X} \to \mathbb{P}^1$ identify to closed subschemes of $X \times X$. Further, the fiber $\pi^{-1}(1)$ equals $\mathrm{diag}(X)$. In fact, $\pi^{-1}(\mathbb{C}^*)$ identifies to $\mathrm{diag}(X) \times \mathbb{C}^*$ via $(x, y, t) \mapsto (x, \lambda(t^{-1})x, t)$, and this identifies the restriction of π to the projection $\mathrm{diag}(X) \times \mathbb{C}^* \to \mathbb{C}^*$.

3.1.2. Theorem. *We have equalities of subschemes of* $X \times X$*:*

$$\pi^{-1}(0) = \bigcup_{w \in W} X_w \times X^w \quad \textit{and} \quad \pi^{-1}(\infty) = \bigcup_{w \in W} X^w \times X_w.$$

Proof. By symmetry, it suffices to prove the first equality. We begin by showing the inclusion $\bigcup_{w\in W} X_w \times X^w \subseteq \pi^{-1}(0)$. Equivalently, we claim that $C_w \times C^w \subset \pi^{-1}(0)$ for all $w \in W$.

For this, we analyze the structure of $X \times X$ in a neighborhood of the base point (F_w, F_w) of $C_w \times C^w$ (recall that F_w denotes the image under w of the standard flag F). By Proposition 1.3.5, wC^{id} is a T-invariant open neighborhood of F_w in X, isomorphic to wU^-w^{-1}. Further, $C_w = UF_w \cong (wU^-w^{-1} \cap U)F_w$ identifies via this isomorphism to the subgroup $wU^-w^{-1} \cap U$. Likewise, C^w identifies to the subgroup $wU^-w^{-1} \cap U^-$, and the product map in the group wU^-w^{-1}

$$(wU^-w^{-1} \cap U) \times (wU^-w^{-1} \cap U^-) \to wU^-w^{-1}$$

is an isomorphism. Further, each factor is isomorphic to an affine space.

The group $\mathbb{C}^*$ acts on wU^-w^{-1} via its homomorphism $t \mapsto (t^{a_1}, \dots, t^{a_n})$ to T and the action of T on wU^-w^{-1} by conjugation. In fact, this action is linear, and hence wU^-w^{-1} decomposes into a direct sum of weight subspaces. Using the assumption that $a_1 > \cdots > a_n$, one checks that the sum of all the positive weight subspaces is $wU^-w^{-1} \cap U = C_w$; likewise, the sum of all the negative weight subspaces is C^w. In other words,

$$C_w = \{x \in wU^-w^{-1} \mid \lim_{t\to 0} \lambda(t)x = \mathrm{id}\}, \quad C^w = \{y \in wU^-w^{-1} \mid \lim_{t\to\infty} \lambda(t)y = \mathrm{id}\}.$$

Now identify our neighborhood $wC^{\mathrm{id}} \times wC^{\mathrm{id}}$ to $C_w \times C^w \times C_w \times C^w$. Take arbitrary $x \in C_w$ and $y \in C^w$, then

$$(x, \lambda(t)^{-1}y, \lambda(t)x, y) \to (x, \mathrm{id}, \mathrm{id}, y) \quad \text{as} \quad t \to 0.$$

By the definition of $\mathcal{X}$, it follows that $\pi^{-1}(0)$ contains the point $(x, \mathrm{id}, \mathrm{id}, y)$, identified to $(x, y) \in X \times X$. This proves the claim.

From this claim, it follows that $\pi^{-1}(0)$ contains $\bigcup_{w\in W} X_w \times X^w$ (as schemes). On the other hand, the cohomology class of $\pi^{-1}(0)$ equals that of $\pi^{-1}(1)$, i.e., $\sum_{w\in W}[X_w \times X^w]$ by Lemma 3.1.1. Further, the cohomology class of any nonempty subvariety of $X \times X$ is a positive integer combination of classes $[X_w \times X^v]$ by Proposition 1.3.6. It follows that the irreducible components of $\pi^{-1}(0)$ are exactly the $X_w \times X^w$, and that the corresponding multiplicities are all 1. Thus, the scheme $\pi^{-1}(0)$ is generically reduced.

To complete the proof, it suffices to show that $\pi^{-1}(0)$ is reduced. Since π may be regarded as a regular function on $\mathcal{X}$, it suffices in turn to show that $\mathcal{X}$ is normal. In the next subsection, this will be deduced from a general normality criterion for varieties with group action. □

To apply Theorem 3.1.2, we will also need to analyze the structure sheaf of the special fiber $\pi^{-1}(0)$. This is the content of the following statement.

3.1.3. Proposition. *The sheaf $\mathcal{O}_{\pi^{-1}(0)}$ admits a filtration with associated graded*

$$\bigoplus_{w\in W} \mathcal{O}_{X_w} \otimes \mathcal{O}_{X^w}(-\partial X^w).$$

Proof. We may index the finite partially ordered set $W = \{w_1, \dots, w_N\}$ so that $i \leq j$ whenever $w_i \leq w_j$ (then $w_N = w_o$). Put

$$Z_i := X_{w_i} \times X^{w_i} \quad \text{and} \quad Z_{\geq i} := \bigcup_{j \geq i} Z_j,$$

where $1 \leq i \leq N$. Then $Z_{\geq 1} = \pi^{-1}(0)$ and $Z_{\geq N} = X_{w_o} \times X^{w_o} = X \times \{w_o F\}$. Further, the $Z_{\geq i}$ form a decreasing filtration of $\pi^{-1}(0)$. This yields exact sequences

$$0 \to \mathcal{I}_i \to \mathcal{O}_{Z_{\geq i}} \to \mathcal{O}_{Z_{\geq i+1}} \to 0,$$

where $\mathcal{I}_i$ denotes the ideal sheaf of $Z_{\geq i+1}$ in $Z_{\geq i}$. In turn, these exact sequences yield an increasing filtration of the sheaf $\mathcal{O}_{\pi^{-1}(0)}$ with associated graded $\bigoplus_i \mathcal{I}_i$. Since $Z_{\geq i} = Z_{\geq i+1} \cup Z_i$, we may identify $\mathcal{I}_i$ to the ideal sheaf of $Z_i \cap Z_{\geq i+1}$ in $Z_i = X_{w_i} \times X^{w_i}$. To complete the proof, it suffices to show that

$$Z_i \cap Z_{\geq i+1} = X_{w_i} \times \partial X^{w_i}.$$

We first check the inclusion "$\subseteq$". Note that $Z_i \cap Z_{\geq i+1}$ is invariant under $B \times B^-$, and hence is a union of products $X_u \times X^v$ for certain $u, v \in W$. We must have $u \leq w_i \leq v$ (since $X_u \times X^v \subseteq Z_i$) and $w_i \neq v$ (since $X_u \times X^v \subseteq Z_{\geq i+1}$). Thus, $X_u \times X^v \subseteq X_{w_i} \times \partial X^{w_i}$. To check the opposite inclusion, note that if $X^v \subseteq \partial X^{w_i}$ then $v > w_i$, so that $v = w_j$ with $j > i$. Thus, $X_{w_i} \times X^v \subset X_{w_j} \times X^{w_j} \subseteq Z_{\geq i+1}$. □

3.2. A normality criterion

Let G be a connected linear algebraic group acting on an algebraic variety Z. Let $Y \subset Z$ be a subvariety, invariant under the action of a Borel subgroup $B \subseteq G$, and let $P \supset B$ be a parabolic subgroup of G. Then, as in Subsection 2.1, we may define the "induced" variety $P \times^B Y$. It is equipped with a P-action and with P-equivariant maps $\pi : P \times^B Y \to Z$ (a proper morphism with image PY), and $f : P \times^B Y \to P/B$ (a locally trivial fibration with fiber Y). If, in addition, P is a minimal parabolic subgroup (i.e., $P/B \cong \mathbb{P}^1$), and if $PY \neq Y$, then $\dim(PY) = \dim(Y) + 1$, and the morphism π is generically finite over its image PY.

We say that Y is *multiplicity-free* if it satisfies the following conditions:

(i) $GY = Z$.
(ii) Either $Y = Z$, or Z contains no G-orbit.
(iii) For all minimal parabolic subgroups $P \supset B$ such that $PY \neq Y$, the morphism $\pi : P \times^B Y \to PY$ is birational, and the variety PY is multiplicity-free.

(This defines indeed the class of multiplicity-free subvarieties by induction on the codimension, starting with Z).

For example, Schubert varieties are multiplicity-free. Further, the proof of their normality given in Subsection 2.1 readily adapts to show the following

3.2.1. Theorem. *Let Y be a B-invariant subvariety of a G-variety Z. If Z is normal and Y is multiplicity-free, then Y is normal.*

Next we obtain a criterion for multiplicity-freeness of any B-stable subvariety of $Z := G$, where G acts by left multiplication. Note that the B-stable subvarieties $Y \subseteq G$ correspond to the subvarieties V of the full flag variety G/B, by taking $V := \{g^{-1}B \mid g \in Y\}$.

3.2.2. Lemma. *With the preceding notation, Y is multiplicity-free if and only if $[V]$ is a multiplicity-free combination of Schubert classes, i.e., the coefficients of $[V]$ in the basis $\{[X_w]\}$ are either 0 or 1. Equivalently, $\langle [V], [X^w] \rangle \leq 1$ for all w.*

Proof. Clearly, Y satisfies conditions (i) and (ii) of multiplicity-freeness. For condition (iii), consider a minimal parabolic subgroup $P \supset B$ and the natural map $f : G/B \to G/P$. Then the subvariety of G associated with $f^{-1}f(V)$ is PY. As a consequence, $PY \neq Y$ if and only if the restriction $f|_V : V \to f(V)$ is generically finite. Further, the fibers of $f|_V$ identify to those of the natural map $\pi : P \times^B Y \to PY$; in particular, both morphisms have the same degree d. Note that $d = 1$ if and only if π (or, equivalently, $f|_V$) is birational.

Let $X_w \subseteq G/B$ be a Schubert variety of positive dimension. We may write $X_w = P_1 \cdots P_\ell/B$, where $(P_1, \ldots, P_\ell)$ is a sequence of minimal parabolic subgroups, and $\ell = \dim(X_w)$. Put $P := P_\ell$ and $X_v := P_1 \cdots P_{\ell-1}/B$. Then $X_w = f^{-1}f(X_w)$, and the restriction $X_v \to f(X_v) = f(X_w)$ is birational. We thus obtain the equalities of intersection numbers

$$\begin{aligned}\langle [V], [X_w] \rangle_{G/B} &= \langle [V], f^{-1}[f(X_w)] \rangle_{G/B} = \langle f_*[V], [f(X_w)] \rangle_{G/P} \\ &= d\, \langle [f(V)], [f(X_w)] \rangle_{G/P} = d\, \langle [f^{-1}f(V), [X_v] \rangle_{G/B},\end{aligned}$$

as follows from the projection formula and from the equalities $f_*[V] = d\,[f(V)]$, $f_*[X_v] = [f(X_v)] = [f(X_w)]$. From these equalities, it follows that $[V]$ is a multiplicity-free combination of Schubert classes if and only if: $d = 1$ and $[f^{-1}f(V)]$ is a multiplicity-free combination of Schubert classes, for any minimal parabolic subgroup P such that $PY \neq Y$. Now the proof is completed by induction on $\operatorname{codim}_{G/B}(V) = \operatorname{codim}_G(Y)$. □

We may now complete the proof of Theorem 3.1.2 by showing that $\mathcal{X}$ is normal. Consider first the group $G \times G$, the Borel subgroup $B \times B$, and the variety $Z := G \times G$, where $G \times G$ acts by left multiplication. Then the subvariety $Y := (B \times B)\operatorname{diag}(G)$ is multiplicity-free. (Indeed, Y corresponds to the variety $V = \operatorname{diag}(X) \subset X \times X$, where $X = G/B$. By Lemma 3.1.1, the coefficients of $[\operatorname{diag}(X)]$ in the basis of Schubert classes are either 0 or 1, so that Lemma 3.2.2 applies.)

Next consider the same group $G \times G$ and take $Z := G \times G \times \mathbb{P}^1$, where $G \times G$ acts via left multiplication on the factor $G \times G$. Let Y be the preimage in Z of the subvariety $\mathcal{X} \subset X \times X \times \mathbb{P}^1$ under the natural map $G \times G \times \mathbb{P}^1 \to X \times X \times \mathbb{P}^1$ (a locally trivial fibration). Clearly, Y satisfies conditions (i), (ii) of multiplicity-freeness. Further, condition (iii) follows from the fact that Y contains an open subset isomorphic to $(B \times B)\operatorname{diag}(G) \times \mathbb{C}^*$, together with the multiplicity-freeness of $(B \times B)\operatorname{diag}(G)$. Since Z is nonsingular, it follows that Y is normal by Theorem 3.2.1. Hence, $\mathcal{X}$ is normal as well.

3.3. The Grothendieck group

For any scheme X, the *Grothendieck group of coherent sheaves on* X is the abelian group $K_0(X)$ generated by symbols $[\mathcal{F}]$, where $\mathcal{F}$ is a coherent sheaf on X, subject to the relations $[\mathcal{F}] = [\mathcal{F}_1]+[\mathcal{F}_2]$ whenever there exists an exact sequence of sheaves $0 \to \mathcal{F}_1 \to \mathcal{F} \to \mathcal{F}_2 \to 0$. (In particular, $[\mathcal{F}]$ only depends on the isomorphism class of $\mathcal{F}$.) For example, any closed subscheme $Y \subseteq X$ yields a class $[\mathcal{O}_Y]$ in $K(X)$.

Likewise, we have the *Grothendieck group* $K^0(X)$ *of vector bundles on* X, generated by symbols $[E]$, where E is a vector bundle on X, subject to the relations $[E] = [E_1] + [E_2]$ whenever there exists an exact sequence of vector bundles $0 \to E_1 \to E \to E_2 \to 0$. The tensor product of vector bundles yields a commutative, associative multiplication law on $K^0(X)$ denoted by $(\alpha, \beta) \mapsto \alpha \cdot \beta$. With this multiplication, $K^0(X)$ is a commutative ring, the identity element being the class of the trivial bundle of rank 1.

The duality of vector bundles $E \mapsto E^\vee$ is compatible with the defining relations of $K^0(X)$. Thus, it yields a map $K^0(X) \to K^0(X)$, $\alpha \mapsto \alpha^\vee$, which is an involution of the ring $K^0(X)$: the *duality involution.*

By associating with each vector bundle E its (locally free) sheaf of sections $\mathcal{E}$, we obtain a map

$$\varphi : K^0(X) \to K_0(X).$$

More generally, since tensoring with a locally free sheaf is exact, the ring $K^0(X)$ acts on $K_0(X)$ via

$$[E] \cdot [\mathcal{F}] := [\mathcal{E} \otimes_{\mathcal{O}_X} \mathcal{F}],$$

where E is a vector bundle on X with sheaf of sections $\mathcal{E}$, and $\mathcal{F}$ is a coherent sheaf on X. This makes $K_0(X)$ a module over $K^0(X)$; further, $\varphi(\alpha) = \alpha \cdot [\mathcal{O}_X]$ for any $\alpha \in K^0(X)$.

If Y is another scheme, then the external tensor product of sheaves (resp. vector bundles) yields product maps $K_0(X) \times K_0(Y) \to K_0(X \times Y)$, $K^0(X) \times K^0(Y) \to K^0(X \times Y)$, compatible with the corresponding maps φ. We will denote both product maps by $(\alpha, \beta) \mapsto \alpha \times \beta$.

If X is a nonsingular variety, then φ is an isomorphism. In this case, we identify $K_0(X)$ to $K^0(X)$, and we denote this ring by $K(X)$, the *Grothendieck ring of* X. For any coherent sheaves $\mathcal{F}$, $\mathcal{G}$ on X, we have

$$[\mathcal{F}] \cdot [\mathcal{G}] = \sum_j (-1)^j \, [\mathcal{T}or_j^X(\mathcal{F}, \mathcal{G})].$$

(This formula makes sense because the sheaves $\mathcal{T}or_j^X(\mathcal{F}, \mathcal{G})$ are coherent, and vanish for $j > \dim(X)$.) In particular, $[\mathcal{F}] \cdot [\mathcal{G}] = 0$ if the sheaves $\mathcal{F}$ and $\mathcal{G}$ have disjoint supports. Further,

$$[\mathcal{F}]^\vee = \sum_j (-1)^j \, [\mathcal{E}xt_X^j(\mathcal{F}, \mathcal{O}_X)].$$

In particular, if Y is an equidimensional Cohen-Macaulay subscheme of X, then

$$[\mathcal{O}_Y]^\vee = (-1)^c\,[Ext^c_X(\mathcal{O}_Y, \mathcal{O}_X)] = (-1)^c\,[\omega_{Y/X}] = (-1)^c\,[\omega_Y]\cdot[\omega_X]^\vee,$$

where c denotes the codimension of Y, and $\omega_{Y/X} := \omega_Y \otimes \omega_X^{-1}$ denotes the relative dualizing sheaf of Y in X.

Returning to an arbitrary scheme X, any morphism of schemes $f : X \to Y$ yields a pull-back map

$$f^* : K^0(Y) \to K^0(X), \quad [E] \mapsto [f^*E].$$

If, in addition, f is *flat*, then it defines similarly a pull-back map $f^* : K_0(Y) \to K_0(X)$.

On the other hand, any *proper* morphism $f : X \to Y$ yields a push-forward map

$$f_* : K_0(X) \to K_0(Y), \quad [\mathcal{F}] \mapsto \sum_j (-1)^j\,[R^j f_*(\mathcal{F})].$$

As above, this formula makes sense as the higher direct images $R^j f_*(\mathcal{F})$ are coherent sheaves on Y, which vanish for $j > \dim(X)$. Moreover, we have the *projection formula*

$$f_*((f^*\alpha)\cdot\beta) = \alpha\cdot f_*\beta$$

for all $\alpha \in K^0(Y)$ and $\beta \in K_0(X)$.

In particular, if X is complete then we obtain a map

$$\chi : K_0(X) \to \mathbb{Z}, \quad [\mathcal{F}] \mapsto \chi(\mathcal{F}) = \sum_j (-1)^j\,h^j(\mathcal{F}),$$

where $h^j(\mathcal{F})$ denotes the dimension of the jth cohomology group of $\mathcal{F}$, and χ stands for the *Euler-Poincaré characteristic.*

We will repeatedly use the following result of "homotopy invariance" in the Grothendieck group.

3.3.1. Lemma. *Let X be a variety and let $\mathcal{X}$ be a subvariety of $X \times \mathbb{P}^1$ with projections $\pi : \mathcal{X} \to \mathbb{P}^1$ and $p : \mathcal{X} \to X$. Then the class $[\mathcal{O}_{p(\pi^{-1}(z))}] \in K_0(X)$ is independent of $z \in \mathbb{P}^1$, if π is dominant.*

Proof. The exact sequence $0 \to \mathcal{O}_{\mathbb{P}^1}(-1) \to \mathcal{O}_{\mathbb{P}^1} \to \mathcal{O}_z \to 0$ of sheaves on $\mathbb{P}^1$ shows that the class $[\mathcal{O}_z] \in K_0(\mathbb{P}^1)$ is independent of z. Since π is flat, it follows that the class $\pi^*[\mathcal{O}_z] = [\mathcal{O}_{\pi^{-1}(z)}] \in K_0(\mathcal{X})$ is also independent of z, and the same holds for $p_*[\mathcal{O}_{\pi^{-1}(z)}] \in K_0(X)$ since p is proper. But $p_*[\mathcal{O}_{\pi^{-1}(z)}] = [\mathcal{O}_{p(\pi^{-1}(z))}]$, since p restricts to an isomorphism $\pi^{-1}(z) \to p(\pi^{-1}(z))$. □

Finally, we present a relation of $K_0(X)$ to the Chow group $A_*(X)$ of rational equivalence classes of algebraic cycles on X (graded by the dimension), see [22] Example 15.1.5. Define the *topological filtration* on $K_0(X)$ by letting $F_jK_0(X)$ to be the subgroup generated by coherent sheaves whose support has dimension at most j. Let $\operatorname{Gr} K_0(X)$ be the associated graded group. Then assigning to any subvariety $Y \subseteq X$ the class $[\mathcal{O}_Y]$ passes to rational equivalence (as follows from

Lemma 3.3.1) and hence defines a morphism $A_*(X) \to \operatorname{Gr} K_0(X)$ of graded abelian groups. This morphism is surjective ; it is an isomorphism over the rationals if, in addition, X is nonsingular (see [22] Example 15.2.16).

3.4. The Grothendieck group of the flag variety

The Chow group of the full flag variety X is isomorphic to its cohomology group and, in particular, is torsion-free. It follows that the associated graded of the Grothendieck group (for the topological filtration) is isomorphic to the cohomology group; this isomorphism maps the image of the structure sheaf $\mathcal{O}_Y$ of any subvariety, to the cohomology class $[Y]$. Thus, the following result may be viewed as a refinement in $K(X \times X)$ of the equality $[\operatorname{diag}(X)] = \sum_{w\in W}[X_w \times X^w]$ in $H^*(X \times X)$.

3.4.1. Theorem.

(i) *In* $K(X \times X)$ *holds*

$$[\mathcal{O}_{\operatorname{diag}(X)}] = \sum_{w\in W} [\mathcal{O}_{X_w}] \times [\mathcal{O}_{X^w}(-\partial X^w)].$$

(ii) *The bilinear map*

$$K(X) \times K(X) \to \mathbb{Z}, \quad (\alpha, \beta) \mapsto \chi(\alpha \cdot \beta)$$

is a nondegenerate pairing. Further, $\{[\mathcal{O}_{X_w}]\}$, $\{[\mathcal{O}_{X^w}(-\partial X^w)]\}$ *are bases of the abelian group* $K(X)$*, dual for this pairing.*

Proof. (i) By Theorem 3.1.2 and Lemma 3.3.1, we have

$$[\mathcal{O}_{\operatorname{diag}(X)}] = [\mathcal{O}_{\bigcup_{w\in W} X_w \times X^w}].$$

Further, $[\mathcal{O}_{\bigcup_{w\in W} X_w \times X^w}] = \sum_{w\in W}[\mathcal{O}_{X_w}] \times [\mathcal{O}_{X^w}(-\partial X^w)]$ by Proposition 3.1.3.

(ii) Let $p_1, p_2 : X \times X \to X$ be the projections. Let $\mathcal{E}$ be a locally free sheaf on X. Then we have by (i):

$$\begin{aligned}[\mathcal{E}|_{\operatorname{diag}(X)}] &= [p_2^*\mathcal{E}] \cdot [\mathcal{O}_{\operatorname{diag}(X)}]\\ &= \sum_{w\in W} [p_2^*\mathcal{E}] \cdot [p_1^*\mathcal{O}_{X_w} \otimes p_2^*\mathcal{O}_{X^w}(-\partial X^w)] = \sum_{w\in W} [p_1^*\mathcal{O}_{X_w} \otimes p_2^*\mathcal{E}|_{X^w}(-\partial X^w)].\end{aligned}$$

Applying $(p_1)_*$ to both sides and using the projection formula yields

$$[\mathcal{E}] = \sum_{w\in W} \chi(\mathcal{E}|_{X^w}(-\partial X^w))\, [\mathcal{O}_{X_w}].$$

Since the group $K(X)$ is generated by classes of locally free sheaves, it follows that

$$\alpha = \sum_{w\in W} \chi(\alpha \cdot [\mathcal{O}_{X^w}(-\partial X^w)])\, [\mathcal{O}_{X_w}]$$

for all $\alpha \in K(X)$. Thus, the classes $[\mathcal{O}_{X_w}]$ generate the group $K(X)$.

To complete the proof, it suffices to show that these classes are linearly independent. If $\sum_{w\in W} n_w\, [\mathcal{O}_{X_w}] = 0$ is a non-trivial relation in $K(X)$, then we may choose $v \in W$ maximal such that $n_v \neq 0$. Now a product $[\mathcal{O}_{X_w}] \cdot [\mathcal{O}_{X^v}]$ is non-zero only if $X_w \cap X^v$ is non-empty, i.e., $v \leq w$. Thus, we have by maximality of v:

$$0 = \sum_{w \in W} n_w [\mathcal{O}_{X_w}] \cdot [\mathcal{O}_{X^v}] = n_v [\mathcal{O}_{X_v}] \cdot [\mathcal{O}_{X^v}].$$

Further, we have $[\mathcal{O}_{X_v}] \cdot [\mathcal{O}_{X^v}] = [\mathcal{O}_{vF}]$. (Indeed, X_v and X^v meet transversally at the unique point vF; see Lemma 4.1.1 below for a more general result.) Further, $[\mathcal{O}_{vF}]$ is non-zero since $\chi(\mathcal{O}_{vF}) = 1$; a contradiction. □

We put for simplicity

$$\mathcal{O}_w := [\mathcal{O}_{X_w}] \quad \text{and} \quad \mathcal{I}_w := [\mathcal{O}_{X_w}(-\partial X_w)].$$

The $\mathcal{O}_w$ are the *Schubert classes* in $K(X)$. Further, $\mathcal{I}_w = [\mathcal{O}_{X_w}] - [\mathcal{O}_{\partial X_w}]$ by the exact sequence $0 \to \mathcal{O}_{X_w}(-\partial X_w) \to \mathcal{O}_{X_w} \to \mathcal{O}_{\partial X_w} \to 0$. We will express the $\mathcal{I}_w$ in terms of the $\mathcal{O}_w$, and vice versa, in Proposition 4.3.2 below.

Define likewise

$$\mathcal{O}^w := [\mathcal{O}_{X^w}] \quad \text{and} \quad \mathcal{I}^w := [\mathcal{O}_{X^w}(-\partial X^w)].$$

In other words, $\mathcal{O}^w = [\mathcal{O}_{w_0 X_{w_0 w}}]$ and $\mathcal{I}^w = [\mathcal{O}_{w_0 X_{w_0 w}}(-w_0 \partial X_{w_0 w})]$. But $[\mathcal{O}_{gY}] = [\mathcal{O}_Y]$ for any $g \in G$ and any subvariety $Y \subseteq X$. Indeed, this follows from Lemma 3.3.1 together with the existence of a connected chain of rational curves in G joining g to id (since the group G is generated by images of algebraic group homomorphisms $\mathbb{C} \to G$ and $\mathbb{C}^* \to G$). Thus,

$$\mathcal{O}^w = \mathcal{O}_{w_0 w} \quad \text{and} \quad \mathcal{I}^w = \mathcal{I}_{w_0 w}.$$

Now Theorem 3.4.1(ii) yields the equalities

$$\alpha = \sum_{w \in W} \chi(\alpha \cdot \mathcal{I}^w)\, \mathcal{O}_w = \sum_{w \in W} \chi(\alpha \cdot \mathcal{O}_w)\, \mathcal{I}^w,$$

for any $\alpha \in K(X)$.

3.4.2. Remarks.

1) Theorem 3.4.1 and the isomorphism $\operatorname{Gr} K(X) \cong H^*(X)$ imply that the classes $\mathcal{O}_w$ ($w \in W$, $\ell(w) \leq j$) form a basis of $F_j K(X)$; another basis of this group consists of the $\mathcal{I}_w$ ($w \in W$, $\ell(w) \leq j$).
2) All the results of this section extend to an arbitrary flag variety G/P by replacing W with the set W^P of minimal representatives.

3.4.3. Examples. 1) Consider the case where X is the projective space $\mathbb{P}^n$. Then the Schubert varieties are the linear subspaces $\mathbb{P}^j$, $0 \leq j \leq n$, and the corresponding opposite Schubert varieties are the $\mathbb{P}^{n-j}$. Further, $\partial \mathbb{P}^j = \mathbb{P}^{j-1}$ so that

$$[\mathcal{O}_{\mathbb{P}^j}(-\partial \mathbb{P}^j)] = [\mathcal{O}_{\mathbb{P}^j}] - [\mathcal{O}_{\mathbb{P}^{j-1}}] = [\mathcal{O}_{\mathbb{P}^j}(-1)].$$

Thus, $\{[\mathcal{O}_{\mathbb{P}^j}]\}$ is a basis of $K(\mathbb{P}^n)$ with dual basis $\{[\mathcal{O}_{\mathbb{P}^{n-j}}(-1)]\}$.

The group $K(\mathbb{P}^n)$ may be described more concretely in terms of polynomials, as follows. For each coherent sheaf $\mathcal{F}$ on $\mathbb{P}^n$, the function $\mathbb{Z} \to \mathbb{Z}$, $k \mapsto \chi(\mathcal{F}(k))$ is polynomial of degree equal to the dimension of the support of $\mathcal{F}$; this defines the

Hilbert polynomial $P_{\mathcal{F}}(t) \in \mathbb{Q}[t]$. Clearly, $P_{\mathcal{F}}(t) = P_{\mathcal{F}_1}(t) + P_{\mathcal{F}_2}(t)$ for any exact sequence $0 \to \mathcal{F}_1 \to \mathcal{F} \to \mathcal{F}_2 \to 0$. Thus, the Hilbert polynomial yields an additive map

$$P : K(\mathbb{P}^{n-1}) \to \mathbb{Q}[t], \quad [\mathcal{F}] \mapsto P_{\mathcal{F}}(t).$$

Since $\chi(\mathcal{O}_{\mathbb{P}^j}(k)) = \binom{k+j}{j}$, it follows that P maps the basis $\{[\mathcal{O}_{\mathbb{P}^j}]\}$ to the linearly independent polynomials $\{\binom{t+j}{j}\}$. Thus, P identifies $K(\mathbb{P}^{n-1})$ to the additive group of polynomials of degree $\leq n$ in one variable which take integral values at all integers. Note that P takes non-zero values at classes of non-trivial sheaves.

2) More generally, consider the case where X is a Grassmannian. Let L be the ample generator of $\operatorname{Pic}(X)$, then the boundary of each Schubert variety X_I (regarded as a reduced Weil divisor on X_I) is the divisor of the section p_I of $L|_{X_I}$; see Remark 1.4.6.3. Thus, we have an exact sequence

$$0 \to L^{-1}|_{X_I} \to \mathcal{O}_{X_I} \to \mathcal{O}_{\partial X_I} \to 0,$$

where the map on the left is the multiplication by p_I. It follows that

$$[\mathcal{O}_{X_I}(-\partial X_I)] = [L^{-1}|_{X_I}].$$

Thus, the dual basis of the basis of Schubert classes $\{\mathcal{O}_{X_I} := \mathcal{O}_I\}$ is the basis $\{[L^{-1}] \cdot \mathcal{O}^I\}$.

Notes. The cohomology class of the diagonal is discussed in [23] Appendix G, in a relative situation which yields a generalization of Lemma 3.1.1.

Our degeneration of the diagonal of a flag variety was first constructed in [5], by using canonical compactifications of adjoint semisimple groups; see [7] for further developments realizing these compactifications as irreducible components of Hilbert schemes. The direct construction of 3.1 follows [9] with some simplifications. In [loc.cit.], this degeneration was combined with vanishing theorems for unions of Richardson varieties, to obtain a geometric approach to standard monomial theory. Conversely, this theory also yields the degeneration of the diagonal presented here, see [41].

The normality criterion in 3.2 appears first in [8]. It is also proved there that a B-invariant multiplicity-free subvariety Y of a G-variety Z is normal and Cohen-Macaulay (resp. has rational singularities), if Y is normal and Cohen-Macaulay (resp. has rational singularities). This yields an alternative proof for the rationality of singularities of Schubert varieties.

The exposition in 3.3 is based on [3] regarding fundamental results on the Grothendieck ring $K(X)$, where X is any nonsingular variety, and on [22] regarding the relation of this ring to intersection theory on X. The reader will find another overview of K-theory in [11] together with several developments concerning degeneracy loci. In particular, a combinatorial expression for the structure constants of the Grothendieck ring of Grassmannians is presented there, after [10]. This yields another proof of the result in Example 3.4.3(ii); see the proof of Corollary 1 in [11].

The dual bases of the K-theory of the flag manifold presented in 3.4 appear in [43] for the variety of complete flags. In the general framework of T-equivariant

K-theory of flag varieties, they were constructed by Kostant and Kumar [38]. In fact, our approach fits into this framework. Indeed, T acts on $X \times X \times \mathbb{P}^1$ via $t(x,y,z) = (tx,ty,z)$. This action commutes with the $\mathbb{C}^*$-action via λ and leaves $\mathcal{X}$ invariant; clearly, the morphism $\pi : \mathcal{X} \to \mathbb{P}^1$ is T-invariant as well. Thus, π is a degeneration of T-varieties. Further, the filtration of $\mathcal{O}_{\pi^{-1}}(0)$ constructed in Proposition 3.1.3 is also T-invariant. So Theorem 3.4.1 extends readily to the T-equivariant Grothendieck group.

The idea of determining the (equivariant) class of a subvariety by an (equivariant) degeneration to a union of simpler subvarieties plays an essential role in the articles of Graham [25] on the structure constants of the equivariant cohomology ring of flag varieties, and of Knutson and Miller [36] on Schubert polynomials. These polynomials are special representatives of Schubert classes in the cohomology ring of the variety of complete flags. They were introduced by Lascoux and Schützenberger [42], [44] and given geometric interpretations in [24], [36]. Likewise, the Grothendieck polynomials are special representatives of Schubert classes in the Grothendieck ring of the complete flag variety, see [43] and [11].

It would be very interesting to have further examples of varieties with a torus action, where the diagonal admits an equivariant degeneration to a reduced union of products of subvarieties. The Bott-Samelson varieties should provide such examples; their T-equivariant Grothendieck ring has been described by Willems [69] with applications to equivariant Schubert calculus that generalize results of Duan [17].

4. Positivity in the Grothendieck group of the flag variety

Let Y be a subvariety of the full flag variety $X = G/B$. By the results of Section 3, we may write in the Grothendieck group $K(X)$:

$$[\mathcal{O}_Y] = \sum_{w \in W} c^w(Y)\, \mathcal{O}_w,$$

where the $\mathcal{O}_w = [\mathcal{O}_{X_w}]$ are the Schubert classes. Further, $c^w(Y) = 0$ unless $\dim(Y) \geq \dim(X_w) = \ell(w)$, and we have in the cohomology group $H^*(X)$:

$$[Y] = \sum_{w \in W,\, \ell(w) = \dim(Y)} c^w(Y)\, [X_w].$$

By Proposition 1.3.6, it follows that $c^w(Y) = \#(Y \cap gX^w)$ for general $g \in G$, if $\ell(w) = \dim(Y)$; in particular, $c^w(Y) \geq 0$ in this case.

One may ask for the signs of the integers $c^w(Y)$, where w is arbitrary. In this section, we show that these signs are *alternating*, i.e.,

$$(-1)^{\dim(Y) - \ell(w)} c^w(Y) \geq 0,$$

whenever Y has rational singularities (but not for arbitrary Y, see Remark 4.1.4.2).

We also show that the Richardson varieties have rational singularities, and we generalize to these varieties the results of Section 2 for cohomology groups of homogeneous line bundles on Schubert varieties. From this, we deduce that the structure constants of the ring $K(X)$ in its basis of Schubert classes have alternating signs as well, and we present several related positivity results.

Finally, we obtain a version in $K(X)$ of the Chevalley formula, that is, we decompose the product $[L_\lambda] \cdot \mathcal{O}_w$ in the basis of Schubert classes, where λ is any dominant weight, and X_w is any Schubert variety.

4.1. The class of a subvariety

In this subsection, we sketch a proof of the alternation of signs for the coefficients $c^w(Y)$. By Theorem 3.4.1, we have

$$c^w(Y) = \chi([\mathcal{O}_Y] \cdot [\mathcal{O}_{X^w}(-\partial X^w)]) = \chi([\mathcal{O}_Y] \cdot [\mathcal{O}_{X^w}]) - \chi([\mathcal{O}_Y] \cdot [\mathcal{O}_{\partial X^w}]).$$

Our first aim is to obtain a more tractable formula for $c^w(Y)$. For this, we need the following version of a lemma of Fulton and Pragacz (see [23] p. 108).

4.1.1. Lemma. *Let Y, Z be equidimensional Cohen-Macaulay subschemes of a nonsingular variety X. If Y meets Z properly in X, then the scheme-theoretic intersection $Y \cap Z$ is equidimensional and Cohen-Macaulay, of dimension* $\dim(Y) + \dim(Z) - \dim(X)$. *Further,*

$$Tor_i^X(\mathcal{O}_Y, \mathcal{O}_Z) = 0 = Tor_i^X(\omega_Y, \omega_Z)$$

for any $j \geq 1$, and $\omega_{Y \cap Z} = \omega_Y \otimes \omega_Z \otimes \omega_X^{-1}$.

Thus, we have in $K(X)$:

$$[\mathcal{O}_{Y \cap Z}] = [\mathcal{O}_Y] \cdot [\mathcal{O}_Z] \quad \textit{and} \quad [\omega_{Y \cap Z}] = [\omega_Y] \cdot [\omega_Z] \cdot [\omega_X^{-1}].$$

We also need another variant of Kleiman's transversality theorem (Lemma 1.3.1):

4.1.2. Lemma. *Let Y be a Cohen-Macaulay subscheme of the flag variety X and let $w \in W$. Then Y meets properly gX^w for general $g \in G$; further, $Y \cap gX^w$ is equidimensional and Cohen-Macaulay.*

If, in addition, Y is a variety with rational singularities, then $Y \cap gX^w$ is a disjoint union of varieties with rational singularities (again, for general $g \in G$).

We refer to [8] p. 142–144 for the proof of these results. Together with the fact that the boundary of any Schubert variety is Cohen-Macaulay (Corollary 2.2.7), they imply that

$$\begin{aligned} c^w(Y) &= \chi(\mathcal{O}_{Y \cap gX^w}) - \chi(\mathcal{O}_{Y \cap g\partial X^w}) \\ &= \chi(\mathcal{O}_{Y \cap gX^w}(-Y \cap g\partial X^w)) = \sum_{j=0}^{\dim(Y \cap gX^w)} (-1)^j \, h^j(\mathcal{O}_{Y \cap gX^w}(-Y \cap g\partial X^w)). \end{aligned}$$

Further, $\dim(Y \cap gX^w) = \dim(Y) + \dim(X^w) - \dim(X) = \dim(Y) - \ell(w)$. Thus, the assertion on the sign of $c^w(Y)$ will result from the following vanishing theorem, which holds in fact for any partial flag variety X.

4.1.3. Theorem. *Let $Y \subseteq X$ be a subvariety with rational singularities and let $Z \subseteq X$ be a Schubert variety. Then we have for general $g \in G$:*

$$H^j(Y \cap gZ, \mathcal{O}_{Y \cap gZ}(-Y \cap g\partial Z)) = 0 \quad \textit{whenever} \quad j < \dim(Y) + \dim(Z) - \dim(X).$$

Proof. First we present the argument in the simplest case, where $X = \mathbb{P}^n$ and Y is nonsingular. Then $Z = \mathbb{P}^j$ and $\mathcal{O}_Z(-\partial Z) = \mathcal{O}_{\mathbb{P}^j}(-1)$, see Example 3.4.3.1. Thus, $Y \cap gZ =: V$ is a general linear section of Y. By Bertini's theorem, V is nonsingular (and irreducible if its dimension is positive). Further, $\mathcal{O}_{Y \cap gZ}(-Y \cap g\partial Z) = \mathcal{O}_V(-1)$. Thus, we are reduced to showing the vanishing of $H^j(V, \mathcal{O}(-1))$ for $j < \dim(V)$, where V is a nonsingular subvariety of $\mathbb{P}^n$. But this follows from the Kodaira vanishing theorem.

Next we consider the case where X is a Grassmannian, and Y is allowed to have rational singularities. Let L be the ample generator of $\mathrm{Pic}(X)$ and recall that $\mathcal{O}_Z(-\partial Z) = L^{-1}|_Z$. It follows that $\mathcal{O}_{Y \cap gZ}(-Y \cap g\partial Z) = L^{-1}|_{Y \cap gZ}$. Further, by Lemma 4.1.2, $Y \cap gZ$ is a disjoint union of varieties with rational singularities, of dimension $\dim(Y) + \dim(Z) - \dim(X)$. Thus, it suffices to show that $H^j(V, L^{-1}) = 0$ whenever V is a variety with rational singularities, L is an ample line bundle on V, and $j < \dim(V)$. Let $\pi : \tilde{V} \to V$ be a desingularization and put $\tilde{L} := \pi^* L$. Since $R^i \pi_* \mathcal{O}_{\tilde{V}} = 0$ for any $i \geq 1$, we obtain isomorphisms $H^j(V, L^{-1}) \cong H^j(\tilde{V}, \tilde{L}^{-1})$ for all j. Thus, the Grauert-Riemenschneider theorem (see [19] Corollary 5.6) yields the desired vanishing.

The proof for arbitrary flag varieties goes along similar lines, but is much more technical. Like in the proof of Theorem 2.3.1, one applies the Kawamata-Viehweg theorem to a desingularization of $Y \cap gZ$; see [8] p. 153–156 for details. □

4.1.4. Remarks. 1) As a consequence of Theorem 4.1.3, we have

$$c^w(Y) = (-1)^{\dim(Y) - \ell(w)} \, h^{\dim(Y) - \ell(w)}(\mathcal{O}_{Y \cap gX^w}(-Y \cap g\partial X^w)).$$

By using Serre duality on $Y \cap gX^w$, it follows that

$$c^w(Y) = (-1)^{\dim(Y) - \ell(w)} \, h^0(Y \cap gX^w, L_\rho \otimes \omega_Y).$$

2) The property of alternation of signs for the coefficients of $[\mathcal{O}_Y]$ on Schubert varieties fails for certain (highly singular) subvarieties Y of a flag variety X. Indeed, there exist surfaces $Y \subset X = \mathbb{P}^4$ such that the coefficient of $[\mathcal{O}_Y]$ on $[\mathcal{O}_x]$ (where x is any point of $\mathbb{P}^4$) is arbitrarily negative.

Specifically, let $d \geq 3$ be an integer and let C be the image of the morphism $\mathbb{P}^1 \to \mathbb{P}^3$, $(x, y) \mapsto (x^d, x^{d-1}y, xy^{d-1}, y^d)$ (a closed immersion). Then C is a nonsingular rational curve of degree d in $\mathbb{P}^3$. Regarding C as a curve in $\mathbb{P}^4 \supset \mathbb{P}^3$, choose $x \in \mathbb{P}^4 \setminus \mathbb{P}^3$ and denote by $Y \subset \mathbb{P}^4$ the projective cone over C with vertex x, that is, the union of all projective lines containing x and meeting C. Then Y is a surface, so that we have by Example 3.4.3.1:

$$[\mathcal{O}_Y] = c_2(Y)\,[\mathcal{O}_{\mathbb{P}^2}] + c_1(Y)\,[\mathcal{O}_{\mathbb{P}^1}] + c_0(Y)\,[\mathcal{O}_x].$$

We claim that $c_0(Y) \leq 3 - d$.

To see this, first notice that $c_0(Y) = \chi(\mathcal{O}_Y(-1))$, as $\chi(\mathcal{O}_{\mathbb{P}^j}(-1)) = 0$ for all $j \geq 1$. Thus,

$$c_0(Y) = \chi(\mathcal{O}_Y) - \chi(\mathcal{O}_{Y\cap\mathbb{P}^3}) = \chi(\mathcal{O}_Y) - \chi(\mathcal{O}_C) = \chi(\mathcal{O}_Y) - 1.$$

To compute $\chi(\mathcal{O}_Y)$, consider the desingularization $\pi : Z \to Y$, where Z is the total space of the projective line bundle $\mathbb{P}(\mathcal{O}_C \oplus \mathcal{O}_C(-1))$ on C (that is, the blow-up of x in Y). Then we have an exact sequence

$$0 \to \mathcal{O}_Y \to \pi_*\mathcal{O}_Z \to \mathcal{F} \to 0,$$

where the sheaf $\mathcal{F}$ is supported at x. Further, $R^i\pi_*\mathcal{O}_Z = 0$ for all $i \geq 1$. (Indeed, since the affine cone $Y_0 := Y \setminus C$ is an affine neighborhood of x in Y, it suffices to show that $H^i(Z_0, \mathcal{O}_{Z_0}) = 0$ for $i \geq 1$, where $Z_0 := \pi^{-1}(Y_0)$. Now Z_0 is the total space of the line bundle $\mathcal{O}_C(-1) \cong \mathcal{O}_{\mathbb{P}^1}(-d)$ on $C \cong \mathbb{P}^1$, whence

$$H^i(Z_0, \mathcal{O}_{Z_0}) \cong \bigoplus_{n=0}^{\infty} H^i(\mathbb{P}^1, \mathcal{O}_{\mathbb{P}^1}(nd))$$

for any $i \geq 0$.) Thus, we obtain $\chi(\mathcal{O}_Y) = \chi(\mathcal{O}_Z) - \chi(\mathcal{F}) = 1 - h^0(\mathcal{F})$, so that $c_0(Y) = -h^0(\mathcal{F})$. Further, $\mathcal{F}$ identifies with the quotient $(\pi_*\mathcal{O}_{Z_0})/\mathcal{O}_{Y_0}$. Since $Y_0 \subset \mathbb{C}^4$ is the affine cone over $C \subset \mathbb{P}^3$, this quotient is a graded vector space with component of degree 1 being $H^0(\mathcal{O}_{\mathbb{P}^1}(d))/H^0(\mathcal{O}_{\mathbb{P}^3}(1))$, of dimension $d-3$. Thus, $h^0(\mathcal{F}) \geq d-3$. This completes the proof of the claim.

On the other hand, for any surface $Y \subset \mathbb{P}^n$, the coefficient $c_2(Y)$ is the degree of Y, a positive integer. Further, one checks that

$$c_1(Y) = \chi(\mathcal{O}_Y(-1)) - \chi(\mathcal{O}_Y(-2)) = \chi(\mathcal{O}_{Y\cap\mathbb{P}^{n-1}}(-1)) = -h^1(\mathcal{O}_{Y\cap\mathbb{P}^{n-1}}(-1))$$

for any hyperplane $\mathbb{P}^{n-1}$ which does not contain Y. Thus, $c_1(Y) \leq 0$.

Likewise, one may check that the property of alternation of signs holds for any curve in any flag variety. In other words, the preceding counterexample has the smallest dimension.

4.2. More on Richardson varieties

We begin with a vanishing theorem for these varieties that generalizes Theorem 2.3.1. Let v, w in W such that $v \leq w$ and let X_w^v be the corresponding Richardson variety. Then X_w^v has two kinds of boundaries, namely

$$(\partial X_w)^v := (\partial X_w) \cap X^v \quad \text{and} \quad (\partial X^v)_w := (\partial X^v) \cap X_w,$$

where $\partial X^v = X^v \setminus C^v = \bigcup_{u>v} X^u$ denotes the boundary of the opposite Schubert variety X^v. Define the total boundary by

$$\partial X_w^v := (\partial X_w)^v \cup (\partial X^v)_w,$$

this is a closed subset of pure codimension 1 in X_w^v.

We may now state

4.2.1. Theorem.

(i) *The Richardson variety* X_w^v *has rational singularities, and its dualizing sheaf equals* $\mathcal{O}_{X_w^v}(-\partial X_w^v)$. *Further, we have in* $K(X)$:

$$[\mathcal{O}_{X_w^v}] = \mathcal{O}_w \cdot \mathcal{O}^v = \mathcal{O}_w \cdot \mathcal{O}_{w_o v}.$$

(ii) $H^j(X_w^v, L_\lambda) = 0$ *for any* $j \geq 1$ *and any dominant weight* λ.

(iii) $H^j(X_w^v, L_\lambda(-(\partial X^v)_w)) = 0$ *for any* $j \geq 1$ *and any dominant weight* λ.

(iv) $H^j(X_w^v, L_\lambda(-\partial X_w^v)) = 0$ *for any* $j \geq 1$ *and any regular dominant weight* λ.

Proof. (i) follows from the rationality of singularities of Schubert varieties and the structure of their dualizing sheaves, together with Lemmas 4.1.1 and 4.1.2.

(ii) We adapt the proof of Theorem 2.3.1 to this setting. Choose a reduced decomposition $\underline{w}$ of w and let $Z_{\underline{w}}$ be the associated Bott-Samelson variety with morphism

$$\pi_{\underline{w}} : Z_{\underline{w}} \to X_w.$$

Likewise, a reduced decomposition $\underline{v}$ of v yields an opposite Bott-Samelson variety $Z^{\underline{v}}$ (defined via the opposite Borel subgroup B^- and the corresponding minimal parabolic subgroups) together with a morphism

$$\pi^{\underline{v}} : Z^{\underline{v}} \to X^v.$$

Now consider the fibered product

$$Z = Z_{\underline{w}}^{\underline{v}} := Z_{\underline{w}} \times_X Z^{\underline{v}}$$

with projection $\pi = \pi_{\underline{w}}^{\underline{v}} : Z_{\underline{w}}^{\underline{v}} \to X_w \cap X^v = X_w^v$. Using Kleiman's transversality theorem, one checks that $Z_{\underline{w}}^{\underline{v}}$ is a nonsingular variety and π is a desingularization of X_w^v. Let ∂Z be the union of the boundaries

$$(\partial Z_{\underline{w}})^{\underline{v}} := \partial Z_{\underline{w}} \times_X Z^{\underline{v}}, \quad (\partial Z^{\underline{v}})_{\underline{w}} := Z_{\underline{w}} \times_X \partial Z^{\underline{v}}.$$

This is a union of irreducible nonsingular divisors intersecting transversally, and one checks that $\omega_Z \cong \mathcal{O}_Z(-\partial Z)$.

Since X_w^v has rational singularities, it suffices to show that $H^j(Z, \pi^*\mathcal{L}_\lambda) = 0$ for $j \geq 1$. By Lemma 2.3.2 and the fact that Z is a subvariety of $Z_{\underline{w}} \times Z^{\underline{v}}$, the boundary ∂Z is the support of an effective ample divisor E on Z. Applying the Kawamata-Viehweg theorem with $D := N\partial Z - E$, where N is a large integer, and $\mathcal{L} := (\pi^*\mathcal{L}_\lambda)(\partial Z)$, we obtain the desired vanishing as in the proof of Theorem 2.3.1.

(iii) is checked similarly: let now E be the pull-back on Z of an effective ample divisor on $Z_{\underline{w}}$ with support $\partial Z_{\underline{w}}$. Let N be a large integer, and put $\mathcal{L} := (\pi^*\mathcal{L}_\lambda)((\partial Z_{\underline{w}})^{\underline{v}})$. Then the assumptions of the Kawamata-Viehweg theorem are still verified, since the projection $Z \to Z_{\underline{w}}$ is generically injective. Thus, we obtain

$$H^j(Z, (\pi^*\mathcal{L}_\lambda)(-(\partial Z^{\underline{v}})_{\underline{w}})) = 0 \quad \text{for} \quad j \geq 1.$$

This implies in turn that

$$R^j\pi_*\mathcal{O}_Z(-(\partial Z^{\underline{v}})_{\underline{w}}) = 0 \quad \text{for} \quad j \geq 1.$$

Together with the isomorphism

$$\pi_*\mathcal{O}_Z(-(\partial Z^{\underline{v}})_{\underline{w}}) = \mathcal{O}_{X_w^v}(-(\partial X^v)_w)$$

and a Leray spectral sequence argument, this completes the proof.

Likewise, (iv) follows from the vanishing of $H^j(Z, (\pi^*L_\lambda) \otimes \omega_Z)$ for $j \geq 1$. In turn, this is a consequence of the Grauert-Riemenschneider theorem, since L_λ is ample on X_w^v. □

4.2.2. Remarks.

1) One may also show that the restriction $H^0(\lambda) \to H^0(X_w^v, L_\lambda)$ is surjective for any dominant weight λ. As in Corollary 2.3.3, it follows that the affine cone over X_w^v has rational singularities in the projective embedding given by any ample line bundle on X. In particular, X_w^v is projectively normal in any such embedding.
2) Theorem 4.2.1 (iv) does not extend to all the dominant weights λ. Indeed, for $\lambda = 0$ we obtain
$$H^j(X_w^v, \mathcal{O}_{X_w^v}(-\partial X_w^v)) = H^j(X_w^v, \omega_{X_w^v}).$$
By Serre duality, this equals $H^{\ell(w)-\ell(v)-j}(X_w^v, \mathcal{O}_{X_w^v})$; i.e., $\mathbb{C}$ if $j = \ell(w)-\ell(v)$, and 0 otherwise, by Theorem 4.2.1 (iii).

Next we adapt the construction of Section 3 to obtain a degeneration of the diagonal of any Richardson variety X_w^v. Let $\lambda : \mathbb{C}^* \to T$ be as in Subsection 3.1 and let $\mathcal{X}_w^v$ be the closure in $X \times X \times \mathbb{P}^1$ of the subset

$$\{(x, \lambda(t)x, t) \mid x \in X_w^v,\ t \in \mathbb{C}^*\} \subseteq X \times X \times \mathbb{C}^*.$$

We still denote by $\pi : \mathcal{X}_w^v \to \mathbb{P}^1$ the projection, then $\pi^{-1}(\mathbb{C}^*)$ identifies again to the product $\mathrm{diag}(X_w^v) \times \mathbb{C}^*$ above $\mathbb{C}^*$. Further, we have the following analogues of Theorem 3.1.2 and Proposition 3.1.3.

4.2.3. Proposition.

(i) *With the preceding notation, we have equalities of subschemes of $X \times X$:*
$$\pi^{-1}(0) = \bigcup_{x\in W,\ v\leq x\leq w} X_x^v \times X_w^x \quad \textit{and} \quad \pi^{-1}(\infty) = \bigcup_{x\in W,\ v\leq x\leq w} X_w^x \times X_x^v.$$
(ii) *The sheaf $\mathcal{O}_{\pi^{-1}(0)}$ admits a filtration with associated graded*
$$\bigoplus_{x\in W,\ v\leq x\leq w} \mathcal{O}_{X_x^v} \otimes \mathcal{O}_{X_w^x}(-(\partial X^x)_w).$$
Therefore, we have in $K(X \times X)$:
$$[\mathcal{O}_{\mathrm{diag}(X_w^v)}] = \sum_{x\in W,\ v\leq x\leq w} [\mathcal{O}_{X_x^v}] \times [\mathcal{O}_{X_w^x}(-(\partial X^x)_w)].$$

Proof. Put

$$Y_w^v := \bigcup_{x \in W} X_x^v \times X_w^x.$$

By the argument of the proof of Theorem 3.1.2, we obtain the inclusion $Y_w^v \subseteq \pi^{-1}(0)$. Further, the proof of Proposition 3.1.3 shows that the structure sheaf $\mathcal{O}_{Y_w^v}$ admits a filtration with associated graded given by (ii).

On the other hand, Lemma 3.3.1 implies the equality $[\mathcal{O}_{\pi^{-1}(0)}] = [\mathcal{O}_{\mathrm{diag}(X_w^v)}]$ in $K(X \times X)$. Further, we have

$$[\mathcal{O}_{\mathrm{diag}(X_w^v)}] = [\mathcal{O}_{\mathrm{diag}(X)}] \cdot [\mathcal{O}_{X^v \times X_w}]$$

by Lemma 4.1.1, since $\mathrm{diag}(X)$ and $X^v \times X_w$ meet properly in $X \times X$ along $\mathrm{diag}(X_w^v)$. Together with Theorem 3.4.1 and Lemma 4.1.1 again, this yields

$$[\mathcal{O}_{\mathrm{diag}(X_w^v)}] = \sum_{x \in W} [\mathcal{O}_{X_x^v}] \times [\mathcal{O}_{X_w^x}(-(\partial X^x)_w)] = [\mathcal{O}_{Y_w^v}].$$

Thus, the structure sheaves of Y_w^v and of $\pi^{-1}(0)$ have the same class in $K(X \times X)$. But we have an exact sequence

$$0 \to \mathcal{F} \to \mathcal{O}_{\pi^{-1}(0)} \to \mathcal{O}_{Y_w^v} \to 0,$$

where $\mathcal{F}$ is a coherent sheaf on $X \times X$. So $[\mathcal{F}] = 0$ in $K(X \times X)$, and it follows that $\mathcal{F} = 0$ (e.g., by Example 3.4.3.1). In other words, $Y_w^v = \pi^{-1}(0)$. This proves (ii) and the first assertion of (i); the second assertion follows by symmetry. □

4.3. Structure constants and bases of the Grothendieck group

Let c_{vw}^x be the structure constants of the Grothendieck ring $K(X)$ in its basis $\{\mathcal{O}_w\}$ of Schubert classes, that is, we have in $K(X)$:

$$\mathcal{O}_v \cdot \mathcal{O}_w = \sum_{x \in W} c_{vw}^x \, \mathcal{O}_x.$$

Then Theorem 4.2.1 (i) yields the equality $c_{vw}^x = c^x(X_w^{w_o v})$. Together with Theorem 4.1.3, this implies a solution to Buch's conjecture:

4.3.1. Theorem. *The structure constants c_{vw}^x satisfy*

$$(-1)^{\ell(v)+\ell(w)+\ell(x)+\ell(w_o)} \, c_{vw}^x \geq 0.$$

Another consequence of Theorem 4.2.1 is the following relation between the bases $\{\mathcal{O}_w\}$ and $\{\mathcal{I}_w\}$ of the group $K(X)$ introduced in 3.4.

4.3.2. Proposition. *We have in $K(X)$*

$$\mathcal{O}_w = \sum_{v \in W,\, v \leq w} \mathcal{I}_v \quad \textit{and} \quad \mathcal{I}_w = \sum_{v \in W,\, v \leq w} (-1)^{\ell(w)-\ell(v)} \, \mathcal{O}_v.$$

Proof. By Theorem 3.4.1, we have

$$\mathcal{O}_w = \sum_{v \in W} \chi(\mathcal{O}_w \cdot \mathcal{O}^v) \, \mathcal{I}_v.$$

Further,

$$\chi(\mathcal{O}_w \cdot \mathcal{O}^v) = \chi(\mathcal{O}_{X_w^v}) = \sum_j (-1)^j\, h^j(\mathcal{O}_{X_w^v})$$

equals 1 if $v = w$, and 0 otherwise, by Theorem 4.2.1.

Likewise, we obtain

$$\mathcal{I}_w = \sum_{v\in W} \chi(\mathcal{I}_w \cdot \mathcal{I}^v)\, \mathcal{O}_v \quad \text{and} \quad \chi(\mathcal{I}_w \cdot \mathcal{I}^v) = \chi(\mathcal{O}_{X_w^v}(-\partial X_w^v)) = \chi(\omega_{X_w^v})$$

by using the equalities $\mathcal{I}_w = [\mathcal{O}_{X_w}] - [\mathcal{O}_{\partial X_w}]$, $\mathcal{I}^v = [\mathcal{O}_{X^v}] - [\mathcal{O}_{\partial X^v}]$, together with Lemma 4.1.2 and Cohen-Macaulayness of Schubert varieties and their boundaries. Further, we have by Serre duality and Theorem 4.2.1:

$$\chi(\omega_{X_w^v}) = (-1)^{\dim(X_w^v)}\, \chi(\mathcal{O}_{X_w^v}) = (-1)^{\ell(w)-\ell(v)}. \qquad \square$$

4.3.3. Remark. The preceding proposition implies that the Möbius function of the Bruhat order on W maps $(v,w) \in W \times W$ to $(-1)^{\ell(w)-\ell(v)}$ if $v \le w$, and to 0 otherwise. We refer to [15] for a direct proof of this combinatorial fact.

Using the duality involution $\alpha \mapsto \alpha^\vee$ of $K(X)$, we now introduce another natural basis of this group for which the structure constants become positive.

4.3.4. Proposition.

(i) *We have in* $K(X)$

$$[L_\rho|_{X_w}(-\partial X_w)] = (-1)^{\ell(w_o)-\ell(w)}\, \mathcal{O}_w^\vee.$$

In particular, the classes

$$\mathcal{I}_w(\rho) := [L_\rho|_{X_w}(-\partial X_w)] = [L_\rho]\cdot \mathcal{I}_w$$

form a basis of the Grothendieck group $K(X)$.

(ii) *For any Cohen-Macaulay subscheme* Y *of* X *with relative dualizing sheaf* $\omega_{Y/X} = \omega_Y \otimes \omega_X^{-1}$, *we have*

$$[\omega_{Y/X}] = \sum_{w\in W} (-1)^{\dim(Y)-\ell(w)}\, c^w(Y)\, \mathcal{I}_w(\rho).$$

Thus, if Y *is a variety with rational singularities, then the coordinates of* $\omega_{Y/X}$ *in the basis* $\{\mathcal{I}_w(\rho)\}$ *are the absolute values of the* $c^w(Y)$.

(iii) *The structure constants of* $K(X)$ *in the basis* $\{\mathcal{I}_w(\rho)\}$ *are the absolute values of the structure constants* c_{vw}^x.

Proof. We obtain

$$[\mathcal{O}_{X_w}]^\vee = (-1)^{\operatorname{codim}(X_w)}\, [\omega_{X_w}]\cdot[\omega_X^{-1}]$$
$$= (-1)^{\operatorname{codim}(X_w)}\, [L_{-\rho}|_{X_w}(-\partial X_w)]\cdot[L_{2\rho}] = (-1)^{\ell(w_o)-\ell(w)}\, \mathcal{I}_w(\rho).$$

This proves (i). The assertions (ii), (iii) follow by applying the duality involution to Theorems 4.1.3 and 4.3.1. $\square$

By similar arguments, we obtain the following relations between the bases $\{\mathcal{I}_w(\rho)\}$ and $\{\mathcal{O}_w\}$.

4.3.5. Proposition.

$$\mathcal{I}_w(\rho) = \sum_{v \in W} h_w^v \, \mathcal{O}_v, \quad \text{where} \quad h_w^v := h^0(X_w^v, L_\rho(-\partial X_w^v)).$$

In particular, $h_w^v \neq 0$ *only if* $v \leq w$. *Further,*

$$\mathcal{O}_w = \sum_{v \in W,\, v \leq w} (-1)^{\ell(w)-\ell(v)} h_w^v \, \mathcal{I}_v(\rho).$$

Next we consider the decomposition of the products $[L_\lambda] \cdot \mathcal{O}_w$ in the basis $\{\mathcal{O}_v\}$, where λ is a dominant weight. These products also determine the multiplication in $K(X)$. Indeed, by [52], this ring is generated by the classes of line bundles (using [22], Example 15.2.16, it follows that the cohomology ring is generated over the rationals by the Chern classes of line bundles). Since any weight is the difference of two dominant weights, it follows that the ring $K(X)$ is generated by the classes $[L_\lambda]$, where λ is dominant. This motivates the following:

4.3.6. Theorem. *For any dominant weight* λ *and any* $w \in W$, *we have in* $K(X)$

$$[L_\lambda] \cdot \mathcal{O}_w = [L_\lambda|_{X_w}] = \sum_{v \in W,\, v \leq w} h^0(X_w^v, L_\lambda(-(\partial X^v)_w)) \, \mathcal{O}_v.$$

In particular, the coefficients of $[L_\lambda] \cdot \mathcal{O}_w$ *in the basis of Schubert classes are non-negative.*

Proof. By Theorem 3.4.1, we have

$$[L_\lambda] \cdot \mathcal{O}_w = \sum_{v \in W} \chi([L_\lambda] \cdot \mathcal{O}_w \cdot \mathcal{I}^v) \, \mathcal{O}_v.$$

Further, as in the proof of Proposition 4.3.2, we obtain

$$\chi([L_\lambda] \cdot \mathcal{O}_w \cdot \mathcal{I}^v) = \chi(X_w^v, L_\lambda(-(\partial X^v)_w)).$$

The latter equals $h^0(X_w^v, L_\lambda(-(\partial X^v)_w))$ by Theorem 4.2.1. □

Next let σ be a non-zero section of L_λ on X. Then the structure sheaf of the zero subscheme $Z(\sigma) \subset X$ fits into an exact sequence

$$0 \to L_{-\lambda} \to \mathcal{O}_X \to \mathcal{O}_{Z(\sigma)} \to 0.$$

Thus, the class $[\mathcal{O}_{Z(\sigma)}] = 1 - [L_{-\lambda}]$ depends only on λ; we denote this class by $\mathcal{O}_\lambda$. Note that the image of $\mathcal{O}_\lambda$ in the associated graded $\operatorname{Gr} K(X) \cong H^*(X)$ is the class of the divisor of σ, i.e., the Chern class $c_1(L_\lambda)$. We now decompose the products $\mathcal{O}_\lambda \cdot \mathcal{O}_w$ in the basis of Schubert classes.

4.3.7. Proposition. *For any dominant weight* λ *and any* $w \in W$, *we have in* $K(X)$

$$\mathcal{O}_\lambda \cdot \mathcal{O}_w = \sum_{v \in W,\, v < w} (-1)^{\ell(w)-\ell(v)-1} \, h^0(X_w^v, L_\lambda(-(\partial X_w)^v)) \, \mathcal{O}_v.$$

Proof. We begin by decomposing the product $[L_\lambda]\cdot\mathcal{I}_w$ in the basis $\{\mathcal{I}_v\}$. As in the proof of Theorem 4.3.6, we obtain

$$\begin{aligned}[L_\lambda]\cdot\mathcal{I}_w &= \sum_{v\in W}\chi([L_\lambda]\cdot\mathcal{I}_w\cdot\mathcal{O}^v)\,\mathcal{I}_v\\ &= \sum_{v\in W,\, v\leq w}\chi(X_w^v, L_\lambda(-(\partial X_w)^v))\,\mathcal{I}_v = \sum_{v\in W,\, v\leq w} h^0(X_w^v, L_\lambda(-(\partial X_w)^v))\,\mathcal{I}_v.\end{aligned}$$

Applying the duality involution and using the equality

$$\mathcal{I}_w^\vee = (-1)^{\ell(w_o)-\ell(w)}\,[L_\rho]\cdot\mathcal{O}_w,$$

we obtain

$$[L_{-\lambda}]\cdot\mathcal{O}_w = \sum_{v\in W, v\leq w}(-1)^{\ell(w)-\ell(v)}\,h^0(X_w^v, L_\lambda(-(\partial X_w)^v))\,\mathcal{O}_v.$$

Further, $[L_{-\lambda}] = 1-\mathcal{O}_\lambda$. Substituting in the previous equality completes the proof. □

4.3.8. Remarks.

1) In the case of a fundamental weight χ_d, the divisor of the section $p_{w_o\chi_d}$ equals $[X_{w_os_d}]$, and hence $\mathcal{O}_{w_o\chi_d}$ is the class of the Schubert divisor $X_{w_os_d}$. Thus, Proposition 4.3.7 expresses the structure constants arising from the product of the classes of Schubert divisors by arbitrary Schubert classes. These structure constants have alternating signs as predicted by Theorem 4.3.1.

2) Proposition 4.3.7 gives back the Chevalley formula in $H^*(X)$ obtained in Proposition 1.4.3. Indeed, going to the associated graded $\operatorname{Gr} K(X)\cong H^*(X)$ yields

$$c_1(L_\lambda)\cup[X_w] = \sum_v h^0(X_w^v, L_\lambda(-(\partial X_w)^v))\,[X_v],$$

the sum over the $v\in W$ such that $v\leq w$ and $\ell(v)=\ell(w)-1$. For any such v, we know that the Richardson variety X_w^v is isomorphic to $\mathbb{P}^1$, identifying the restriction of L_λ to $\mathcal{O}_{\mathbb{P}^1}(\lambda_i-\lambda_j)$, where $v = ws_{ij}$ and $i<j$. Further, $(\partial X_w)^v$ is just the point vF, so that $L_\lambda|_{X_w^v}(-(\partial X_w)^v)$ identifies to $\mathcal{O}_{\mathbb{P}^1}(\lambda_i-\lambda_j-1)$. Thus, $h^0(X_w^v, L_\lambda(-(\partial X_w)^v)) = \lambda_i-\lambda_j$.

3) The results of this subsection adapt to any partial flag variety $X = G/P$. In particular, if X is the Grassmannian $\operatorname{Gr}(d,n)$ and L is the ample generator of $\operatorname{Pic}(X)$, then we have $L_{X_I}(-\partial X_I)\cong\mathcal{O}_{X_I}$, so that Theorem 4.3.6 yields the very simple formula

$$[L|_{X_I}] = [L]\cdot\mathcal{O}_I = \sum_{J,\, J\leq I}\mathcal{O}_J.$$

In particular, $[L]=\sum_I\mathcal{O}_I$ (sum over all the multi-indices I).

By Möbius inversion, it follows that $[L^{-1}] \cdot \mathcal{O}_I = \sum_{J,\, J \leq I} (-1)^{|I|-|J|} \mathcal{O}_J$. This yields

$$\mathcal{O}_{\omega_d} \cdot \mathcal{O}_I = \sum_{J,\, J<I} (-1)^{|I|-|J|-1} \mathcal{O}_J,$$

where $\mathcal{O}_{\omega_d}$ is the class of the Schubert divisor.

Notes. A general reference for this section is [6], from which much of the exposition is taken.

Stronger versions of Theorem 4.2.1 were obtained in [9] by the techniques of Frobenius splitting, and Proposition 4.2.3 was also proved there. These results also follow from standard monomial theory by work of Kreiman and Lakshmibai for Grassmannians [40], Lakshmibai and Littelmann in general [41].

Propositions 4.3.2, 4.3.4 (i) and 4.3.5 are due to Kostant and Kumar [38] in the framework of T-equivariant K-theory; again, the present approach is also valid in this framework.

Theorem 4.3.6 also extends readily to T-equivariant K-theory. In this form, it is due to Fulton and Lascoux [20] in the case of the general linear group. Then the general case was settled by Pittie and Ram [57], Mathieu [54], Littelmann and Seshadri [48], via very different methods. The latter authors obtained a more precise version by using standard monomial theory. Specifically, they constructed a B-stable filtration of the sheaf $L_\lambda|_{X_w}(-\partial X_w)$ with associated graded sheaf being the direct sum of structure sheaves of Schubert varieties (with twists by characters). This was generalized to Richardson varieties by Lakshmibai and Littelmann [41], again by using standard monomial theory.

This theory constructs bases for spaces of sections of line bundles over flag varieties, consisting of T-eigenvectors which satisfy very strong compatibility properties to Schubert and opposite Schubert varieties. It was completed by Littelmann [46], [47] after a series of important contributions of Lakshmibai, Musili, and Seshadri. Littelmann's approach is based on methods from combinatorics (the path model in representation theory) and algebra (quantum groups at roots of unity). It would be highly desirable to obtain a completely geometric derivation of standard monomial theory; some steps in this direction are taken in [9].

Another open problem is to obtain a positivity result for the structure constants of the T-equivariant Grothendieck ring. Such a result would imply both Theorem 4.3.1 and Graham's positivity theorem [25] for the structure constants in the T-equivariant cohomology ring. A precise conjecture in this direction is formulated in [27], where a combinatorial approach to T-equivariant K-theory of flag varieties is developed.

References

[1] S. Billey and V. Lakshmibai: *Singular loci of Schubert varieties*, Progr. Math. **182**, Birkhäuser, Boston (2000).

[2] S. Billey and G. Warrington: Maximal singular loci of Schubert varieties in $\mathrm{SL}(n)/B$, *Trans. Amer. Math. Soc.* **335** (2003), 3915–3945.

[3] A. Borel and J.-P. Serre: Le théorème de Riemann-Roch, *Bull. Soc. Math. France* **86** (1958), 97–136.

[4] R. Bott and H. Samelson: Application of the theory of Morse to symmetric spaces, *Amer. J. Math.* **80** (1958), 964–1029.

[5] M. Brion and P. Polo: Large Schubert varieties, *Represent. Theory* **4** (2000), 97–126.

[6] M. Brion: Positivity in the Grothendieck group of complex flag varieties, *J. Algebra* **258** (2002), 137–159.

[7] M. Brion: Group completions via Hilbert schemes, *J. Algebraic Geom.* **12** (2003), 605–626.

[8] M. Brion: Multiplicity-free subvarieties of flag varieties, in: *Commutative algebra, interactions with algebraic geometry*, Contemporary Math. **331**, AMS, Providence (2003), 13–24.

[9] M. Brion and V. Lakshmibai: A geometric approach to standard monomial theory, *Represent. Theory* **7** (2003), 651–680.

[10] A.S. Buch: A Littlewood-Richardson rule for the K-theory of Grassmannians, *Acta Math.* **189** (2002), 37–78.

[11] A.S. Buch: Combinatorial K-theory, *this volume.*

[12] C. Chevalley: Sur les décompositions cellulaires des espaces G/B, in: *Proceedings of Symposia in Pure Mathematics* **56**, Amer. Math. Soc., Providence (1994), 1–25.

[13] A. Cortez: Singularités génériques et quasi-résolutions des variétés de Schubert pour le groupe linéaire, *Adv. Math.* **178** (2003), 396–445.

[14] M. Demazure: Désingularisations des variétés de Schubert généralisées, *Ann. Sci. Éc. Norm. Supér.* **7** (1974), 53–88.

[15] V.V. Deodhar: Some characterizations of Bruhat ordering on a Coxeter group and determination of the relative Möbius function, *Invent. math.* **39** (1977), 187–198.

[16] V.V. Deodhar: On some geometric aspects of Bruhat orderings. I. A finer decomposition of Bruhat cells, *Invent. math.* **79** (1985), 499–511.

[17] H. Duan: Multiplicative rule of Schubert classes, *preprint available on* arXiv: math.AG/0306227.

[18] H. Duan: Morse functions and cohomology of homogeneous spaces, *this volume.*

[19] H. Esnault and E. Viehweg: *Lectures on vanishing theorems*, DMV Seminar Band **13**, Birkhäuser (1992).

[20] W. Fulton and A. Lascoux: A Pieri formula in the Grothendieck ring of a flag bundle, *Duke Math. J.* **76** (1994), 711–729.

[21] W. Fulton: *Young tableaux with applications to representation theory and geometry*, London Mathematical Society Student Texts **35**, Cambridge University Press (1997).

[22] W. Fulton: *Intersection theory* (second edition), Springer-Verlag (1998).

[23] W. Fulton and P. Pragacz: *Schubert varieties and degeneracy loci*, Lecture Notes in Mathematics **1689**, Springer-Verlag (1998).

[24] R. Goldin: The cohomology ring of weight varieties and polygon spaces, *Adv. Math.* **160** (2001), 175–204.

[25] W. Graham: Positivity in equivariant Schubert calculus, *Duke Math. J.* **109** (2001), 599-614.

[26] M. Greenberg and J. Harper: *Algebraic topology. A first course*, Benjamin/Cummings Publishing Co. (1981).

[27] S. Griffeth and A. Ram: Affine Hecke algebras and the Schubert calculus, *preprint available on* arXiv: math.RT/0405333.

[28] A. Grothendieck: Sur quelques points d'algèbre homologique, *Tohoku Math. J.* **9** (1957), 119–221.

[29] H.L. Hansen: On cycles in flag manifolds, *Math. Scand.* **33** (1973), 269–274.

[30] R. Hartshorne: *Algebraic Geometry*, Graduate Texts in Math. **52**, Springer-Verlag (1977).

[31] H. Hiller and B. Boe: Pieri formula for $\mathrm{SO}_{2n+1}/\mathrm{U}_n$ and $\mathrm{Sp}_n/\mathrm{U}_n$, *Adv. in Math.* **62** (1986), 49–67.

[32] W.V.D. Hodge: The intersection formulae for a Grassmannian variety, *J. London Math. Soc.* **17** (1942), 48–64.

[33] C. Kassel, A. Lascoux and C. Reutenauer: The singular locus of a Schubert variety, *J. Algebra* **269** (2003), 74–108.

[34] G. Kempf, F. Knudsen, D. Mumford and B. Saint-Donat: *Toroidal embeddings*, Lecture Notes in Math. **339**, Springer-Verlag (1973).

[35] S. Kleiman: The transversality of a general translate, *Compos. Math.* **28** (1974), 287–297.

[36] A. Knutson and E. Miller: Gröbner geometry of Schubert polynomials, *preprint available on* arXiv: math.AG/011058.

[37] B. Kostant and S. Kumar: The nil Hecke ring and cohomology of G/P for a Kac-Moody group G, *Adv. Math.* **62** (1986), 187–237.

[38] B. Kostant and S. Kumar: T-equivariant K-theory of generalized flag varieties, *J. Differential Geom.* **32** (1990), 549–603.

[39] S. Kumar: *Kac-Moody groups, their flag varieties and representation theory*, Progr. Math. **204**, Birkhäuser (2002).

[40] V. Kreiman and V. Lakshmibai: Richardson varieties in the Grassmannian, in: *Contributions to Automorphic Forms, Geometry and Number Theory: Shalikafest 2002*, Johns Hopkins University Press (2003), 573–597.

[41] V. Lakshmibai and P. Littelmann: Richardson varieties and equivariant K-theory, *J. Algebra* **260** (2003), 230–260.

[42] A. Lascoux and M.-P. Schützenberger: Polynômes de Schubert, *C. R. Acad. Sci. Paris Sér. 1 Math.* **294** (1982), 447–450.

[43] A. Lascoux and M.-P. Schützenberger: Symmetry and flag manifolds, in: *Invariant Theory (Proceedings, Montecatini 1982)*, Lecture Note in Math. **996**, Springer-Verlag (1983), 118–144.

[44] A. Lascoux and M.-P. Schützenberger: Interpolation de Newton à plusieurs variables, in: *Séminaire d'algèbre Paul Dubreil et Marie-Paule Malliavin*, Lecture Notes in Math. **1146**, Springer-Verlag (1985), 161–175.

[45] N. Lauritzen and J.F. Thomsen: Line bundles on Bott-Samelson varieties, *J. Algebraic Geom.* **13** (2004), 461–473.

[46] P. Littelmann: Contracting modules and standard monomial theory for symmetrizable Kac-Moody algebras, *J. Amer. Math. Soc.* **11** (1998), 551–567.

[47] P. Littelmann: The path model, the quantum Frobenius map and standard monomial theory, in: *Algebraic groups and their representations (Cambridge, 1997)*, Kluwer Acad. Publ. (1998), 175–212.

[48] P. Littelmann and C.S. Seshadri: A Pieri-Chevalley type formula for $K(G/B)$ and standard monomial theory, in: *Studies in memory of Issai Schur (Chevaleret/Rehovot, 2000)*, Progr. Math. **210**, Birkhäuser (2003), 155–176.

[49] L. Manivel: *Fonctions symétriques, polynômes de Schubert et lieux de dégénérescence*, Cours Spécialisés **3**, Soc. Math. France (1998).

[50] L. Manivel: Le lieu singulier des variétés de Schubert, *Internat. Math. Res. Notices* **16** (2001), 849–871.

[51] L. Manivel: Generic singularities of Schubert varieties, *preprint available on* arXiv: math.AG/0105239.

[52] R. Marlin: Anneaux de Grothendieck des variétés de drapeaux, *Bull. Soc. Math. France* **104** (1976), 337–348.

[53] O. Mathieu: Formules de caractères pour les algèbres de Kac-Moody générales, *Astérisque* **159–160** (1988).

[54] O. Mathieu: Positivity of some intersections in $K_0(G/B)$, *J. Pure Appl. Algebra* **152** (2000), 231–243.

[55] V. Mehta and A. Ramanathan: Frobenius splitting and cohomology vanishing for Schubert varieties, *Ann. of Math. (2)* **122** (1985), 27–40.

[56] D. Monk: The geometry of flag manifolds, *Proc. London Math. Soc.* **9** (1959), 253–286.

[57] H. Pittie and A. Ram: A Pieri-Chevalley formula in the K-theory of a G/B-bundle, *Electron. Res. Announc. Amer. Math. Soc.* **5** (1999), 102–107.

[58] P. Pragacz and J. Ratajski: Pieri type formula for isotropic Grassmannians; the operator approach, *Manuscripta Math.* **79** (1993), 127–151.

[59] P. Pragacz and J. Ratajski: A Pieri-type theorem for Lagrangian and odd orthogonal Grassmannians, *J. Reine Angew. Math.* **476** (1996), 143–189.

[60] P. Pragacz and J. Ratajski: A Pieri-type formula for even orthogonal Grassmannians, *Fund. Math.* **178** (2003), 49–96.

[61] P. Pragacz: Multiplying Schubert classes, *this volume.*

[62] S. Ramanan and A. Ramanathan: Projective normality of flag varieties and Schubert varieties, *Invent. Math.* **79** (1985), 217–224.

[63] A. Ramanathan: Equations defining Schubert varieties and Frobenius splitting of diagonals, *Pub. Math. IHES* **65** (1987), 61–90.

[64] R.W. Richardson: Intersections of double cosets in algebraic groups, *Indag. Math. (N.S.)* **3** (1992), 69–77.

[65] C.S. Seshadri: Line bundles over Schubert varieties, in: *Vector Bundles on Algebraic Varieties, Bombay Colloquium 1984*, Oxford University Press (1987), 499–528.

[66] T.A. Springer: *Linear algebraic groups* (second edition), Progr. Math. **9**, Birkhäuser (1998).

[67] T.A. Springer: Schubert varieties and generalizations, in: *Representation theories and algebraic geometry (Montreal, PQ, 1997)*, Kluwer Acad. Publ. (1998), 413–440.

[68] H. Tamvakis: Gromov-Witten invariants and quantum cohomology of Grassmannians, *this volume.*

[69] M. Willems: Cohomologie et K-théorie équivariantes des tours de Bott et des variétés de drapeaux, *Bull. Soc. Math. France* **132** (2004), 569–589.

Michel Brion
Institut Fourier, BP 74
F-38402 Saint-Martin d'Hères Cedex
France
e-mail: `Michel.Brion@ujf-grenoble.fr`

Trends in Mathematics:
Topics in Cohomological Studies of Algebraic Varieties, 87–104

Combinatorial K-theory

Anders Skovsted Buch

1. Introduction

Let X be an algebraic variety. The *Grothendieck ring* $K(X)$ of X is defined as the free abelian group generated by isomorphism classes $[F]$ of algebraic vector bundles on X, modulo the relations $[F] = [F'] + [F'']$ whenever there exists a short exact sequence $0 \to F' \to F \to F'' \to 0$. Multiplication is defined by the tensor product, $[F_1] \cdot [F_2] = [F_1 \otimes F_2]$. Any morphism $f : X \to X'$ of varieties defines a pullback ring homomorphism $f^* : K(X') \to K(X)$ given by $f^*([E]) = [f^* E]$.

In the following, we will assume that X is non-singular and quasi-projective. This implies that any coherent $\mathcal{O}_X$-module $\mathcal{G}$ has a resolution

$$0 \to F_n \to F_{n-1} \to \cdots \to F_1 \to F_0 \to \mathcal{G} \to 0$$

by locally free sheaves on X. One can therefore define the class of $\mathcal{G}$ in the Grothendieck ring by

$$[\mathcal{G}] = \sum_{i=0}^{n} (-1)^i [F_i] \in K(X) .$$

In particular, if $Y \subset X$ is a closed subvariety (or subscheme), we define the *Grothendieck class* of Y to be the class $[\mathcal{O}_Y]$ of its structure sheaf.

Suppose Z is another closed subvariety of X. In intersection theory one studies the geometry of the intersection $Y \cap Z$. This can be done using (Chow) cohomology by examining the product $[Y] \cdot [Z] \in H^*(X)$, or with K-theory by studying the product of Grothendieck classes $[\mathcal{O}_Y] \cdot [\mathcal{O}_Z] \in K(X)$. When Y and Z meet sufficiently transversally, these products give the respective classes of the intersection, in the sense that $[Y \cap Z] = [Y] \cdot [Z] \in H^*(X)$ and $[\mathcal{O}_{Y \cap Z}] = [\mathcal{O}_Y] \cdot [\mathcal{O}_Z] \in K(X)$. For example, one can show that these identities hold when Y and Z are Cohen-Macaulay, and all components of $Y \cap Z$ have dimension $\dim(Y) + \dim(Z) - \dim(X)$.

The Grothendieck ring has a *topological filtration* $K(X) = \mathcal{F}_0 \supset \mathcal{F}_1 \supset \cdots$ by ideals. The ideal $\mathcal{F}_i$ is generated by all classes $[\mathcal{O}_Y]$ for which $Y \subset X$ has codimension at least i. There is a group homomorphism $H^*(X) \to \operatorname{gr}(K(X)) = \bigoplus_{i \geq 0} \mathcal{F}_i / \mathcal{F}_{i+1}$ from the Chow ring to the associated graded ring of $K(X)$ given by

$[Y] \mapsto [\mathcal{O}_Y]$, and this map becomes an isomorphism of rings after tensoring with $\mathbb{Q}$ (see [11, Ex. 15.2.16]). The geometric meaning of this is that intersection theory in the cohomology ring only captures the lowest graded piece of information available in the Grothendieck ring.

Example 1. Let $C \subset \mathbb{P}^2$ be a curve of degree 3. Since C can be degenerated into a union of three lines, its cohomology class is $[C] = 3[\text{line}]$. The Grothendieck class is finer and can "see" that these lines overlap in three points. Thus $[\mathcal{O}_C] = 3[\mathcal{O}_{\text{line}}] - 3[\mathcal{O}_{\text{point}}]$. This example generalizes easily to a hypersurface of any degree in $\mathbb{P}^n$.

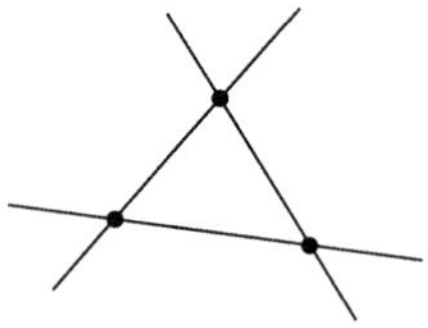

The goal of these notes is to explain the combinatorics that arises in the study of K-theoretic intersection theory of some concrete varieties and degeneracy loci. The material is mostly taken from the papers [5, 4], as well as from joint work [8] with A. Kresch, H. Tamvakis, and A. Yong.

Our notes are based on a lecture series presented at the *Mini-School on Schubert Varieties* at the Banach Center in the Spring of 2003. We thank Piotr Pragacz, Andrzej Weber, and the Banach Center for their kind invitation and hospitality. We also thank R. Vakil for private communication concerning a symmetry property of K-theoretic Schubert structure constants.

2. K-theory of Grassmannians

Let $X = \text{Gr}(d, n)$ be the Grassmann variety of d-dimensional subspaces of $\mathbb{C}^n$. For each partition $\lambda = (\lambda_1 \geq \cdots \geq \lambda_d \geq 0)$ with $\lambda_1 \leq n-d$, there is a Schubert variety

$$X_\lambda = \{V \in X \mid \dim(V \cap \mathbb{C}^{n-d+i-\lambda_i}) \geq i \ \forall 1 \leq i \leq d\}\,.$$

Here $\mathbb{C}^k \subset \mathbb{C}^n$ denotes the subspace of vectors where the last $n-k$ coordinates are zero. The codimension of X_λ in X is equal to the weight $|\lambda| = \sum_{i=1}^d \lambda_i$ of λ. We will identify the partition λ with its *Young diagram* of boxes, which has λ_1 boxes in the top row, λ_2 boxes in the second row, etc. For example, we write $(3,1,0) =$ 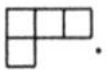. If we let $R = (n-d)^d$ denote a rectangular partition with d rows and $n-d$ columns, then the Schubert varieties in X correspond to the Young diagrams λ contained in R.

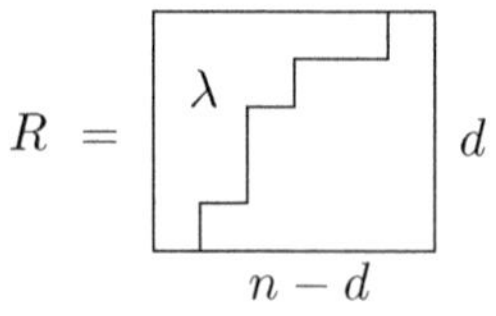

We will write $\mathcal{O}_\lambda = [\mathcal{O}_{X_\lambda}]$ for the Grothendieck class of the Schubert variety X_λ. In the notation of Brion's lectures we have $\mathcal{O}_\lambda = \mathcal{O}^I$ for $I = \{\lambda_d + 1, \dots, \lambda_1 + d\}$ (see [2, Ex. 3.4.3]). These classes form a basis for the Grothendieck ring of X:

$$K(X) = \bigoplus_{\lambda \subset R} \mathbb{Z} \cdot \mathcal{O}_\lambda \,.$$

As a first example of the multiplicative structure in this ring, we compute the Grothendieck class of the intersection of two general translates of the divisor $X_{(1)}$.

Example 2. Set $Y = X_{(1)} = X_{\square}$ and $Z = \{V \in X \mid V \cap M \neq 0\}$, where $M = \mathbb{C}^{n-d-1} \oplus 0 \oplus \mathbb{C} \oplus 0^{d-1} \subset \mathbb{C}^n$ is the sum of $\mathbb{C}^{n-d-1}$ and the line spanned by the $(n-d+1)$st standard basis vector in $\mathbb{C}^n$. Since Z is a translate of Y by the action of $\mathrm{GL}_n(\mathbb{C})$ on X, it follows that $[\mathcal{O}_Y] = [\mathcal{O}_Z] = \mathcal{O}_{\square}$. By the definitions we have $Y \cap Z = X_{\square\square} \cup X_{\substack{\square\\\square}}$. Since we have a reduced intersection $X_{\square\square} \cap X_{\substack{\square\\\square}} = X_{\substack{\square\square\\\square}}$, we obtain a short exact sequence

$$0 \to \mathcal{O}_{Y \cap Z} \to \mathcal{O}_{X_{\square\square}} \oplus \mathcal{O}_{X_{\substack{\square\\\square}}} \to \mathcal{O}_{X_{\substack{\square\square\\\square}}} \to 0 \,,$$

where the first map is given by $f \mapsto (f, f)$ and the second by $(f, g) \mapsto f - g$. Since Y and Z are Cohen-Macaulay, it follows that $\mathcal{O}_{\square} \cdot \mathcal{O}_{\square} = [\mathcal{O}_Y] \cdot [\mathcal{O}_Z] = [\mathcal{O}_{Y \cap Z}] = \mathcal{O}_{\square\square} + \mathcal{O}_{\substack{\square\\\square}} - \mathcal{O}_{\substack{\square\square\\\square}}$ in $K(X)$. We note that the geometric construction in this example is a special case of the degeneration techniques used by Vakil to compute cohomology products on Grassmannians [27].

More generally, given two partitions $\lambda, \mu \subset R$, we can write

$$\mathcal{O}_\lambda \cdot \mathcal{O}_\mu = \sum_\nu c^\nu_{\lambda\mu} \mathcal{O}_\nu$$

for unique structure constants $c^\nu_{\lambda\mu} \in \mathbb{Z}$. These structure constants depend only on the partitions λ, μ, ν, and not on our choice of Grassmann variety $X = \mathrm{Gr}(d, \mathbb{C}^n)$. Our first goal will be to describe these constants, which determine the K-theoretic intersection theory on X. Using the isomorphism $H^*(X) \cong \mathrm{gr}(K(X))$ it follows that the integer $c^\nu_{\lambda\mu}$ is non-zero only when $|\nu| \geq |\lambda| + |\mu|$. When we have equality $|\nu| = |\lambda| + |\mu|$, the coefficient $c^\nu_{\lambda\mu}$ is a structure constant in the cohomology ring. These cohomology constants are called *Littlewood-Richardson coefficients* and play an important role in numerous fields, including representation theory, geometry, and combinatorics, see, e.g., [13]. Lenart [22] has proved that the K-theoretic Pieri coefficients $c^\nu_{(p),\mu}$ can be expressed as plus or minus a binomial coefficient.

Set-valued tableaux

We will formulate a general rule for the structure constants $c^\nu_{\lambda\mu}$, and for this purpose we introduce some notation. Define a *set-valued tableau* to be a labeling of the boxes of a skew diagram with finite non-empty sets of positive integers, so that the numbers in the diagram increase weakly along rows and increase strictly down columns. More precisely, the maximum number of any box must be less than or equal to the minimum of the box to the right of it, and strictly smaller than the minimum of the box below it. For example

$T =$ 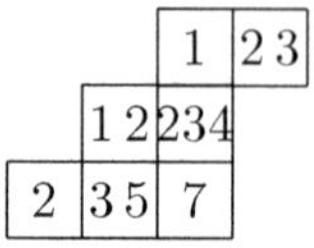

is a set-valued tableau whose shape is the skew diagram $(4,3,3)/(2,1)$. Define the *content* of a set-valued tableau to be the sequence $(c_1, c_2, \dots)$, where c_i is the number of boxes containing the integer i. We also define the *word* of a set-valued tableau to be the sequence of integers in its boxes when read from left to right then bottom to top; the integers of a single box are arranged in increasing order. The above tableau T has content $(2,4,3,1,1,0,1)$ and word $(2,3,5,7,1,2,2,3,4,1,2,3)$. A sequence of integers is called a *reverse lattice word* if every integer $i \geq 2$ in the sequence is followed by more $i-1$'s than i's.

Given two partitions λ and μ, we let $\lambda * \mu$ denote the skew shape obtained by attaching the Young diagrams for λ and μ corner to corner as shown.

$\lambda * \mu \quad =$ 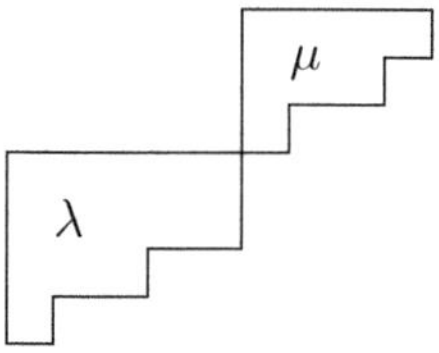

The following theorem from [5] is a K-theoretic generalization of the classical Littlewood-Richardson rule [23].

Theorem 1. *The structure constant $c^{\nu}_{\lambda\mu}$ is equal to $(-1)^{|\nu|-|\lambda|-|\mu|}$ times the number of set-valued tableaux of shape $\lambda * \mu$, such that the word is a reverse lattice word with content ν.*

Example 3. Let $\lambda = \mu = (2,1) =$  and $\nu = (3,3,1) =$ 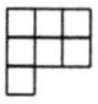. Then there are two set-valued tableaux of shape $\lambda * \mu$ such that the word is a reverse lattice word with content ν.

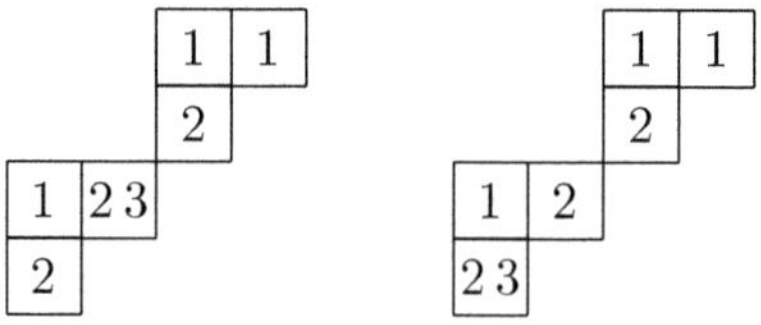

It follows that $c^{\nu}_{\lambda\mu} = -2$. The complete product of $\mathcal{O}_\lambda$ and $\mathcal{O}_\mu$ is given by

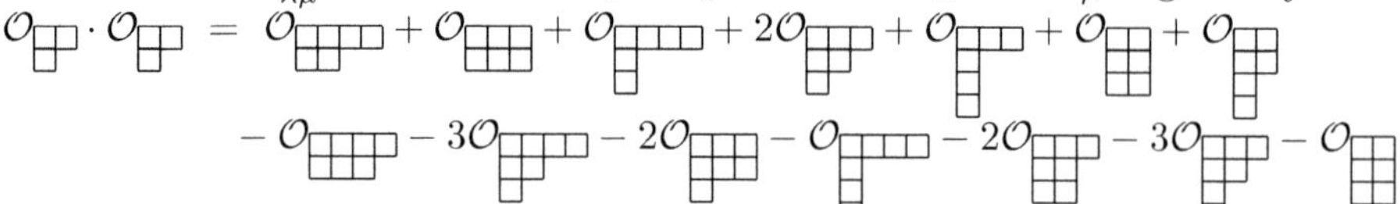

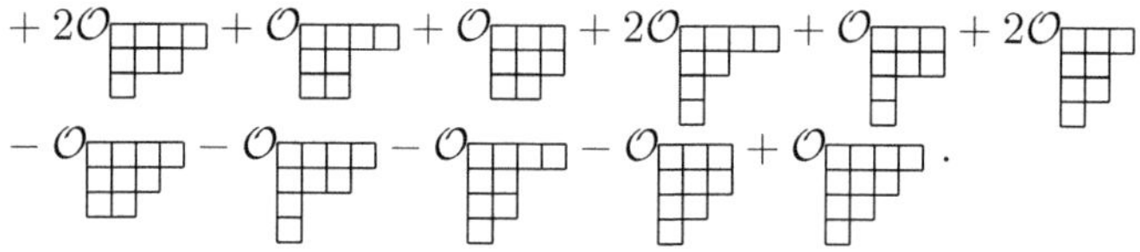

It follows from the theorem that K-theoretic structure constants on Grassmannians have signs that alternate with codimension, that is $(-1)^{|\nu|-|\lambda|-|\mu|} c^{\nu}_{\lambda\mu} \geq 0$. This alternation in sign has in fact been established by Brion for all homogeneous spaces G/P, where G is a simply connected semisimple algebraic group [3]. This is one of the topics discussed in Brion's lectures [2]. One can also derive from Theorem 1 that, if $\mathcal{O}_\nu$ occurs with non-zero coefficient in $\mathcal{O}_\lambda \cdot \mathcal{O}_\mu$, then we have $X_\lambda \cap X_\mu \supset X_\nu \supset \bigcup_\rho X_\rho$, where the last union is over all partitions ρ such that $[X_\rho]$ occurs in the cohomology product $[X_\lambda] \cdot [X_\mu] \in H^*(X)$. In Example 3 above, this says that all shapes occurring in the square of $\mathcal{O}_{(2,1)}$ are contained in the union of the shapes in the first row of terms.

Another consequence of the theorem is the following symmetry property of the K-theoretic Schubert structure constants $c^{\nu}_{\lambda\mu}$. For a partition $\lambda \subset R$ we let $\lambda^\vee = (n-d-\lambda_d, \ldots, n-d-\lambda_1)$ be the dual partition.

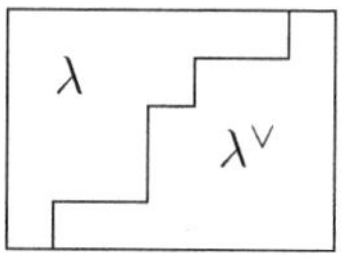

Corollary 1. *For any three partitions $\lambda, \mu, \nu \subset R$ we have $c^{\nu}_{\lambda\mu} = c^{\mu^\vee}_{\lambda\nu^\vee}$.*

Notice that when $|\nu| = |\lambda| + |\mu|$, this is immediate from the identity $c^{\nu}_{\lambda\mu} = \int_X [X_\lambda] \cdot [X_\mu] \cdot [X_{\nu^\vee}]$ for cohomological structure constants.

The proof given below was produced in response to a question of R. Vakil, who had observed that the corollary follows from a conjectured K-theoretic puzzle rule of A. Knutson and T. Tao. We note that this conjecture has now been proved by Knutson and Vakil by translating K-theoretic puzzles to set-valued tableaux [17].

Proof. Let $\rho : X \to \{\text{point}\}$ be a map to a point. The corresponding push-forward map $\rho_* : K(X) \to K(\text{point}) = \mathbb{Z}$ is the linear map given by $\rho_*(\mathcal{O}_\lambda) = 1$ for all $\lambda \subset R$. Define the bilinear form

$$K(X) \otimes K(X) \to \mathbb{Z} \quad ; \quad \alpha \otimes \beta \mapsto \rho_*(\alpha \cdot \beta)\,.$$

We claim that this is a perfect pairing, and that the dual basis element of $\mathcal{O}_\lambda$ equals $t\mathcal{O}_{\lambda^\vee}$, where $t = 1 - \mathcal{O}_{(1)}$. (This class t represents the top exterior power of the tautological subbundle on X.) The corollary is immediate from this claim since $c^{\nu}_{\lambda\mu} = \rho_*(\mathcal{O}_\lambda \cdot \mathcal{O}_\mu \cdot t\mathcal{O}_{\nu^\vee}) = c^{\mu^\vee}_{\lambda\nu^\vee}$.

It follows from Theorem 1 that $t\mathcal{O}_\lambda = \mathcal{O}_\lambda - \mathcal{O}_{(1)} \cdot \mathcal{O}_\lambda = \sum_\nu (-1)^{|\nu/\lambda|} \mathcal{O}_\nu$, where the sum is over all partitions ν such that $\lambda \subset \nu \subset R$ and ν/λ has at most one box in any row or column. This implies that $\rho_*(t\mathcal{O}_\lambda) = \delta_{\lambda,R}$, and we obtain

$$\rho_*(t\mathcal{O}_\lambda \cdot \mathcal{O}_\mu) = \sum_\nu c^\nu_{\lambda\mu}\, \rho_*(t\mathcal{O}_\nu) = c^R_{\lambda\mu}\,.$$

Finally, it is an easy exercise to check from Theorem 1 that $c^R_{\lambda\mu} = \delta_{\lambda,\mu^\vee}$. □

A geometric proof that the bases $\{\mathcal{O}_\lambda\}$ and $\{t\mathcal{O}_\lambda\}$ of $K(X)$ are dual to each other can be found in Brion's lectures [2] (see Thm. 3.4.1 and Ex. 3.4.3). In the notation used there, we have $t = [L^{-1}]$ where L is the ample generator of the Picard group of X.

3. The bialgebra of stable Grothendieck polynomials

For a set-valued tableau T with content $(c_1, c_2, \ldots, c_r)$, we set $x^T = x_1^{c_1} x_2^{c_2} \cdots x_r^{c_r}$ and $|T| = c_1 + c_2 + \cdots + c_r$. For any partition λ we define a *stable Grothendieck polynomial* for this partition by

$$G_\lambda = \sum_{\mathrm{sh}(T)=\lambda} (-1)^{|T|-|\lambda|}\, x^T \in \mathbb{Z}[\![x_1, x_2, \ldots]\!]\,.$$

The sum is over all set-valued tableaux T of shape λ. This is a special case of the stable Grothendieck polynomials studied by Fomin and Kirillov [9] (see Section 6). One can check from the definition that the power series G_λ is symmetric in the variables x_i.

The term of lowest total degree in G_λ is the Schur function s_λ. This power series can be expressed as the determinant $s_\lambda = \det(e_{\lambda'_i+j-i})_{1\le i,j\le \ell(\lambda')}$ where λ' is the conjugate partition of λ, defined so that λ'_i is the number of boxes in the ith column of λ, and e_i is the elementary symmetric function of degree i, i.e., the sum of all monomials which are products of i distinct variables. Given a vector bundle E on a variety X, one defines a characteristic class $s_\lambda(E) = \det(c_{\lambda'_i+j-i}(E)) \in H^*(X)$ by substituting the ith Chern class $c_i(E)$ for e_i. When $X = \mathrm{Gr}(d, \mathbb{C}^n)$ is a Grassmann variety and $S^\vee$ is the dual of the tautological subbundle $S \subset \mathbb{C}^n \times X$, then the classical Giambelli formula says that $s_\lambda(S^\vee)$ is equal to the cohomology class of the Schubert variety X_λ. As a consequence, the cohomology ring of a Grassmann variety is a quotient of the ring of symmetric functions $\Lambda = \bigoplus_\lambda \mathbb{Z} s_\lambda$.

The stable Grothendieck polynomials G_λ provide a K-theoretic analogue of this construction. For a vector bundle which can be written as a direct sum of line bundles, $E = L_1 \oplus \cdots \oplus L_e$, we define

$$G_\lambda(E) = G_\lambda(1 - L_1^{-1}, \ldots, 1 - L_e^{-1}, 0, 0, \ldots) \in K(X)\,.$$

Since G_λ is symmetric, this can be written as a polynomial in the exterior powers of the dual bundle $E^\vee$, so the definition also makes sense when E is not a direct sum of line bundles. With this notation, it follows from [21] or [14, Thm. 3] that $\mathcal{O}_\lambda = G_\lambda(S^\vee)$ in the Grothendieck ring of $\mathrm{Gr}(d, \mathbb{C}^n)$. We explain more about this in Section 6.

Example 4. We compute $G_{(2,1)}(1-a_1, 1-a_2, 1-a_3) = 1 - a_1a_2 - a_1a_3 - a_2a_3 + a_1^2a_2a_3 + a_1a_2^2a_3 + a_1a_2a_3^2 - a_1^2a_2^2a_3^2 = 1 - (a_1a_2 + a_1a_3 + a_2a_3) + (a_1 + a_2 + a_3)(a_1a_2a_3) - (a_1a_2a_3)^2$. It follows that for a vector bundle E of rank 3, we have $G_{(2,1)}(E) = 1 - [\bigwedge^2 E^\vee] + [E^\vee \otimes \bigwedge^3 E^\vee] - [\bigwedge^3 E^\vee]^2 \in K(X)$.

Since the lowest term of G_λ is the Schur function s_λ, it follows that the power series G_λ are linearly independent. We let Γ denote their linear span.

$$\Gamma = \bigoplus_\lambda \mathbb{Z}\, G_\lambda \subset \mathbb{Z}[\![x_1, x_2, \ldots]\!]$$

The following result generalizes Theorem 1.

Theorem 2. *For any partitions λ and μ we have*

$$G_\lambda \cdot G_\mu = \sum c_{\lambda\mu}^\nu \, G_\nu$$

where the sum is over all partitions ν, and the constants $c_{\lambda\mu}^\nu$ are given by Theorem 1.

It follows from this theorem that each product $G_\lambda \cdot G_\mu$ is a finite linear combination of stable Grothendieck polynomials G_ν. In particular, Γ is a subring of the power series ring $\mathbb{Z}[\![x_1, x_2, \ldots]\!]$. This ring provides a combinatorial model for the Grothendieck ring of a Grassmann variety. The topological filtration of the Grothendieck ring $K(X)$ corresponds to ideals $I_j = \bigoplus_{|\lambda| \geq j} \mathbb{Z}G_\lambda$ of Γ, and the ring of symmetric functions is recovered as the associated graded ring $\Lambda = \bigoplus_{j\geq 0} I_j/I_{j+1}$ of Γ.

For partitions λ, μ, and ν, we set $d_{\lambda\mu}^\nu = c_{\nu R}^\rho$, where R is a rectangular partition which is wider than λ and taller than μ, and $\rho = (R + \mu, \lambda)$ is the partition obtained by attaching λ and μ to the sides of R.

$$\rho \;=\; (R + \mu, \lambda) \;=$$

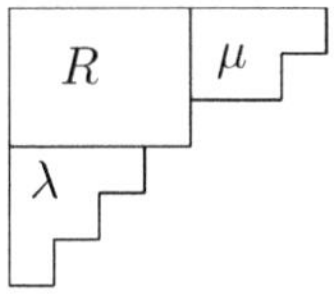

It is not hard to see that $d_{\lambda\mu}^\nu$ is independent of the choice of the rectangle R. Given a finite set of variables $x_1, \ldots, x_n$, we write $G_\lambda(x_1,\ldots,x_n) = G_\lambda(x_1,\ldots,x_n,0,0,\ldots)$. The coefficients $d_{\lambda\mu}^\nu$ have the following interpretation.

Theorem 3. *For any partition ν and integers $0 < m < n$ we have*

$$G_\nu(x_1, \ldots, x_m, x_{m+1}, \ldots, x_n) = \sum_{\lambda,\mu} d_{\lambda\mu}^\nu G_\lambda(x_1, \ldots, x_m) \cdot G_\mu(x_{m+1}, \ldots, x_n)\,.$$

Theorem 3 implies that we can define a coassociative and cocommutative coproduct $\Delta : \Gamma \to \Gamma \otimes \Gamma$ on Γ by setting

$$\Delta(G_\nu) = \sum_{\lambda,\mu} d_{\lambda\mu}^\nu \, G_\lambda \otimes G_\mu\,.$$

Example 5. Let $\nu = (2,1) = \square\!\!\square/\square$, i.e. the Young diagram of $(2,1)$. Using Theorem 2 we compute

$$\begin{aligned} G_{(2,1)} \cdot G_{(2,2)} = {} & G_{(4,3)} + G_{(4,2,1)} + G_{(3,3,1)} + G_{(3,2,2)} + G_{(3,2,1,1)} + G_{(2,2,2,1)} \\ & - 2G_{(4,3,1)} - G_{(4,2,2)} - G_{(3,3,2)} - G_{(4,2,1,1)} - G_{(3,3,1,1)} - 2G_{(3,2,2,1)} \\ & + G_{(4,3,2)} + G_{(4,3,1,1)} + G_{(4,2,2,1)} + G_{(3,3,2,1)} - G_{(4,3,2,1)} . \end{aligned}$$

It follows that

$$\begin{aligned} \Delta(G_{(2,1)}) = {} & 1 \otimes G_{(2,1)} + G_{(1)} \otimes G_{(2)} + G_{(1)} \otimes G_{(1,1)} + G_{(2)} \otimes G_{(1)} + G_{(1,1)} \otimes G_{(1)} + G_{(2,1)} \otimes 1 \\ & - 2G_{(1)} \otimes G_{(2,1)} - G_{(2)} \otimes G_{(2)} - G_{(2)} \otimes G_{(1,1)} - G_{(1,1)} \otimes G_{(2)} - G_{(1,1)} \otimes G_{(1,1)} - 2G_{(2,1)} \otimes G_{(1)} \\ & + G_{(2)} \otimes G_{(2,1)} + G_{(1,1)} \otimes G_{(2,1)} + G_{(2,1)} \otimes G_{(2)} + G_{(2,1)} \otimes G_{(1,1)} - G_{(2,1)} \otimes G_{(2,1)} . \end{aligned}$$

The coproduct Δ corresponds to the pullback along the embedding

$$\iota : \mathrm{Gr}(d_1, n_1) \times \mathrm{Gr}(d_2, n_2) \to \mathrm{Gr}(d_1 + d_2, n_1 + n_2)$$

which maps a pair $(V_1 \subset \mathbb{C}^{n_1}, V_2 \subset \mathbb{C}^{n_2})$ to $V_1 \oplus V_2 \subset \mathbb{C}^{n_1+n_2}$. In fact, if we let S_1, S_2, and S denote the tautological subbundles on $\mathrm{Gr}(d_1, n_1)$, $\mathrm{Gr}(d_2, n_2)$, and $\mathrm{Gr}(d_1 + d_2, n_1 + n_2)$, respectively, then $\iota^*(G_\nu(S)) = G_\nu(\iota^* S) = G_\nu(S_1 \oplus S_2) = \sum_{\lambda,\mu} d^\nu_{\lambda\mu} G_\lambda(S_1) \cdot G_\mu(S_2)$. Here we have identified S_1 and S_2 with their pullbacks to the product $\mathrm{Gr}(d_1, n_1) \times \mathrm{Gr}(d_2, n_2)$.

The coproduct Δ makes Γ into a commutative and cocommutative bialgebra. The linear map $\Gamma \to \mathbb{Z}$ which sends $1 = G_{(0)}$ to one and G_λ to zero for $|\lambda| > 0$ furthermore gives a counit. The bialgebra Γ also has an involution $\Gamma \to \Gamma$ defined by $G_\lambda \mapsto G_{\lambda'}$. This involution corresponds to the duality isomorphism $\mathrm{Gr}(d, n) \cong \mathrm{Gr}(n - d, n)$ of Grassmannians. Equivalently, the structure constants of Γ satisfy the identities $c^\nu_{\lambda\mu} = c^{\nu'}_{\lambda'\mu'}$ and $d^\nu_{\lambda\mu} = d^{\nu'}_{\lambda'\mu'}$.

4. Geometric specializations of stable Grothendieck polynomials

Given two vector bundles E and F over the variety X and a partition ν, we define

$$G_\nu(F - E) \;=\; \sum_{\lambda,\mu} d^\nu_{\lambda\mu}\, G_\lambda(F) \cdot G_{\mu'}(E^\vee) \;\in\; K(X)$$

where $E^\vee$ is the dual bundle of E and μ' is the conjugate partition of μ. It is proved in [5] that $G_\lambda(F - E)$ is a specialization of the double stable Grothendieck polynomials of Fomin and Kirillov [9]. It therefore follows from a super symmetry property of the latter polynomials that, for any vector bundle H we have

$$G_\lambda(F \oplus H - E \oplus H) = G_\lambda(F - E)\,. \tag{1}$$

In other words, G_λ gives a well-defined linear map $G_\lambda : K(X) \to K(X)$. Since the coproduct Δ is coassociative, it follows from this that

$$G_\nu(F-E) = \sum_{\lambda,\mu} d^\nu_{\lambda\mu}\, G_\lambda(F-H)\cdot G_\mu(H-E)\,. \tag{2}$$

Another useful identity is $G_\lambda(F-E) = G_{\lambda'}(E^\vee - F^\vee)$, which was first proved by Fomin (see [5, Lemma 3.4].)

If the bundles E and F have ranks e and f, and if λ and μ are partitions such that $\ell(\lambda) \leq f$ and $\mu_1 \leq e$, we also have the factorization formula

$$G_{(e^f)+\lambda,\mu}(F-E) = G_\lambda(F)\, G_{(e^f)}(F-E)\, G_\mu(-E) \tag{3}$$

where $(e^f)+\lambda,\mu$ denotes the partition $(e+\lambda_1,\dots,e+\lambda_f,\mu_1,\mu_2,\dots)$.

Given a proper morphism $\pi : X \to X'$ of varieties, there is a corresponding push-forward homomorphism $\pi_* : K(X) \to K(X')$, which is defined by $\pi_*([\mathcal{G}]) = \sum_{j\geq 0}(-1)^j[R^j\pi_*(\mathcal{G})]$ for any coherent sheaf $\mathcal{G}$ on X. The factorization formula (3) is a special case of the following Gysin formula from [4]. It generalizes Pragacz's cohomological Gysin formula from [25] and [15, (4.4)] to K-theory.

Theorem 4. *Let E and F be vector bundles of ranks e and f over X, and let $\pi : \mathrm{Gr}(d,F) \to X$ denote the Grassmann bundle of subbundles of F of rank d, with universal exact sequence $0 \to S \to \pi^*F \to Q \to 0$. Set $q = f-d$. Let λ and μ be partitions such that $\ell(\lambda) \leq q$, and $\lambda_q \geq \max(e, \mu_1+d)$. Then we have in $K(X)$ that*

$$\pi_*(G_\lambda(Q-\pi^*E)\cdot G_\mu(S-\pi^*E)) = G_{(\lambda_1-d,\dots,\lambda_q-d,\mu_1,\mu_2,\dots)}(F-E)\,.$$

Like Pragacz's cohomological Gysin formula, this result remains true with slightly weaker conditions. Namely, one may drop the requirement that $\lambda_q - d \geq \mu_1$. This means that the sequence $I = (\lambda_1-d,\dots,\lambda_q-d,\mu_1,\mu_2,\dots)$ is not necessarily a partition; the definition of G_I when I is an arbitrary sequence of integers is given in [4, §3], but will not be discussed here. It is curious to note that the same formula is also true in cohomology when λ and μ are arbitrary partitions such that $\ell(\lambda) \leq q$ and $\ell(\mu) \leq d$ [25], [15, (4.3)], but this case of the formula is false in K-theory. For example, if $F = \mathcal{O}_X^{\oplus 2}$ is a trivial bundle of rank 2 and $d = q = 1$, then one can check that $\pi_*(G_2(Q-\pi^*E)G_1(S-\pi^*E)) = G_{(1,1)}(-E) + G_{(3)}(-E) - G_{(3,1)}(-E)$, and this agrees with $G_{(1,1)}(F-E)$ only when E has rank at most two.

5. Degeneracy loci

Let $\phi : E \to F$ be a map of vector bundles of ranks e and f over X. Given an integer $t \leq \min(e,f)$, there is a degeneracy locus

$$\Omega_t = \Omega_t(\phi) = \{x \in X \mid \mathrm{rank}(\phi_x : E(x) \to F(x)) \leq t\}\,.$$

This locus has a natural structure of subscheme of X, given as the zero-section of the map $\wedge^{t+1}\phi : \wedge^{t+1}E \to \wedge^{t+1}F$. The expected (and maximal possible) codimension of Ω_t is the number $(e-t)(f-t)$. When this codimension is attained, the classical Thom-Porteous formula expresses the cohomology class of Ω_t as a Schur determinant (see [11, § 14.4]). This formula generalizes to K-theory as follows.

Theorem 5. *If Ω_t has codimension $(e-t)(f-t)$ in X then its Grothendieck class is given by*

$$[\mathcal{O}_{\Omega_t}] = G_{(e-t)^{f-t}}(F-E) \in K(X)\,.$$

The Thom-Porteous degeneracy locus Ω_t can be generalized as follows. Let $E_0 \to E_1 \to \cdots \to E_n$ be a sequence of vector bundles and bundle maps over X. A set of *rank conditions* for this sequence is a collection $r = \{r_{ij}\}$ of non-negative integers, for $0 \leq i < j \leq n$. This data defines the *quiver variety*

$$\Omega_r = \Omega_r(E_\bullet) = \{x \in X \mid \operatorname{rank}(E_i(x) \to E_j(x)) \leq r_{ij} \ \forall i < j\}\,.$$

We will demand that the rank conditions can occur, i.e., there exists a sequence of vector spaces and linear maps $V_0 \to V_1 \to \cdots \to V_n$ such that $\dim(V_i) = \operatorname{rank}(E_i)$ and $\operatorname{rank}(V_i \to V_j) = r_{ij}$ for all $i < j$. If we set $r_{ii} = \operatorname{rank}(E_i)$ for each i, then this condition is equivalent to demanding that $r_{ij} \leq \min\{r_{i,j-1}, r_{i+1,j}\}$ for all $0 \leq i < j \leq n$ and $r_{ij} + r_{i-1,j+1} \geq r_{i-1,j} + r_{i,j+1}$ for all $0 < i \leq j < n$ [1]. In this case, the expected codimension of the quiver variety Ω_r is equal to

$$d(r) = \sum_{i<j} (r_{i,j-1} - r_{ij})(r_{i+1,j} - r_{ij})\,.$$

In work with Fulton [7], we proved a formula for the cohomology class of a quiver variety Ω_r. The following theorem from [4] generalizes this formula to K-theory.

Theorem 6. *Let $E_0 \to E_1 \to \cdots \to E_n$ be a sequence of vector bundle maps such that the quiver variety $\Omega_r(E_\bullet)$ has codimension $d(r)$ in X. Then the Grothendieck class of this quiver variety is given by*

$$[\mathcal{O}_{\Omega_r}] = \sum c_\mu(r)\, G_{\mu_1}(E_1 - E_0) \cdot G_{\mu_2}(E_2 - E_1) \cdots G_{\mu_n}(E_n - E_{n-1}) \in K(X)$$

where the sum is over sequences $\mu = (\mu_1, \ldots, \mu_n)$ of partitions μ_i, and the coefficients $c_\mu(r)$ are integers depending on the rank conditions.

The coefficients $c_\mu(r)$ of this formula are called *quiver coefficients.* They are uniquely determined by the fact that the theorem is true for all varieties X, as well as the requirement that $c_\mu(r)$ is non-zero only if $\ell(\mu_i) \leq r_{ii}$ for all i. These coefficients can furthermore be computed by an explicit combinatorial algorithm based on the operations of the bialgebra Γ, see [4, §4].

A quiver coefficient $c_\mu(r)$ is non-zero only when $\sum |\mu_i| \geq d(r)$. When $\sum |\mu_i| = d(r)$, the coefficient $c_\mu(r)$ also appears in the formula for the cohomology class of Ω_r; these coefficients are called *cohomological quiver coefficients.* It was conjectured in [7] and proved by Knutson, Miller, and Shimozono in [16] that cohomological quiver coefficients are non-negative. The K-theoretic quiver coefficients were conjectured to have alternating signs, in the following sense.[1]

[1] After these lectures were presented, this conjecture from [4] has been proved in the papers [6] and [24].

Conjecture 1. *For any rank conditions r and sequence of partitions μ, we have*

$$(-1)^{\sum |\mu_i| - d(r)}\, c_\mu(r) \geq 0\,.$$

We explain how to prove Theorem 6 for a sequence of three vector bundles $E \to F \to H$ of ranks e, f, and h. In this case the quiver variety is defined as

$$\Omega = \{x \in X \mid \operatorname{rank}(E \to F) \leq r, \operatorname{rank}(F \to H) \leq s, \operatorname{rank}(E \to H) \leq t\}$$

for some non-negative integers r, s, and t. The requirement that these rank conditions can occur says that $r \leq \min(e, f)$, $s \leq \min(f, h)$, $t \leq \min(r, s)$, and $f + t \geq r + s$.

To make sure that the bundle maps are maximally generic, we start by replacing X with the bundle $\operatorname{Hom}(E, F) \oplus \operatorname{Hom}(F, H)$ and the bundle maps with the universal maps $E \to F \to H$ on this bundle. (In the rest of this section we avoid explicit notation for pullbacks of bundles.)

Form the product $Y = \operatorname{Gr}(r, F) \times_X \operatorname{Gr}(s, H)$ of Grassmann bundles with projection $\pi : Y \to X$, and let $F' \subset F$ and $H' \subset H$ denote the tautological subbundles on Y. On the subset $Z = Z(E \to F/F') \cap Z(F \to H/H') \subset Y$, the bundle map from E to F factors through F', and the map $F \to H$ factors through H'.

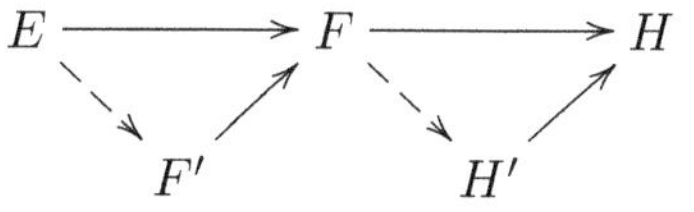

This gives a map $F' \to F \to H'$ on Z, and we let $\Omega_t = \Omega_t(F' \to H') \subset Z$ denote the corresponding Thom-Porteous locus.

Since all the rank conditions are satisfied on Ω_t, it follows that $\pi(\Omega_t) \subset \Omega$. In fact, by using that the bundle maps are universal, one can show that π maps Ω_t birationally onto Ω (see [7, Lemma 1].) Since quiver varieties have rational singularities [18] and π is proper, it follows that $\pi_*(\mathcal{O}_{\Omega_t}) = \mathcal{O}_\Omega$ and $R^j\pi_*(\mathcal{O}_{\Omega_t}) = 0$ for all $j \geq 1$, so we get

$$\pi_*([\mathcal{O}_{\Omega_t}]) \;=\; [\mathcal{O}_\Omega] \;\in K(X)\,.$$

Let $R = (r-t)^{s-t}$ be a rectangle with $s-t$ rows and $r-t$ columns. By Theorem 5 and equations (2) and (3) we have

$$\begin{aligned}
[\mathcal{O}_{\Omega_t}] &= G_{(e)^{f-r}}(F/F' - E)\, G_{(f)^{h-s}}(H/H' - F)\, G_R(H' - F') \\
&= G_{(e)^{f-r}}(F/F' - E)\, G_{(f)^{h-s}}(H/H' - F) \sum_{\lambda,\mu} d^R_{\lambda\mu}\, G_\lambda(F/F')\, G_\mu(H' - F) \\
&= \sum_{\lambda,\mu} d^R_{\lambda\mu}\, G_{(e)^{f-r}+\lambda}(F/F' - E)\, G_{(f)^{h-s}}(H/H' - F)\, G_\mu(H' - F)
\end{aligned}$$

in the Grothendieck ring of Y. Using Theorem 4 we therefore obtain

$$[\mathcal{O}_\Omega] = \pi_*([\mathcal{O}_{\Omega_t}]) = \sum_{\lambda,\mu} d^R_{\lambda\mu}\, G_{(e-r)^{f-r}+\lambda}(F - E) \cdot G_{(f-s)^{h-s},\mu}(H - F)$$

in $K(X)$ as required. Notice that the quiver coefficients for a sequence of three bundles are just the coproduct coefficients $d^R_{\lambda,\mu}$. In particular, they have alternating signs.

Example 6. Suppose $\operatorname{rank}(E) = \operatorname{rank}(F) = 5$, $\operatorname{rank}(H) = 4$, $r = 3$, $s = 2$, and $t = 1$. Then we have $(e-r)^{f-r} = (2,2)$, $(f-s)^{h-s} = (3,3)$, and $R = (r-t)^{s-t} = (2)$. Since $\Delta(G_R) = G_{(2)} \otimes 1 + G_{(1)} \otimes G_{(1)} + 1 \otimes G_{(2)} - G_{(2)} \otimes G_{(1)} - G_{(1)} \otimes G_{(2)}$, we obtain

$$\begin{aligned}[\mathcal{O}_\Omega] = {} & G_{(4,2)}(F-E)\, G_{(3,3)}(H-F) + G_{(3,2)}(F-E)\, G_{(3,3,1)}(H-F) \\ & + G_{(2,2)}(F-E)\, G_{(3,3,2)}(H-F) - G_{(4,2)}(F-E)\, G_{(3,3,1)}(H-F) \\ & - G_{(3,2)}(F-E)\, G_{(3,3,2)}(H-F)\,.\end{aligned}$$

6. Grothendieck polynomials

For each permutation $w \in S_{n+1}$, there is a *double Grothendieck polynomial*

$$\mathfrak{G}_w(x;y) = \mathfrak{G}_w(x_1,\dots,x_n;y_1,\dots,y_n)$$

defined as follows. If $w = w_0$ is the longest permutation in S_{n+1}, then we set

$$\mathfrak{G}_{w_0}(x;y) = \prod_{i+j\leq n+1} (x_i + y_j - x_i y_j)\,.$$

Otherwise there is a simple transposition $s_i \in S_{n+1}$ such that $\ell(ws_i) = \ell(w) + 1$. In this case we set

$$\mathfrak{G}_w(x;y) = \pi_i(\mathfrak{G}_{ws_i})$$

where the *Demazure operator* π_i is defined by

$$\pi_i(f) = \frac{(1-x_{i+1})f(x_1,\dots,x_i,x_{i+1},\dots,x_n) - (1-x_i)f(x_1,\dots,x_{i+1},x_i,\dots,x_n)}{x_i - x_{i+1}}\,.$$

Since the Demazure operators satisfy the Coxeter relations $\pi_i\pi_{i+1}\pi_i = \pi_{i+1}\pi_i\pi_{i+1}$ and $\pi_i\pi_j = \pi_j\pi_i$ for $|i-j| \geq 2$, it follows that this definition is independent of the choice of s_i. The specialization $\mathfrak{G}_w(x_1,\dots,x_n) = \mathfrak{G}_w(x_1,\dots,x_n;0,\dots,0)$ is called the single Grothendieck polynomial for w.

Grothendieck polynomials were introduced by Lascoux and Schützenberger [21, 19] as representatives for the Grothendieck classes of Schubert varieties in a flag variety. More generally, the double Grothendieck polynomials have the following interpretation. Let $F_1 \subset F_2 \subset \cdots \subset F_n \to H_n \twoheadrightarrow \cdots \twoheadrightarrow H_2 \twoheadrightarrow H_1$ be a sequence of bundles on X consisting of a full flag $F_\bullet$ with a general map to a dual full flag $H_\bullet$, such that $\operatorname{rank}(F_i) = \operatorname{rank}(H_i) = i$ for all i. Given a permutation $w \in S_{n+1}$, Fulton [12] defined the degeneracy locus

$$\Omega_w = \Omega_w(F_\bullet \to H_\bullet) = \{x \in X \mid \operatorname{rank}(F_q(x) \to H_p(x)) \leq r_w(p,q)\ \forall p,q\}\,,$$

where $r_w(p,q) = \#\{i \leq p \mid w(i) \leq q\}$. Fulton proved that the cohomology class of Ω_w is given by a double Schubert polynomial [12]. This was generalized to the following formula for the Grothendieck class in [4, Thm. 2.1] as an application of [14, Thm. 3].

Theorem 7. *If the codimension of Ω_w equals the length of w, then*

$$[\mathcal{O}_{\Omega_w}] = \mathfrak{G}_w(x;y)$$

where $x_i = 1 - L_i^{-1} \in K(X)$ for $L_i = \ker(H_i \to H_{i-1})$, and $y_i = 1 - M_i \in K(X)$ for $M_i = F_i/F_{i-1}$.

Fulton's degeneracy locus Ω_w is a special case of a quiver variety, that is $\Omega_w = \Omega_r(F_\bullet \to H_\bullet)$ where $r = \{r_{ij}\}$ is the set of rank conditions defined by

$$r_{ij} = \begin{cases} r_w(2n+1-j, i) & \text{if } i \le n < j \\ i & \text{if } j \le n \\ 2n+1-j & \text{if } i \ge n+1. \end{cases}$$

It follows from this that $\mathfrak{G}_w(x;y) = [\mathcal{O}_{\Omega_r}]$, so

$$\mathfrak{G}_w(x;y) = \sum_\mu c_{w,\mu}\, G_{\mu_1}(F_2 - F_1) \cdots G_{\mu_n}(H_n - F_n) \cdots G_{\mu_{2n-1}}(H_1 - H_2) \qquad (4)$$

where the constants $c_{w,\mu} = c_\mu(r)$ are quiver coefficients. Note that $G_{\mu_i}(F_{i+1} - F_i) = G_{\mu_i}(M_{i+1})$ is either zero or a power of $1 - M_{i+1}^{-1} = 1 - (1 - y_{i+1})^{-1}$, which can be expressed as a formal power series in y_{i+1}. Similarly $G_{\mu_{2n-i}}(H_i - H_{i+1}) = G_{\mu_{2n-i}}(-L_{i+1})$ is a power of $1 - (1 - x_{i+1})^{-1}$. In Section 7 we describe a proof that the coefficients $c_{w,\mu}$ have alternating signs.

Let $Y = \mathrm{Fl}(n+1)$ be the variety of complete flags in $\mathbb{C}^{n+1}$, and $S_1 \subset \cdots \subset S_n \subset \mathbb{C}^{n+1} \times Y$ the tautological flag of subbundles on Y. If we let $H_\bullet$ be the dualized flag $S_n^\vee \twoheadrightarrow S_{n-1}^\vee \twoheadrightarrow \cdots \twoheadrightarrow S_1^\vee$ and let $F_\bullet \subset \mathbb{C}^{n+1} \times Y$ be the trivial flag given by $F_p = (0^{n+1-p} \oplus \mathbb{C}^p) \times Y$, then the degeneracy locus $\Omega_w(F_\bullet \to H_\bullet) \subset Y$ is identical to the Schubert variety denoted by $Y_{w_0 w}$ in Brion's lectures (see [2, Def. 1.2.2]). It follows that the single Grothendieck polynomial $\mathfrak{G}_w(x_1, \ldots, x_n)$ gives a formula for the Grothendieck class of this Schubert variety in $K(Y)$ [21].

The double Grothendieck polynomials are related to stable Grothendieck polynomials as follows. Given a permutation $w \in S_{n+1}$, we let $1^m \times w \in S_{m+n+1}$ denote the shifted permutation, which acts as the identity on the set $\{1, 2, \ldots, m\}$ and maps $m+i$ to $m + w(i)$ for $i \ge 1$. The *stable double Grothendieck polynomial* $G_w(x;y)$ of Fomin and Kirillov [9] is defined by

$$G_w(x_1, \ldots, x_m; y_1, \ldots, y_m) = \mathfrak{G}_{1^m \times w}(x_1, \ldots, x_m, 0, \ldots, 0; y_1, \ldots, y_m, 0, \ldots, 0)\,.$$

Let $\lambda = (\lambda_1 \ge \cdots \ge \lambda_d \ge 0)$ be a partition. The corresponding Grassmannian permutation w_λ with descent at position d is defined by $w_\lambda(i) = i + \lambda_{d+1-i}$ for $i \le d$ and $w_\lambda(i) < w_\lambda(i+1)$ for $i \ne d$. With the notation of Section 3, we have $G_\lambda(x) = G_{w_\lambda}(x;0)$ and

$$G_{w_\lambda}(x;y) = G_\lambda(x;y) = \sum_{\sigma,\tau} d^\lambda_{\sigma\tau}\, G_\sigma(x) G_{\tau'}(y)\,.$$

One can also show that the single Grothendieck polynomial for w_λ is given by $\mathfrak{G}_{w_\lambda}(x) = G_\lambda(x_1, \ldots, x_d)$, but no such identity holds for the double Grothendieck polynomial $\mathfrak{G}_{w_\lambda}(x; y)$.

Let $X = \mathrm{Gr}(d, \mathbb{C}^{n+1})$ and let $\rho : Y = \mathrm{Fl}(n+1) \to X$ be the projection that maps a complete flag $V_\bullet \subset \mathbb{C}^{n+1}$ to its subspace V_d of dimension d. Then we have $\rho^{-1}(X_\lambda) = \Omega_{w_\lambda} = \Omega_{w_\lambda}(F_\bullet \to H_\bullet)$. Since $X_\lambda \subset X$ and $\Omega_{w_\lambda} \subset Y$ have the same codimension $|\lambda| = \ell(w_\lambda)$, and since Schubert varieties are Cohen-Macaulay [26], it follows from this that $\rho^*(\mathcal{O}_\lambda) = [\mathcal{O}_{\Omega_{w_\lambda}}] = \mathfrak{G}_{w_\lambda}(x_1, \ldots, x_n)$. Since the Grothendieck classes of the Schubert varieties in Y are linearly independent, this implies that the pullback map $\rho^* : K(X) \to K(Y)$ is injective. If $S \subset \mathbb{C}^{n+1} \times X$ is the tautological subbundle on X, then we also have $\rho^*(G_\lambda(S^\vee)) = G_\lambda(\rho^* S^\vee) = G_\lambda(H_d) = G_\lambda(x_1, \ldots, x_d) = \mathfrak{G}_{w_\lambda}(x_1, \ldots, x_n)$, which shows that $\mathcal{O}_\lambda = G_\lambda(S^\vee)$ in the Grothendieck ring of X.

Every stable Grothendieck polynomial $G_w(x; y)$ can be written as a finite linear combination of the polynomials $G_\lambda(x; y)$ given by partitions [5]. We will describe a formula of Lascoux for the coefficients of this linear combination, which shows that they have alternating signs.

Given a permutation w, let r be the last descent position of w, i.e., r is maximal such that $w(r) > w(r+1)$. Set $w' = w\tau_{rk}$ where $k > r$ is maximal such that $w(r) > w(k)$. We also set $I(w) = \{i < r \mid \ell(w'\tau_{ir}) = \ell(w)\}$. Define a relation $\rhd$ on the set of all permutations as follows. If $I(w) = \emptyset$ we write $w \rhd v$ if and only if $v = 1 \times w$. Otherwise we write $w \rhd v$ if and only if there exist elements $i_1 < \cdots < i_p$ of $I(w)$, $p \geq 1$, such that $v = w'\tau_{i_1 r} \ldots \tau_{i_p r}$. The following is an immediate consequence of [20, Thm. 4].

Theorem 8 (Lascoux). *For any permutation w we have*

$$G_w = \sum_\lambda a_{w,\lambda}\, G_\lambda$$

where the sum is over all partitions λ, and $a_{w,\lambda}$ is equal to $(-1)^{|\lambda| - \ell(w)}$ times the number of sequences $w = w_1 \rhd w_2 \rhd \cdots \rhd w_m$ such that $w_m = w_\lambda$ is a Grassmannian permutation for λ and w_i is not Grassmannian for $i < m$.

Example 7. For the permutation $w = 2\,1\,4\,3$, we get $r = 3$, $k = 4$, $w' = 2\,1\,3\,4$, and $I(w) = \{1, 2\}$. The sequences of permutations of Theorem 8 are $w \rhd 3\,1\,2\,4 = w_{(2)}$, $w \rhd 2\,3\,1\,4 = w_{(1,1)}$, and $w \rhd 3\,2\,1\,4 \rhd 1\,4\,3\,2\,5 \rhd 2\,4\,1\,3\,5 = w_{(2,1)}$. It follows that $G_w = G_{\square\!\square} + G_{\begin{smallmatrix}\square\\\square\end{smallmatrix}} - G_{\begin{smallmatrix}\square\!\square\\\square\;\;\end{smallmatrix}}$.

7. Alternating signs of the coefficients $c_{w,\mu}$

In this section we outline a proof that the quiver coefficients $c_{w,\mu}$ of (4) have alternating signs, based on our joint paper [8] with Kresch, Tamvakis, and Yong.

Suppose we are given a sequence of vector bundles

$$F_1 \subset F_2 \subset \cdots \subset F_n \subset V \twoheadrightarrow H_n \twoheadrightarrow \cdots \twoheadrightarrow H_2 \twoheadrightarrow H_1$$

consisting of a full flag $F_\bullet$ of a bundle V of rank $n+1$, followed by a dual full flag $H_\bullet$ of V. Set $F_i' = V/F_i$ and $H_i' = \ker(V \twoheadrightarrow H_i)$. We then obtain a sequence

$$H_n' \subset \cdots \subset H_1' \subset V \twoheadrightarrow F_1' \twoheadrightarrow \cdots \twoheadrightarrow F_n' ,$$

and it is an easy exercise to show that $\Omega_w(F_\bullet \to H_\bullet) = \Omega_{\widehat{w}}(H_\bullet' \to F_\bullet')$ as subschemes of X, where $\widehat{w} = w_0 w^{-1} w_0$. It follows that $\mathfrak{G}_w(x;y) = [\mathcal{O}_{\Omega_{\widehat{w}}}]$, so we obtain

$$\begin{aligned}\mathfrak{G}_w(x;y) &= \sum_\mu c_{\widehat{w},\mu} G_{\mu_1}(H_{n-1}' - H_n') \cdots G_{\mu_n}(F_1' - H_1') \cdots G_{\mu_{2n-1}}(F_n' - F_{n-1}') \\ &= \sum_\mu c_{\widehat{w},\mu} G_{\mu_1}(x_n) \cdots G_{\mu_n}(x_1; y_1) \cdots G_{\mu_{2n-1}}(0; y_n)\end{aligned}$$

where x_i and y_i are defined as in Theorem 7. Notice that $G_\lambda(x_i)$ equals x_i^p when $\lambda = (p)$ is a single row with p boxes, and is zero otherwise. Similarly $G_\lambda(0; y_i)$ is a power of y_i or zero, and furthermore $G_\lambda(x_1; y_1) = \sum_{\sigma,\tau} d_{\sigma\tau}^\lambda G_\sigma(x_1) G_{\tau'}(y_1)$.

Corollary 2. *The monomial coefficients of Grothendieck polynomials are special cases of the quiver coefficients $c_{w,\mu}$.*

The *degenerate Hecke algebra* is the free $\mathbb{Z}$-algebra $\mathcal{H}$ generated by symbols $s_1, s_2, \ldots$, modulo the relations $s_i^2 = s_i$ and $s_i s_{i+1} s_i = s_{i+1} s_i s_{i+1}$ for all i, and $s_i s_j = s_j s_i$ for $|i-j| \geq 2$. This algebra has a basis of permutations, corresponding to reduced expressions in the generators.

Now define the *universal Grothendieck polynomial* for the permutation $w \in S_n$ to be the element

$$P_w = \sum_\mu c_{w,\mu}\, G_{\mu_1} \otimes \cdots \otimes G_{\mu_{2n-1}} \in \Gamma^{\otimes 2n-1} .$$

The following theorem gives an explicit formula for these polynomials.

Theorem 9. *For $w \in S_{n+1}$ we have*

$$P_w = \sum (-1)^{\sum \ell(u_i) - \ell(w)}\, G_{u_1} \otimes G_{u_2} \otimes \cdots \otimes G_{u_{2n-1}} \in \Gamma^{\otimes 2n-1} ,$$

where the sum is over all factorizations $w = u_1 \cdot u_2 \cdots u_{2n-1}$ in the degenerate Hecke algebra $\mathcal{H}$ such that $u_i \in S_{\min(i,2n-i)+1}$ for each i.

This theorem combined with Lascoux's formula for the expansion of stable Grothendieck polynomials in the basis of Γ implies the following explicit formula for the quiver coefficients $c_{w,\mu}$.

Corollary 3. *The quiver coefficients $c_{w,\mu}$ of P_w are given by*

$$c_{w,\mu} = (-1)^{\sum |\mu_i| - \ell(w)} \sum \prod_{i=1}^{2n-1} |a_{u_i,\mu_i}| ,$$

where the sum is over all factorizations $w = u_1 \cdot u_2 \cdots u_{2n-1}$ in the degenerate Hecke algebra $\mathcal{H}$ such that $u_i \in S_{\min(i,2n-i)+1}$ for each i, and the constants a_{u_i,μ_i} are given by Theorem 8.

The proof of Theorem 9 is based on some identities of universal Grothendieck polynomials. For $1 \leq i \leq j \leq 2n-1$ we define

$$P_w[i,j] = \sum_{\mu:\mu_k=\emptyset \text{ for } k\notin[i,j]} c_{w,\mu}\, 1\otimes\cdots\otimes 1\otimes G_{\mu_i}\otimes\cdots\otimes G_{\mu_j}\otimes 1\otimes\cdots\otimes 1\,.$$

Using a Cauchy identity for double Grothendieck polynomials of Fomin and Kirillov [10] as well as some geometry of quiver varieties, one obtains the identity

$$P_w = \sum_{u\cdot v=w\in\mathcal{H}} (-1)^{\ell(uvw)} P_u[1,i]\cdot P_v[i+1,2n-1]\,.$$

By iterating this formula, we obtain

$$P_w = \sum_{u_1\cdots u_{2n-1}=w\in\mathcal{H}} (-1)^{\sum \ell(u_i)-\ell(w)}\, P_{u_1}[1,1]\cdot P_{u_2}[2,2]\cdots P_{u_{2n-1}}[2n-1,2n-1]\,.$$

Finally we use the identity

$$P_w[i,i] = \begin{cases} 1^{\otimes i-1}\otimes G_w\otimes 1^{\otimes 2n-1-i} & \text{if } w\in S_{\min(i,2n-i)+1} \\ 0 & \text{otherwise.}\end{cases}$$

This was proved in [4] for $i=n$, and the remaining cases are easy consequences of this. Theorem 9 follows immediately from these identities.

References

[1] S. Abeasis and A. Del Fra, *Degenerations for the representations of an equioriented quiver of type A_m*, Boll. Un. Mat. Ital. Suppl. (1980), no. 2, 157–171. MR **84e:**16019

[2] M. Brion, *Lectures on the geometry of flag varieties*, this volume.

[3] ______, *Positivity in the Grothendieck group of complex flag varieties*, J. Algebra **258** (2002), no. 1, 137–159, Special issue in celebration of Claudio Procesi's 60th birthday. MR **2003m:**14017

[4] A.S. Buch, *Grothendieck classes of quiver varieties*, Duke Math. J. **115** (2002), no. 1, 75–103. MR **2003m:**14018

[5] ______, *A Littlewood-Richardson rule for the K-theory of Grassmannians*, Acta Math. **189** (2002), no. 1, 37–78. MR **2003j:**14062

[6] ______, *Alternating signs of quiver coefficients*, preprint, 2003.

[7] A.S. Buch and W. Fulton, *Chern class formulas for quiver varieties*, Invent. Math. **135** (1999), no. 3, 665–687. MR **2000f:**14087

[8] A.S. Buch, A. Kresch, H. Tamvakis, and A. Yong, *Grothendieck polynomials and quiver formulas*, To appear in Amer. J. Math., 2003.

[9] S. Fomin and A.N. Kirillov, *Grothendieck polynomials and the Yang-Baxter equation*, Proc. Formal Power Series and Alg. Comb. (1994), 183–190.

[10] ______, *The Yang-Baxter equation, symmetric functions, and Schubert polynomials*, Discrete Math. **153** (1996), 123–143.

[11] W. Fulton, *Intersection theory*, Ergebnisse der Mathematik und ihrer Grenzgebiete (3) [Results in Mathematics and Related Areas (3)], vol. 2, Springer-Verlag, Berlin, 1984. MR **85k:**14004

[12] ———, *Flags, Schubert polynomials, degeneracy loci, and determinantal formulas*, Duke Math. J. **65** (1992), no. 3, 381–420. MR **93e:**14007

[13] ———, *Young tableaux*, London Mathematical Society Student Texts, vol. 35, Cambridge University Press, Cambridge, 1997, With applications to representation theory and geometry. MR **99f:**05119

[14] W. Fulton and A. Lascoux, *A Pieri formula in the Grothendieck ring of a flag bundle*, Duke Math. J. **76** (1994), no. 3, 711–729. MR **96j:**14036

[15] W. Fulton and P. Pragacz, *Schubert varieties and degeneracy loci*, Lecture Notes in Mathematics, vol. 1689, Springer-Verlag, Berlin, 1998, Appendix J by the authors in collaboration with I. Ciocan-Fontanine. MR **99m:**14092

[16] A. Knutson, E. Miller, and M. Shimozono, *Four positive formulas for type A quiver polynomials*, preprint, 2003.

[17] A. Knutson and R. Vakil, manuscript in preparation.

[18] V. Lakshmibai and P. Magyar, *Degeneracy schemes, quiver schemes, and Schubert varieties*, Internat. Math. Res. Notices (1998), no. 12, 627–640. MR **99g:**14065

[19] A. Lascoux, *Anneau de Grothendieck de la variété de drapeaux*, The Grothendieck Festschrift, Vol. III, Progr. Math., vol. 88, Birkhäuser Boston, Boston, MA, 1990, pp. 1–34. MR **92j:**14064

[20] ———, *Transition on Grothendieck polynomials*, Physics and combinatorics, 2000 (Nagoya), World Sci. Publishing, River Edge, NJ, 2001, pp. 164–179. MR **2002k:**14082

[21] A. Lascoux and M.-P. Schützenberger, *Structure de Hopf de l'anneau de cohomologie et de l'anneau de Grothendieck d'une variété de drapeaux*, C. R. Acad. Sci. Paris Sér. I Math. **295** (1982), no. 11, 629–633. MR **84b:**14030

[22] C. Lenart, *Combinatorial aspects of the K-theory of Grassmannians*, Ann. Comb. **4** (2000), no. 1, 67–82. MR **2001j:**05124

[23] D.E. Littlewood and A.R. Richardson, *Group characters and algebra*, Phil. Trans. R. Soc., A **233** (1934), 99–141.

[24] E. Miller, *Alternating formulae for K-theoretic quiver polynomials*, preprint, 2003.

[25] P. Pragacz, *Enumerative geometry of degeneracy loci*, Ann. Sci. École Norm. Sup. (4) **21** (1988), no. 3, 413–454. MR **90e:**14004

[26] A. Ramanathan, *Schubert varieties are arithmetically Cohen-Macaulay*, Invent. Math. **80** (1985), no. 2, 283–294. MR **MR788411 (87d:**14044)

[27] R. Vakil, *A geometric Littlewood-Richardson rule*, preprint, 2003.

Anders Skovsted Buch
Matmatisk Institut, Århus Universitet
Ny Munkegade
DK-8000 Århus C, Denmark
e-mail: abuch@imf.au.dk

Trends in Mathematics:
Topics in Cohomological Studies of Algebraic Varieties, 105–133

Morse Functions and Cohomology of Homogeneous Spaces

Haibao Duan

This article arose from a series of three lectures given at the Banach Center, Warsaw, during period of 24 March to 13 April, 2003.

Morse functions are useful tool in revealing the geometric formation of its domain manifolds M. They define handle decompositions of M from which the additive homologies $H_*(M)$ may be constructed. In these lectures two further questions were emphasized.

(1) How to find a Morse function on a given manifold?
(2) From Morse functions can one derive the multiplicative cohomology in addition to the additive homology?

It is not our intention here to make detailed studies of these questions. Instead, we will illustrate by examples solutions to them for some classical manifolds as homogeneous spaces.

I am very grateful to Piotr Pragacz for the opportunity to speak of the wonder that I have experienced with Morse functions, and for his hospitality during my stay in Warsaw. Thanks are also due to Marek Szyjewski for taking the lecture notes from which the present article was initiated, and to Maciej Borodzik for many improvements on the earlier version of the note.

1. Computing homology: a classical method

There are many ways to introduce Morse Theory. However, I would like to present it in the effective computation of homology (cohomology) of a manifold.

Homology (cohomology) theory is a bridge between geometry and algebra in the sense that it assigns to a manifold M a graded abelian group $H_*(M)$ (graded ring $H^*(M)$), assigns to a map $f: M \to N$ between manifolds the induced homomorphism

$$f_*: H_*(M) \to H_*(N) \quad (\text{resp.} \quad f^*: H^*(N) \to H^*(M)).$$

Supported by Polish KBN grant No.2 P03A 024 23.

During the past century this idea has been widely applied to translate geometric problems concerning manifolds and maps to problems about groups (or rings) and homomorphisms, so that by solving the latter in the well-developed framework of algebra, one obtains solutions to the problems initiated from geometry.

The first problem one encounters when working with homology theory is the following one.

Problem 1. *Given a manifold M, compute $H_*(M)$ (as a graded abelian group) and $H^*(M)$ (as a graded ring).*

We begin by recalling a classical method to approach the additive homology of manifolds.

1.1. Homology of a cell complex

The simplest geometric object in dimension n, $n \geq 0$, is the unit ball $D^n = \{x \in \mathbb{R}^n \mid \|x\|^2 \leq 1\}$ in the Euclidean n-space $\mathbb{R}^n = \{x = (x_1, \dots, x_n) \mid x_i \in \mathbb{R}\}$, which will be called the *n-dimensional disk* (or *cell*). Its boundary presents us the simplest closed $(n-1)$-dimensional manifold, the $(n-1)$ *sphere*: $S^{n-1} = \partial D^n = \{x \in \mathbb{R}^n \mid \|x\|^2 = 1\}$.

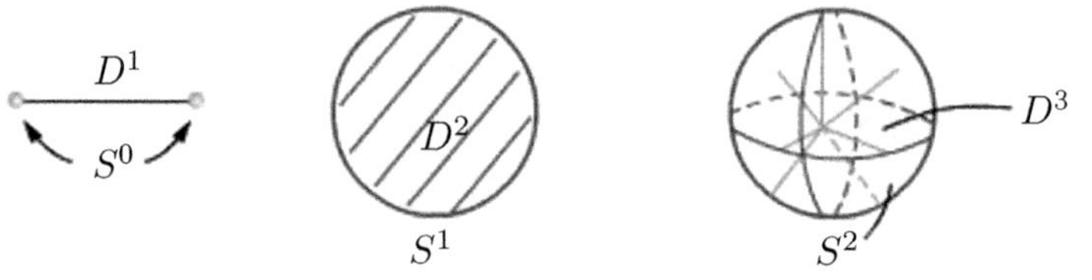

Let $f : S^{r-1} \to X$ be a continuous map from S^{r-1} to a topological space X. From f one gets

(1) an adjunction space

$$X_f = X \cup_f D^r = X \sqcup D^r / y \in S^{r-1} \sim f(y) \in X,$$

called *the space obtained from X by attaching an n-cell using f.*

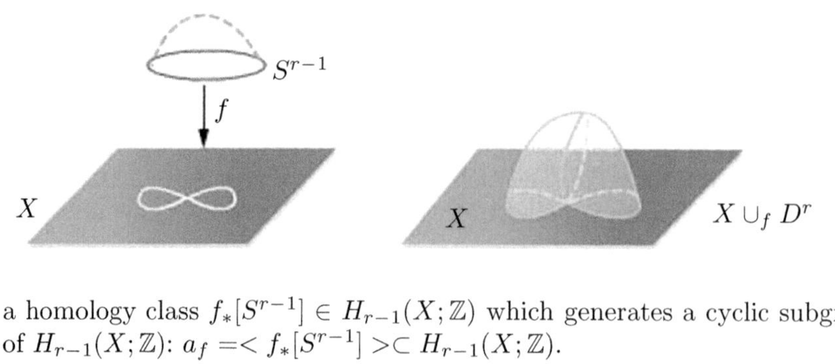

(2) a homology class $f_*[S^{r-1}] \in H_{r-1}(X;\mathbb{Z})$ which generates a cyclic subgroup of $H_{r-1}(X;\mathbb{Z})$: $a_f =< f_*[S^{r-1}] >\subset H_{r-1}(X;\mathbb{Z})$.

We observe that the integral homology of the new space $X \cup_f D^r$ can be computed in terms of $H_*(X;\mathbb{Z})$ and its subgroup a_f.

Theorem 1. *Let $X_f = X \cup_f D^r$. Then the inclusion $i : X \to X_f$*

1) *induces isomorphisms $H_k(X;\mathbb{Z}) \to H_k(X_f;\mathbb{Z})$ for all $k \neq r, r-1$;*
2) *fits into the short exact sequences*

$$0 \to a_f \to H_{r-1}(X;\mathbb{Z}) \xrightarrow{i_*} H_{r-1}(X_f;\mathbb{Z}) \to 0$$

$$0 \to H_r(X;\mathbb{Z}) \xrightarrow{i_*} H_r(X_f;\mathbb{Z}) \to \begin{cases} 0 \text{ if } |a_f| = \infty \\ \mathbb{Z} \to 0 \text{ if } |a_f| < \infty. \end{cases}$$

Proof. Substituting in the homology exact sequence of the pair (X_f, X)

$$H_k(X_f, X;\mathbb{Z}) = \begin{cases} 0 \text{ if } k \neq r; \\ \mathbb{Z} \text{ if } k = r \end{cases}$$

(note that the boundary operator maps the generator of $H_r(X_f, X;\mathbb{Z}) = \mathbb{Z}$ to $f_*[S^{r-1}]$), one obtains (1) and (2) of the theorem. □

Definition 1.1. *Let X be a topological space. A cell-decomposition of X is a sequence of subspaces $X_0 \subset X_1 \subset \cdots \subset X_{m-1} \subset X_m = X$ so that*

a) *X_0 consists of finite many points $X_0 = \{p_1, \ldots, p_l\}$; and*
b) *$X_k = X_{k-1} \cup_{f_i} D^{r_k}$, where $f_i : \partial D^{r_k} = S^{r_k - 1} \to X_{k-1}$ is a continuous map.*

Moreover, X is called a cell complex if a cell-decomposition of X exists.

Two comments are ready for the notion of cell-complex X.

(1) It can be built up using the simplest geometric objects D^n, $n = 1, 2, \ldots$ by repeated application of the same construction as "attaching cell";
(2) Its homology can be computed by repeated application of the single algorithm (i.e., Theorem 1).

The concept of cell-complex was initiated by Ehresmann in 1933–1934. Suggested by the classical work of H. Schubert in algebraic geometry in 1879 [Sch], he found a cell decomposition for the complex Grassmannian manifolds from which the homology of these manifolds were computed [Eh]. The cells involved are currently known as *Schubert cells* (*varieties*) [MS].

In 1944, Whitehead [Wh] described a cell decomposition for the real Stiefel manifolds (including all real orthogonal groups) in order to compute the homotopy groups of these manifolds, where the cells were called *the normal cells* by Steenrod [St] or *Schubert cells* by Dieudonné [D, p. 226]. In terms of this cell decompositions the homologies of these manifolds were computed by C. Miller in 1951 [M]. We refer the reader to Steenrod [St] for the corresponding computation for complex and quaternionic Stiefel manifolds.

Historically, finding a cell decomposition of a manifold was a classical approach to computing its homology. It should be noted that it is generally a difficult and tedious task to find (or to describe) a cell-decomposition for a given manifold. We are looking for simpler methods.

1.2. Attaching handles (Construction in manifolds)

"Attaching cells" is a geometric procedure to construct topological spaces by using the elementary geometric objects D^r, $r \geq 0$. The corresponding construction in manifolds are known as "attaching handles"or more intuitively, "*attaching thickened cells*".

Let M be an n-manifold with boundary $N = \partial M$, and let $f : S^{r-1} \to N$ be a smooth embedding of an $(r-1)$-sphere whose tubular neighborhood in N is trivial: $T(S^{r-1}) = S^{r-1} \times D^{n-r}$. Of course, as in the previous section, one may form a new topological space $M_f = M \cup_f D^r$ by attaching an r-cell to M by using f. However, the space M_f is in general not a manifold!

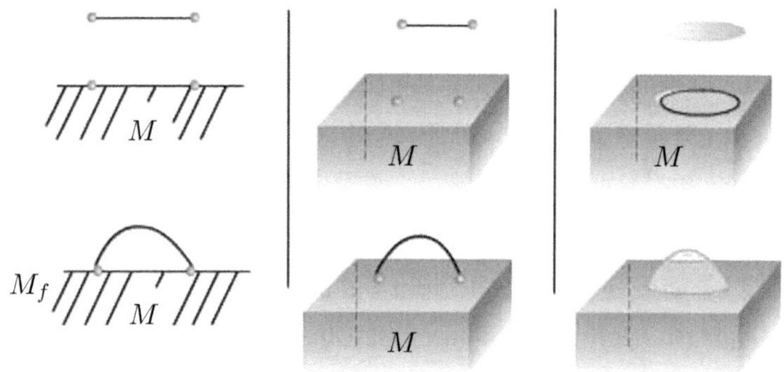

Nevertheless, one may construct a new manifold M' which contains the space M_f as a "*strong deformation retract*" by the procedure below.

Step 1. To match the dimension of M, thicken the r-disc D^r by taking product with D^{n-r}

$$D^r \times 0 \subset D^r \times D^{n-r} \quad \text{(a thickened } r\text{-disc)}$$

and note that $\partial(D^r \times D^{n-r}) = S^{r-1} \times D^{n-r} \cup D^r \times S^{n-r-1}$.

Step 2. Choose a diffeomorphism

$$S^{r-1} \times D^{n-r} (\subset D^r \times D^{n-r}) \xrightarrow{\varphi} T(S^r) \subset M$$

that extends f in the sense that $\varphi \mid S^{r-1} \times \{0\} = f$;

Step 3. Gluing $D^r \times D^{n-r}$ to M by using φ to obtain $M' = M \cup_\varphi D^r \times D^{n-r}$.

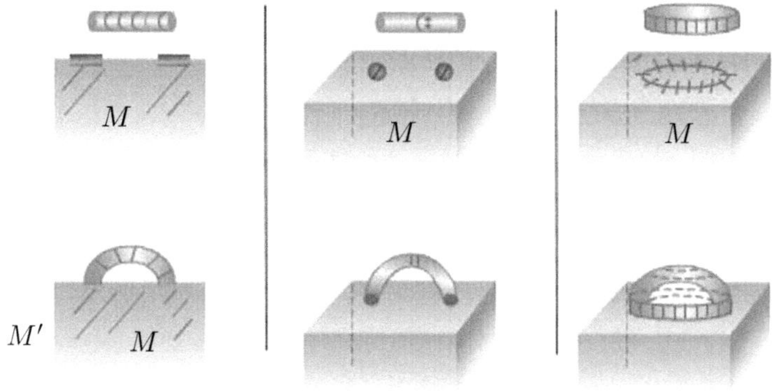

Step 4. Smoothing the angles [M3].

Definition 1.2. *M' is called the manifold obtained from M by adding a thickened r-cell with core M_f.*

Remark. The homotopy type (hence the homology) of M' depends on the homotopy class $[f] \in \pi_{r-1}(M)$ of f.

The diffeomorphism type of M' depends on the isotopy class of the embedding f (with trivial normal bundle), and a choice of $\varphi \in \pi_r(SO(n-r))$.

Inside $M' = M \cup_\varphi D^r \times D^{n-r}$ one finds the submanifold $M \subset M'$ as well as the subspace $M_f = M \cup_f D^r \times \{0\} \subset M' = M \cup_\varphi D^r \times D^{n-r}$ in which the inclusion $j : M_f \to M'$ is a homotopy equivalence. In particular, j induces isomorphisms in every dimension

$$H_k(M_f, \mathbb{Z}) \to H_k(M'; \mathbb{Z}), \quad k \geq 0.$$

Consequently, the integral cohomology of the new manifold M' can be expressed in terms of that of M together with the class $f_*[S^{r-1}] \in H_{r-1}(M; \mathbb{Z})$ by Theorem 1.

Corollary. *Let M' be the manifold obtained from M by adding a thickened r-cell with core M_f. Then the inclusion $i : M \to M'$*

1) *induces isomorphisms $H_k(M; \mathbb{Z}) \to H_k(M'; \mathbb{Z})$ for all $k \neq r, r-1$;*
2) *fits into the short exact sequences*

$$0 \to a_f \to H_{r-1}(M; \mathbb{Z}) \to H_{r-1}(M'; \mathbb{Z}) \to 0$$

$$0 \to H_r(M; \mathbb{Z}) \to H_r(M; \mathbb{Z}) \to \begin{cases} 0 & \text{if } |a_f| = \infty \\ \mathbb{Z} \to 0 & \text{if } |a_f| < \infty. \end{cases}$$

Definition 1.3. *Let M be a smooth closed n-manifold (with or without boundary). A handle decomposition of M is a filtration of submanifolds $M_1 \subset M_2 \subset \cdots \subset M_{m-1} \subset M_m = M$ so that*

(1) *$M_1 = D^n$;*
(2) *M_{k+1} is a manifold obtained from M_k by attaching a thickened r_k-cell, $r_k \leq n$.*

If M is endowed with a handle decomposition, its homology can be computed by repeated applications of the corollary

$$H_*(M_1) \mapsto H_*(M_2) \mapsto \cdots \mapsto H_*(M).$$

Now, Problem 1 can be stated in geometric terms.

Problem 2. *Let M be a smooth manifold.*

(1) *Does M admit a handle decomposition?*
(2) *If yes, find one.*

2. Elements of Morse theory

Using Morse function we prove, in this section, the following result which answers (1) of Problem 2 affirmatively.

Theorem 2. *Any closed smooth manifold admits a handle decomposition.*

2.1. Study manifolds by using functions: the idea

Let M be a smooth closed manifold of dimension n and let $f : M \to \mathbb{R}$ be a non-constant smooth function on M. Put

$$a = \min\{f(x) \mid x \in M\}, \quad b = \max\{f(x) \mid x \in M\}.$$

Then f is actually a map onto the interval $[a, b]$.

Intuitively, f assigns to each point $x \in M$ a *height* $f(x) \in [a, b]$. For a $c \in (a, b)$, those points on M with the same height c form *the level surface* $L_c = f^{-1}(c)$ of f at level c. It cuts the whole manifold into two parts $M = M_c^- \cup M_c^+$ with

$$M_c^- = \{x \in M \mid f(x) \le c\} \quad \text{(the part below } L_c\text{)}$$
$$M_c^+ = \{x \in M \mid f(x) \ge c\} \quad \text{(the part above } L_c\text{)}$$

and with $L_c = M_c^- \cap M_c^+$.

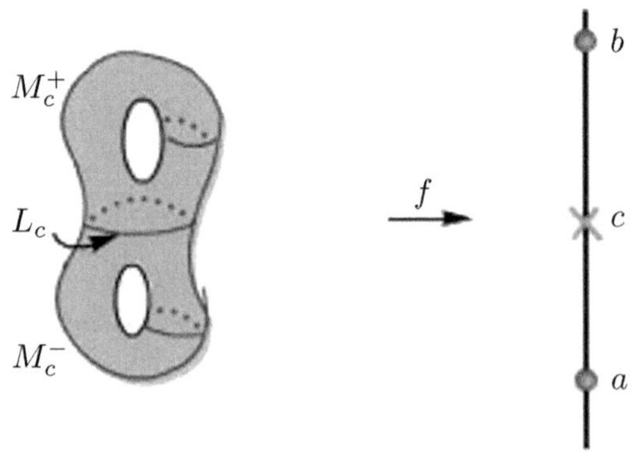

In general, given a sequence of real numbers $a = c_1 < \cdots < c_m = b$, the $m-2$ level surfaces L_{c_i}, $2 \leq i \leq m-1$, define a filtration on M

$$M_1 \subset M_2 \subset \cdots \subset M_{m-1} \subset M_m = M,$$

with $M_i = M^-_{c_i}$.

Our aim is to understand the geometric construction of M (rather than the functions on M). Naturally, one expects to find a good function f as well as suitable reals $a = c_1 < c_2 < \cdots < c_m = b$ so that

(1) each M_i is a smooth manifold with boundary L_{c_i};
(2) the change in topology between each adjoining pair $M_k \subset M_{k+1}$ is as simple as possible.

If this can be done, we may arrive at a global picture of the construction of M.

Among all smooth functions on M, Morse functions are the most suitable ones for this purpose.

2.2. Morse functions

Let $f : M \to \mathbb{R}$ be a smooth function on an n-dimensional manifold M and let $p \in M$ be a point. In a local coordinate system $(x_1, \ldots, x_n)$ centered at p (i.e., a Euclidean neighborhood around p) the Taylor expansion of f near p reads

$$f(x_1, \ldots, x_n) = a + \sum_{1 \leq i \leq n} b_i x_i + \sum_{1 \leq i,j \leq n} c_{ij} x_i x_j + o(\| x \|^3),$$

in which

$$a = f(0); \quad b_i = \frac{\partial f}{\partial x_i}(0), \quad 1 \leq i \leq n;$$

and

$$c_{ij} = \frac{1}{2} \frac{\partial^2 f}{\partial x_j \partial x_i}(0), \quad 1 \leq i,j \leq n.$$

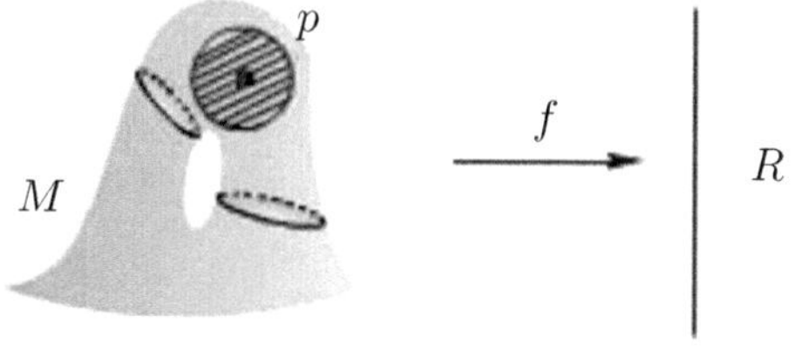

Let T_pM be the tangent space of M at p. The $n \times n$ symmetric matrix,

$$H_p(f) = (c_{ij}) : T_pM \times T_pM \to \mathbb{R} \quad (\text{resp. } T_pM \to T_pM)$$

called the *Hessian form* (resp. Hessian operator) of f at p, can be brought into diagonal form by changing the linear basis $\{\frac{\partial}{\partial x_1}, \ldots, \frac{\partial}{\partial x_n}\}$ of T_pM

$$H_p(f) = (c_{ij}) \sim 0_s \oplus (-I_r) \oplus (I_t), \quad s + r + t = n.$$

Definition 2.1. *$p \in M$ is called a critical point of f if, in a local coordinate system at p, $b_i = 0$ for all $1 \leq i \leq n$. Write Σ_f for the set of all critical points of f.*

A critical point $p \in \Sigma_f$ is called non-degenerate if the form $H_p(f)$ is non-degenerate. In this case the number r is called the index of p (as a non-degenerate critical point of f), and will be denoted by $r = Ind(p)$.

f is said to be a Morse function on M if all its critical points are non-degenerate.

The three items "*critical point*", "*non-degenerate critical point*" as well as the "*index*" of a nondegenerate critical point specified in the above are clearly independent of the choice of a local coordinate system centered at p. Two useful properties of a Morse function are given in the next two lemmas.

Lemma 2.1. *If M is closed and if f is a Morse function on M, then Σ_f is a finite set.*

Proof. The set Σ_f admits an intrinsic description without referring to local coordinate systems.

The tangent map $Tf : TM \to \mathbb{R}$ of f gives rise to a section $\sigma_f : M \to T^*M$ for the cotangent bundle $\pi : T^*M \to M$. Let $\sigma : M \to T^*M$ be the zero section of π. Then $\Sigma_f = \sigma_f^{-1}[\sigma(M)]$. That f is a Morse function is equivalent to the statement that the two embeddings $\sigma_f, \sigma : M \to T^*M$ have transverse intersection. □

Lemma 2.2 (**Morse Lemma**, cf. [H; p.146]). *If $p \in M$ is a non-degenerate critical point of f with index r, there exists a local coordinate system $(x_1, \dots, x_n)$ centered at p so that*

$$f(x_1, \dots, x_n) = f(0) - \sum_{1 \le i \le r} x_i^2 + \sum_{r < i \le n} x_i^2$$

(i.e., the standard nondegenerate quadratic function of index r).

Proof. By a linear coordinate change we may assume that

$$\left(\frac{\partial^2 f}{\partial x_j \partial x_i}(0)\right) = (-I_r) \oplus (I_{n-r}).$$

Applying the fundamental theorem of calculus twice yields the expansion

$$\text{(A)} \qquad f(x_1, \dots, x_n) = f(0) + \sum_{1 \le i,j \le n} x_i x_j b_{ij}(x)$$

in which

$$b_{ij}(x) = \int_0^1 \int_0^1 \frac{\partial^2 \overline{f}}{\partial x_j \partial x_i}(stx_1, \dots, stx_n)dtds.$$

The family of matrix $B(x) = (b_{ij}(x))$, $x \in U$, may be considered as a smooth map

$$B : U \to \mathbb{R}^{\frac{n(n+1)}{2}} \ (= \text{the vector space of all } n \times n \text{ symmetric matrices}).$$

with $B(0) = (-I_r) \oplus (I_{n-r})$, where $U \subset M$ is the Euclidean neighborhood centered at p. It follows that

"there is a smooth map $P : U \to GL(n)$ so that in some neighborhood V of $0 \in U$,

$$B(x) = P(x)\{(-I_r) \oplus (I_{n-r})\}P(x)^\tau \quad \text{and} \quad P(0) = I_n\text{"}.$$

With this we infer from (A) that, for $x = (x_1, \ldots, x_n) \in V$

$$f(x) = f(0) + xB(x)x^\tau = f(0) + xP(x)\{(-I_r) \oplus (I_{n-r})\}P(x)^\tau x^\tau.$$

It implies that if one makes the coordinate change

$$(y_1, \ldots, y_n) = (x_1, \ldots, x_n)P(x)$$

in a neighborhood of $0 \in U$ then one gets

$$f(y_1, \ldots, y_n) = f(0) - \sum_{1 \le i \le r} y_i^2 + \sum_{r < i \le n} y_i^2. \qquad \square$$

2.3. Geometry of gradient flow lines

The first set of information we can derive directly from a Morse function $f : M \to \mathbb{R}$ consists of

(1) the set Σ_f of critical points of f;
(2) the index function $Ind : \Sigma_f \to \mathbb{Z}$.

Equip M with a Riemannian metric so that the gradient field of f

$$v = \text{grad}(f) : M \to TM,$$

is defined. One of the very first things that one learns from the theory of ordinary differential equations is that, for each $x \in M$, there exists a unique smooth curve $\varphi_x : \mathbb{R} \to M$ subject to the following constraints

(1) the initial condition: $\varphi_x(0) = x$;
(2) the ordinary differential equation: $\frac{d\varphi_x(t)}{dt} = v(\varphi_x(t))$;
(3) φ_x varies smoothly with respect to $x \in M$ in the sense that "the map $\varphi : M \times \mathbb{R} \to M$ by $(x, t) \to \varphi_x(t)$ is smooth and, for every $t \in \mathbb{R}$, the restricted function $\varphi : M \times \{t\} \to M$ is a diffeomorphism".

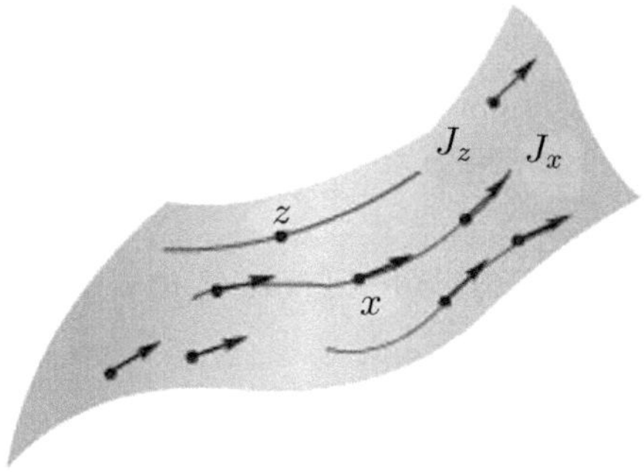

Definition 2.2. *For $x \in M$ let $J_x = \operatorname{Im} \varphi_x \subset M$, and call it the gradient flow line of f through x.*

An alternative description for J_x is the following. It is the image of the parameterized curve $\varphi(t)$ in M that satisfies

1) passing through x at the time $t = 0$;

2) at any point $y \in J_x$, the tangent vector $\frac{d\varphi}{dt}$ to J_x at y agrees with the value of v at y.

We build up the geometric picture of flow lines in the result below.

Lemma 2.3 (Geometry of gradient flow lines).

(1) *$x \in \Sigma_f \Leftrightarrow J_x$ consists of the single point x;*
(2) *$\forall x, y \in M$ we have either $J_x = J_y$ or $J_x \cap J_y = \emptyset$;*
(3) *if $x \notin \Sigma_f$, then J_x meets level surfaces of f transversely; and f is strictly increasing along the directed curve J_x;*
(4) *if $x \notin \Sigma_f$, the two limits $\lim_{t\to\pm\infty} \varphi_x(t)$ exist and belong to Σ_f.*

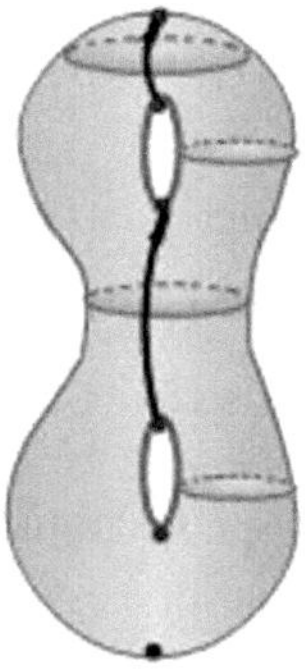

Proof. (2) comes from the fact that $\varphi_{\varphi_x(t)}(s) = \varphi_x(t+s)$.

(3) is verified by

$$\frac{df\varphi_x(t)}{dt} = \left\langle \operatorname{grad} f, \frac{d\varphi_x(t)}{dt} \right\rangle = \mid \operatorname{grad} f \mid^2 > 0.$$

Since the function $f\varphi_x(t)$ is bounded $a \leq f\varphi_x(t) \leq b$ and is monotone in t, the limits $\lim_{t\to\pm\infty} f\varphi_x(t)$ exist. It follows from (3) that $\lim_{t\to\pm\infty} |\operatorname{grad}_{\varphi_x(t)} f \mid^2 = 0$. This shows (4). □

The most important notion subordinate to flow lines is:

Definition 2.3. *For a $p \in \Sigma_f$ we write*

$$S(p) = \bigcup_{\lim_{t\to+\infty} \varphi_x(t) = p} J_x \cup \{p\}; \quad T(p) = \bigcup_{\lim_{t\to-\infty} \varphi_x(t) = p} J_x \cup \{p\}.$$

These will be called respectively the descending cell and the ascending cell of f at the critical point p.

The term "*cell*" appearing in Definition 2.3 is justified by the next result.

Lemma 2.4. *If $p \in \Sigma_f$ with $\operatorname{Ind}(p) = r$, then $(S(p), p) \cong (R^r, 0)$, $(T(p), p) \cong (R^{n-r}, 0)$, and both meet transversely at p.*

Proof. Let $(\mathbb{R}^n, 0) \subset (M, p)$ be a Euclidean neighborhood centered at p so that

$$f(x, y) = f(0) - \mid x \mid^2 + \mid y \mid^2 \quad \text{(cf. Lemma 2.2)},$$

where $(x, y) \in \mathbb{R}^n = \mathbb{R}^r \oplus \mathbb{R}^{n-r}$. We first examine $S(p) \cap \mathbb{R}^n$ and $T(p) \cap \mathbb{R}^n$.

On $\mathbb{R}^n$ the gradient field of f is easily seen to be $\text{grad} f = (-2x, 2y)$. The flow line J_{x_0} through a point $x_0 = (a, b) \in \mathbb{R}^n = \mathbb{R}^r \oplus \mathbb{R}^{n-r}$ is

$$\varphi_{x_0}(t) = (ae^{-2t}, be^{2t}), t \in \mathbb{R}.$$

Now one sees that

$$x_0 \in S(p) \cap \mathbb{R}^n \Longleftrightarrow \lim_{t \to +\infty} \varphi_{x_0}(t) = 0\ (p) \Longleftrightarrow b = 0;$$
$$x_0 \in T(p) \cap \mathbb{R}^n \Longleftrightarrow \lim_{t \to -\infty} \varphi_{x_0}(t) = 0\ (p) \Longleftrightarrow a = 0.$$

It follows that

(B) $\quad S(p) \cap \mathbb{R}^n = \mathbb{R}^r \oplus \{0\} \subset \mathbb{R}^n; T(p) \cap \mathbb{R}^n = \{0\} \oplus \mathbb{R}^{n-r} \subset \mathbb{R}^n$

and both sets meet transversely at $0 = p$.

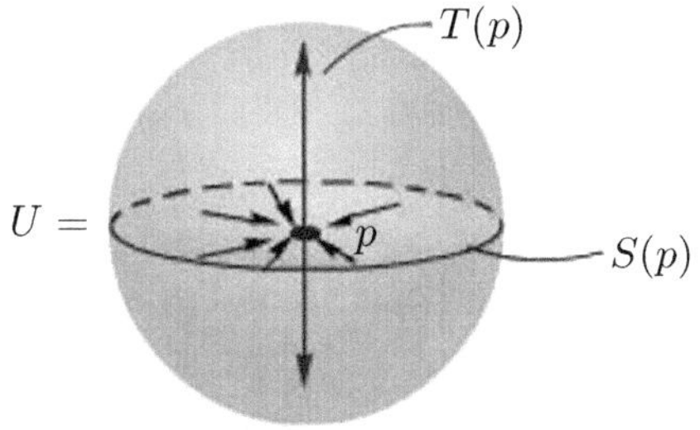

Let S^{n-1} be the unit sphere in $\mathbb{R}^n$ and put

$$S_- = S(p) \cap S^{n-1} \quad (\text{resp. } S_+ = T(p) \cap S^{n-1}).$$

Then (B) implies that $S_- \cong S^{r-1}$ (resp. $S_+ \cong S^{n-r-1}$). Furthermore, (2) of Lemma 2.3 implies that, for any $x \in S(p)$, $J_x = J_v$ for some unique $v \in S_-$ because of $\varphi_x(t) \in S(p) \cap \mathbb{R}^n$ for sufficient large t with $\lim_{t \to +\infty} \varphi_x(t) = p$. Therefore

$$S(p) = \bigcup_{v \in S_-} J_v \cup \{p\} \quad (\text{resp. } T(p) = \bigcup_{v \in S_+} J_v \cup \{p\}).$$

That is, $S(p)$ (resp. $T(p)$) is an open cone over S_- (resp. S_+) with vertex p. □

Summarizing, at a critical point $p \subset \Sigma_f$,

(1) the flow lines that grow to p (as $t \to \infty$) form an open cell of dimension $\text{Ind}(p) = r$ centered at p which lies below the critical level $L_{f(p)}$;
(2) those flow lines that grow out of from p (as $t \to \infty$) form an open cell of dimension $\text{Ind}(p) = n - r$ centered at p which lies above the critical level $L_{f(p)}$.

2.4. Handle decomposition of a manifold

Our proof of Theorem 2 implies that the set of descending cells $\{S(p) \subset M \mid p \in \Sigma_f\}$ of a Morse function on M endows M with the structure of a cell complex.

Proof of Theorem 2. Let $f : M \to [a,b]$ be a Morse function on a closed manifold M with critical set Σ_f and index function $\mathrm{Ind} : \Sigma_f \to \mathbb{Z}$. By Lemma 2.1 the set Σ_f is finite and we can assume that the points in Σ_f are ordered as $\{p_1, \dots, p_m\}$ by its values under f

$$a = f(p_1) < f(p_2) < \cdots < f(p_{m-1}) < f(p_m) = b \quad [M_1, \text{Section 4}].$$

Take a $c_i \in (f(p_i), f(p_{i+1}))$, $i \leq m-1$. Then c_i is a regular value of f. As a result $M_i = f^{-1}[a, c_i] \subset M$ is a smooth submanifold with boundary $\partial M_i = L_{c_i}$. Moreover we get a filtration on M by submanifolds

$$M_1 \subset M_2 \subset \cdots \subset M_{m-1} \subset M_m = M.$$

We establish Theorem 2 by showing that

1) $M_1 = D^n$;
2) For each k there is an embedding $g : S^{r-1} \to \partial M_k$ so that

$$M_k \cup S(p_{k+1}) = M_k \cup_g D^r, \quad r = \mathrm{Ind}(p_{k+1});$$

3) $M_{k+1} = M_k \cup D^r \times D^{n-r}$ with core $M_k \cup_g D^r$.

1) Let $\mathbb{R}^n$ be a Euclidean neighborhood around p_1 so that

$$f(x_1, \dots, x_n) = a + \Sigma x_i^2,$$

here we have made use of the fact $\mathrm{Ind}(p_1) = 0$ (because f attains its absolute minimal value a at p_1) as well as Lemma 2.2. Since $c_1 = a + \varepsilon$ we have

$$f^{-1}[a, c_1] = \{x \in \mathbb{R}^n \mid \parallel x \parallel^2 \leq \varepsilon\} \cong D^n.$$

2) With the notation introduced in the proof of Lemma 2.4 we have

$$\text{(C)} \qquad S(p_{k+1}) = \bigcup_{v \in S_-} J_v \cup \{p_{k+1}\}$$

where $S_- \cong S^{r-1}$, $r = \mathrm{Ind}(p_{k+1})$, and where J_v is the unique flow line $\varphi_v(t)$ with $\varphi_v(0) = v$ and with $\lim\limits_{t \to +\infty} \varphi_v(t) = p_{k+1}$.

For a $v \in S_-$, $\lim\limits_{t \to -\infty} \varphi_v(t) \in \{p_1, \dots, p_k\} \subset Int(M_k)$ by (4) and (3) of Lemma 2.3. So J_v must meet ∂M_k at some unique point. The map $g : S_- \to \partial M_k$ such that $g(v) = J_v \cap \partial M_k$ is now well defined and must be an embedding by (2) of Lemma 2.3. We get

$$M_k \cup S(p_{k+1}) = M_k \cup_g D^r$$

form (C).

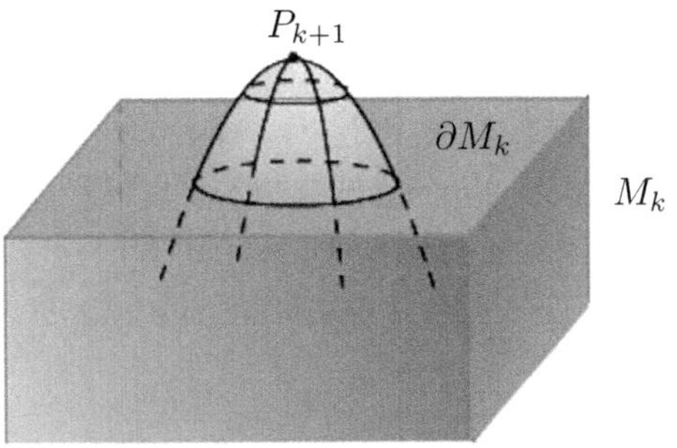

3) In [M₁, p. 33–34], Milnor demonstrated explicitly two deformation retractions

$$r : M_{k+1} \overset{R_1}{\to} M_k \cup D^r \times D^{n-r} \overset{R_2}{\to} M_k \cup S(p_{k+1})$$

where R_1 does not change the diffeomorphism type of M_{k+1} and where $D^r \times D^{n-r}$ is a thickening of the r-cell corresponding to $S(p_{k+1})$. □

3. Morse functions via Euclidean geometry

Our main subject is the effective computation of the additive homology or the multiplicative cohomology of a given manifold M. Recall from Section 1 that if M is endowed with a cell decomposition, the homology $H_*(M)$ can be calculated by repeated application of Theorem 1. Further we have seen in Section 2 that a Morse function f on M may define a cell decomposition on M with each critical point of index r corresponding to an r-cell in the decomposition. The remaining question is

How to find a Morse function on a given manifold?

3.1. Distance function on a Euclidean submanifold

By a classical result of Whitney, every n-dimensional smooth manifold M can be smoothly embedded into Euclidean space of some dimension less than $2n+1$. Therefore, it suffices to assume that M is a submanifold in a Euclidean space E.

A point $a \in E$ gives rise to a function $f_a : M \to \mathbb{R}$ by $f_a(x) = \| x - a \|^2$.

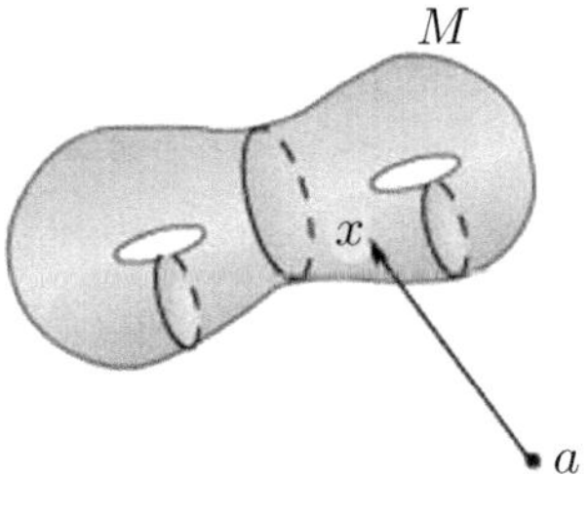

Let Σ_a be the set of all critical points of this function. Two questions are:

(a) How to specify the critical set of f_a?
(b) For which choice of the point $a \in E$, f_a is a Morse function on M?

For a point $x \in M$ let $T_x M \subset E$ be the tangent plane to M at x (an affine plane in E with dimension n). Its orthonormal complement

$$\gamma_x = \{v \in E \mid v \perp M_x\}$$

is called *the normal plane to M at x*. We state the answers to questions (a) and (b) in

Lemma 3.1. *Let $f_a : M \to \mathbb{R}$ be as above.*
(1) $\Sigma_a = \{x \in M \mid a - x \in \gamma_x\}$;
(2) *For almost all $a \in E$, f_a is a Morse function.*

Proof. The function $g_a : E \to \mathbb{R}$ by $x \to \| x - a \|^2$ has gradient field $\mathrm{grad}_x\, g_a = 2(x - a)$. Since $f_a = g_a \mid M$, for a $x \in M$,

$$\mathrm{grad}_x\ f_a = \text{the orthonormal projection of } 2(x - a) \text{ to } T_x M.$$

So $x \in \Sigma_a$ (i.e., $\mathrm{grad}_x\ f_a = 0$) is equivalent to $2(x - a) \perp T_x M$. This shows (1).

Let $\Lambda \subset E$ be the focal set of the submanifold $M \subset E$. It can be shown that f_a is a Morse function if and only if $a \in E \backslash \Lambda$. (2) follows from the fact that Λ has measure 0 in E (cf. [M_2, p. 32–38]). □

3.2. Examples of submanifolds in Euclidean spaces

Many manifolds important in geometry are already sitting in Euclidean spaces in some ready-made fashion. We present such examples.

Let $\mathbb{F}$ be one of $\mathbb{R}$ (the field of reals), $\mathbb{C}$ (the field of complex) or $\mathbb{H}$ (the division algebra of quaternions). Let E be one of the following real vector spaces:

- the space of $n \times n$ matrices over $\mathbb{F}$: $M(n; \mathbb{F})$;
- the space of complex Hermitian matrices:

$$S(n; \mathbb{C}) = \{x \in M(n; \mathbb{C}) \mid x^\tau = x\};$$

- the space of complex symmetric matrices

$$S^+(n; \mathbb{C}) = \{x \in M(n; \mathbb{C}) \mid x^\tau = \overline{x}\};$$

- the space of real skew symmetric matrices:

$$S^-(2n; \mathbb{R}) = \{x \in M(2n; \mathbb{R}) \mid x^\tau = -x\}.$$

Their dimensions as real vector spaces are respectively

$$\dim_{\mathbb{R}} M(n; \mathbb{F}) = \dim_{\mathbb{R}} \mathbb{F} \cdot n^2; \qquad \dim_{\mathbb{R}} S(n; \mathbb{C}) = n(n+1);$$
$$\dim_{\mathbb{R}} S^+(n; \mathbb{C}) = n(n-1); \qquad \dim_{\mathbb{R}} S^-(2n; \mathbb{R}) = n(2n-1).$$

Further, E is a Euclidean space with the metric specified by

$$\langle x, y \rangle = \mathrm{Re}[Tr(x^* y)], \quad x, y \in E,$$

where $*$ means transpose followed by conjugation.

Consider in E the following submanifolds

$$\begin{aligned} O(n;\mathbb{F}) &= \{x \in M(n;\mathbb{F}) \mid x^*x = I_n\} \\ G_{n,k} &= \{x \in S^+(n;\mathbb{C}) \mid x^2 = I_n,\ l(x) = k\}; \\ LG_n &= \{x \in S(n;\mathbb{C}) \mid \overline{x}x = I_n\}; \\ \mathbb{C}S_n &= \{x \in S^-(2n;\mathbb{R}) \mid x^2 = -I_{2n}\}, \end{aligned}$$

where $l(x)$ means "the number of negative eigenvalues of x" and where I_n is the identity matrix. The geometric interests in these manifolds may be illustrated in

$$O(n;\mathbb{F}) = \begin{cases} O(n) \text{ if } \mathbb{F} = \mathbb{R}\text{: the real orthogonal group of rank } n; \\ U(n) \text{ if } \mathbb{F} = \mathbb{C}\text{: the unitary group of rank } n; \\ Sp(n) \text{ if } \mathbb{F} = \mathbb{H}\text{: the symplectic group of rank } n; \end{cases}$$

$G_{n,k}$: the Grassmannian of k-subspaces in $\mathbb{C}^n$;
LG_n: the Grassmannian of Lagrangian subspaces in $\mathbb{C}^n$;
$\mathbb{C}S_n$: the Grassmannian of complex structures on $\mathbb{R}^{2n}$.

3.3. Morse functions via Euclidean geometry

Let $0 < \lambda_1 < \cdots < \lambda_n$ be a sequence of n reals, and let $a \in E$ be the point with

$$a = \begin{cases} \operatorname{diag}\{\lambda_1, \ldots, \lambda_n\} \text{ if } M \neq \mathbb{C}S_n; \\ \lambda_1 J \oplus \cdots \oplus \lambda_n J,\ J = \begin{pmatrix} 0 & -1 \\ 1 & 0 \end{pmatrix}, \text{ if } M = \mathbb{C}S_n. \end{cases}$$

With respect to the metric on E specified in 3.2, the function

$$f_a : M \to \mathbb{R}, \quad f_a(x) = \| x - a \|^2$$

admits a simple-looking expression

$$\begin{aligned} f_a((x_{ij})) &= \langle x, x\rangle + \langle a, a\rangle - 2\langle a, x\rangle \\ &= \text{const} - 2\begin{cases} \Sigma\lambda_i \operatorname{Re}(x_{ii}) \text{ if } M = G_{n,k},\ O(n;\mathbb{F}),\ LG_n; \text{ and} \\ \Sigma\lambda_i x_{2i-1,2i} \text{ if } M = \mathbb{C}S_n. \end{cases} \end{aligned}$$

For a subsequence $I = [i_1, \ldots, i_r] \subseteq [1, \ldots, n]$, denote by $\sigma_I \in E$ the point

$$\sigma_I = \begin{cases} \operatorname{diag}\{\varepsilon_1, \ldots, \varepsilon_n\} \text{ if } M \neq \mathbb{C}S_n; \\ \varepsilon_1 J \oplus \cdots \oplus \varepsilon_n J \text{ if } M = \mathbb{C}S_n, \end{cases}$$

where $\varepsilon_k = -1$ if $k \in I$ and $\varepsilon_k = 1$ otherwise.

Theorem 3. *In each of the above four cases, $f_a : M \to R$ is a Morse function on M. Further,*

(1) *the set of critical points of f_a is*

$$\Sigma_a = \begin{cases} \{\sigma_0,\ \sigma_I \in M \mid I \subseteq [1, \ldots, n]\} \text{ if } M \neq G_{n,k}; \\ \{\sigma_I \in M \mid I \subseteq [1, \ldots, n] \text{ with } |I| = k\} \text{ if } M = G_{n,k}. \end{cases}$$

(2) *the index functions are given respectively by*

$$\mathrm{Ind}(\sigma_{i_1,\dots,i_r}) = \begin{cases} \dim_{\mathbb{R}} \mathbb{F} \cdot (i_1 + \cdots + i_r) - r \text{ if } M = O(n;\mathbb{F}); \\ 2(i_1 + \cdots + i_r - r) \text{ if } M = \mathbb{C}S_n; \\ i_1 + \cdots + i_r \text{ if } M = LG_n; \end{cases}$$

$$\mathrm{Ind}(\sigma_{i_1,\dots,i_k}) = 2 \sum_{1 \le s \le k} (i_s - s) \text{ if } M = G_{n,k}.$$

3.4. Proof of Theorem 3

We conclude Section 3 by a proof of Theorem 3.

Lemma 3.2. *For an* $x \in M$ *one has*

$$T_x M = \begin{cases} \{u \in E \mid xu = -ux\} \text{ for } M = G_{n,k};\ \mathbb{C}S_n \\ \{u \in E \mid x^* u = -u^* x\} \text{ for } M = O(n;\mathbb{F}) \\ \{u \in E \mid \overline{x} u = -\overline{u} x\} \text{ for } M = LG_n. \end{cases}$$

Consequently

$$\gamma_x M = \begin{cases} \{u \in E \mid xu = ux\} \text{ for } M = G_{n,k};\ \mathbb{C}S_n \\ \{u \in E \mid x^* u = u^* x\} \text{ for } M = O(n;\mathbb{F}) \\ \{u \in E \mid \overline{x} u = \overline{u} x\} \text{ for } M = LG_n. \end{cases}$$

Proof. We verify Lemma 3.2 for the case $M = G_{n,k}$ as an example. Consider the map $h : S^+(n;\mathbb{C}) \to S^+(n;\mathbb{C})$ by $x \to x^2$. Then

(1) $h^{-1}(I_n) = \bigsqcup_{1 \le t \le n-1} G_{n,t}$;

(2) the tangent map of h at a point $x \in S^+(n;\mathbb{C})$ is

$$T_x h(u) = \lim_{t \to 0} \frac{h(x+tu) - h(x)}{t} = ux + xu.$$

It follows that, for a $x \in G_{n,k}$,

$$T_x G_{n,k} \subseteq \mathrm{Ker}\, T_x h = \{u \in S^+(n;\mathbb{C}) \mid ux + xu = 0\}.$$

On the other hand $\dim_{\mathbb{C}} \mathrm{Ker}\, T_x h = k(n-k)$ $(= \dim_{\mathbb{C}} T_x G_{n,k})$. So the dimension comparison yields

$$T_x G_{n,k} = \{u \in S^+(n;\mathbb{C}) \mid xu = -ux\}.$$

For any $x \in G_{n,k}$ the ambient space $E = S^+(n;\mathbb{C})$ admits the orthogonal decomposition

$$S^+(n;\mathbb{C}) = \{u \mid xu = -ux\} \oplus \{u \mid xu = ux\}$$

in which the first summand has been identified with $T_x G_{n,k}$ in the above computation. It follows that $\gamma_x G_{n,k} = \{u \mid xu = ux\}$.

The other cases can be verified by the same method. □

Lemma 3.3. *Statement* (1) *of Theorem* 3 *holds true.*

Proof. Consider the case $G_{n,k} \subset S^+(n;\mathbb{C})$.

$$\begin{aligned} x \in \Sigma_a &\Leftrightarrow x - a \in \gamma_x G_{n,k} \text{ (by (1) of Lemma 3.1)} \\ &\Leftrightarrow (x-a)x = x(x-a) \text{ (by Lemma 3.2)} \\ &\Leftrightarrow xa = ax. \end{aligned}$$

Since a is diagonal with the distinguished diagonal entries $\lambda_1 < \cdots < \lambda_n$, x is also diagonal. Since $x^2 = I_n$ with $l(x) = k$, we must have $x = \sigma_{i_1,\ldots,i_k}$ for some $[i_1,\ldots,i_k] \subseteq [1,\ldots,n]$.

Analogous computations verify the other cases. □

To prove Theorem 3 we need examining the Hessian operator $H_{x_0}(f_a) : T_{x_0}M \to T_{x_0}M$ at a critical point $x_0 \in \Sigma_a$. The following formulae will be useful for this purpose.

Lemma 3.4. $$H_{x_0}(f_a)(u) = \begin{cases} (ua - au)x_0 & \textit{for } M = G_{n,k};\ \mathbb{C}S_n; \\ (u^*a - au^*)x_0 & \textit{for } M = O(n;\mathbb{F}); \\ (\overline{u}a - a\overline{u})x_0 & \textit{for } M = LG_n. \end{cases}$$

Proof. As a function on the Euclidean space E, f_a has gradient field $2(x-a)$. However, the gradient field of the restricted function $f_a \mid M$ is the orthogonal projection of $2(x-a)$ to T_xM.

In general, for any $x \in M$, a vector $u \in E$ has the "canonical" decomposition

$$u = \begin{cases} \frac{u - xux}{2} + \frac{u + xux}{2} & \text{if } M = G_{n,k};\ \mathbb{C}S_n; \\ \frac{u - x^*ux}{2} + \frac{u + x^*ux}{2} & \text{if } M = O(n;\mathbb{F}); \\ \frac{u - \overline{x}ux}{2} + \frac{u + \overline{x}ux}{2} & \text{if } M = LG_n. \end{cases}$$

with the first component in the T_xM and the second component in γ_xM by Lemma 3.2. Applying these to $u = 2(x-a)$ yields respectively that

$$\operatorname{grad}_x f_a = \begin{cases} (xax - a) & \text{for } M = G_{n,k};\ \mathbb{C}S_n; \\ (x^*ax - a) & \text{for } M = O(n;\mathbb{F}); \\ (\overline{x}ax - a) & \text{for } M = LG_n. \end{cases}$$

Finally, the Hessian operator can be computed in terms of the gradient as

$$H_{x_0}(f_a)(u) = \lim_{t\to 0} \frac{\operatorname{grad}_{x_0+tu} f_a - \operatorname{grad}_{x_0} f_a}{t},\quad u \in T_xM.$$

As an example we consider the case $M = G_{n,k}$. We have

$$\begin{aligned} &\lim_{t\to 0} \frac{grad_{x_0+tu} f_a - grad_{x_0} f_a}{t} = \lim_{t\to 0} \frac{[(x_0+tu)a(x_0+tu) - a] - [x_0ax_0 - a]}{t} \\ &= uax_0 + x_0au = uax_0 + ax_0u \text{ (because } a \text{ and } x_0 \text{ are diagonal)} \\ &= (ua - au)x_0 \end{aligned}$$

(because vectors in $T_{x_0}G_{n,k}$ anti-commute with x_0 by Lemma 3.2). □

Proof of Theorem 3. In view of Lemma 3.3, Theorem 3 will be completed once we have shown

(a) f_a is non-degenerate at any $x_0 \in \Sigma_a$; and
(b) the index functions on Σ_a is given as that in (2) of Theorem 3.

This can be done by applying Lemma 3.2 and Lemma 3.4. We verify these for the cases $M = G_{n,k}$, $O(n)$ and LG_n in detail, and leave the other cases to the reader.

Case 1. $M = G_{n,k} \subset S^+(n;\mathbb{C})$.

(1) The most convenient vectors that span the real vector space $S^+(n;\mathbb{C})$ are

$$\{b_{s,t} \mid 1 \le s,t \le n\} \sqcup \{c_{s,t} \mid 1 \le s \ne t \le n\},$$

where $b_{s,t}$ has the entry 1 at the places (s,t), (t,s) and 0 otherwise, and where $c_{s,t}$ has the pure imaginary i at (s,t), $-i$ at the (t,s) and 0 otherwise.

(2) For a $x_0 = \sigma_I \in \Sigma_a$, those $b_{s,t}$, $c_{s,t}$ that "*anti-commute*" with x_0 belong to $T_{x_0}G_{n,k}$ by Lemma 3.2, and form a basis for $T_{x_0}G_{n,k}$

$$T_{x_0}G_{n,k} = \{b_{s,t}, c_{s,t} \mid (s,t) \in I \times J\},$$

where J is the complement of I in $[1,\dots,n]$.

(3) Applying the Hessian (Lemma 3.4) to the $b_{s,t}, c_{s,t} \in T_{x_0}G_{n,k}$ yields

$$H_{x_0}(f_a)(b_{s,t}) = (\lambda_t - \lambda_s)b_{s,t};$$
$$H_{x_0}(f_a)(c_{s,t}) = (\lambda_t - \lambda_s)c_{s,t}.$$

That is, the $b_{s,t}, c_{s,t} \in T_{x_0}G_{n,k}$ are precisely the eigenvectors for the operator $H_{x_0}(f_a)$. These indicate that $H_{x_0}(f_a)$ is nondegenerate (since $\lambda_t \ne \lambda_s$ for all $s \ne t$), hence f_a is a Morse function.

(4) It follows from the formulas in (3) that the negative space for $H_{x_0}(f_a)$ is spanned by $\{b_{s,t},\ c_{s,t} \mid (s,t) \in I \times J, t < s\}$. Consequently

$$\mathrm{Ind}(\sigma_I) = 2\#\{(s,t) \in I \times J \mid t < s\} = 2 \sum_{1 \le s \le k} (i_s - s).$$

Case 2. $M = O(n) \subset M(n;\mathbb{R})$.

(1) A natural set of vectors that spans the space $M(n;\mathbb{R})$ is

$$\{b_{s,t} \mid 1 \le s \le t \le n\} \sqcup \{\beta_{s,t} \mid 1 \le s < t \le n\},$$

where $b_{s,t}$ is as case 1, and where $\beta_{s,t}$ is the skew symmetric matrix with entry 1 at the (s,t) place, -1 at the (t,s) place and 0 otherwise;

(2) For a $x_0 = \sigma_I \in \Sigma_a$ those $b_{s,t}$, $\beta_{s,t}$ that "anti-commute" with x_0 yields precisely a basis for

$$T_{x_0}O(n) = \{\beta_{s,t} \mid (s,t) \in I \times I, J \times J, s < t\} \sqcup \{b_{s,t} \mid (s,t) \in I \times J\}$$

by Lemma 3.2, where J is the complement of I in $[1,\dots,n]$.

(3) Applying the Hessian operator (Lemma 3.4) to $b_{s,t}$, $\beta_{s,t} \in T_{x_0}O(n)$ tells

$$H_{x_0}(f_a)(\beta_{s,t}) = \begin{cases} -(\lambda_t + \lambda_s)\beta_{s,t} & \text{if } (s,t) \in I \times I,\ s < t; \\ (\lambda_t + \lambda_s)\beta_{s,t} & \text{if } (s,t) \in J \times J,\ s < t. \end{cases}$$
$$H_{x_0}(f_a)(b_{s,t}) = (\lambda_t - \lambda_s)b_{s,t} \text{ if } (s,t) \in I \times J.$$

This implies that the $b_{s,t}$, $\beta_{s,t} \in T_{x_0}G_{n,k}$ are precisely the eigenvectors for the operator $H_{x_0}(f_a)$, and the f_a is a Morse function.

(4) It follows from the computation in (3) that

$$\begin{aligned} \text{Ind}(\sigma_I) &= \#\{(s,t) \in I \times I \mid s < t\} + \#\{(s,t) \in I \times J \mid t < s\} \\ &= 1 + 2 + \cdots + (r-1) + [(i_1 - 1) + (i_2 - 2) + \cdots + (i_r - r)] \\ &= \Sigma i_s - r. \end{aligned}$$

Case 3. $M = LG_n \subset S(n;\mathbb{C})$.

(1) Over reals, the most natural vectors that span the space $S(n;C)$ are

$$\{b_{s,t} \mid 1 \leq s,t \leq n\} \cup \{ib_{s,t} \mid 1 \leq s,t \leq n\},$$

where $b_{s,t}$ is as that in Case 1 and where i is the pure imaginary.

(2) For a $x_0 = \sigma_I \in \Sigma_a$ those "anti-commute" with x_0 yields precisely a basis for $T_{x_0}LG_n$

$$\begin{aligned} T_{x_0}LG_n = &\{b_{s,t} \mid (s,t) \in I \times J \amalg J \times I\} \\ &\sqcup \{ib_{s,t} \mid (s,t) \in I \times I \sqcup J \times J\} \end{aligned}$$

where J is the complement of I in $[1,\ldots,n]$.

(3) Applying the Hessian to $b_{s,t}$, $ib_{s,t} \in T_{x_0}LG_n$ (cf. Lemma 3.4) tells

$$H_{x_0}(f_a)(ib_{s,t}) = \begin{cases} -(\lambda_t + \lambda_s)ib_{s,t} & \text{if } (s,t) \in I \times I \\ (\lambda_t + \lambda_s)ib_{s,t} & \text{if } (s,t) \in J \times J \end{cases};$$

$$H_{x_0}(f_a)(b_{s,t}) = \begin{cases} (\lambda_t - \lambda_s)b_{s,t} & \text{if } (s,t) \in I \times J \\ (\lambda_s - \lambda_t)b_{s,t} & \text{if } (s,t) \in J \times I. \end{cases}$$

It follows that the $b_{s,t}$, $ib_{s,t} \in T_{x_0}G_{n,k}$ are precisely the eigenvectors for the operator $H_{x_0}(f_a)$, and f_a is a Morse function.

(4) It follows from (2) and (3) that

$$\begin{aligned} \text{Ind}(\sigma_I) &= \#\{(s,t) \in I \times I \mid t \leq s\} + \#\{(s,t) \in I \times J \mid t \leq s\} \\ &= i_1 + \cdots + i_r \end{aligned}$$ □

Remark. Let E be one of the following matrix spaces:

- the space of $n \times k$ matrices over $\mathbb{F}$: $M(n \times k;\mathbb{F})$;
- the space of symmetric matrices $S^+(n;\mathbb{F}) = \{x \in M(n;\mathbb{F}) \mid x^\tau = \overline{x}\}$.

Consider in E the following submanifolds:

$$V_{n,k}(\mathbb{F}) = \{x \in M(n \times k;\mathbb{F}) \mid \overline{x}^\tau x = I_k\};$$
$$G_{n,k}(\mathbb{F}) = \{x \in S^+(n;\mathbb{F}) \mid x^2 = I_n,\ l(x) = k\}.$$

These are known respectively as the *Stiefel manifold* of orthonormal k-frames on $\mathbb{F}^n$ (the n-dimensional $\mathbb{F}$-vector space) and the *Grassmannian* of k-dimensional $\mathbb{F}$-subspaces in $\mathbb{F}^n$. Results analogous to Theorem 3 hold for these two families of manifolds as well [D$_1$], [D$_2$].

Remark. In [VD, Theorem 1.2], the authors proved that the function f_a on $M = G_{n,k}(\mathbb{F})$, LG_n, $\mathbb{C}S_n$ is a perfect Morse function (without specifying the set Σ_a as well as the index function Ind: $\Sigma_a \to \mathbb{Z}$).

4. Morse functions of Bott-Samelson type

We recall the original construction of Bott-Samelson cycles in 4.1 and explain its generalization due to Hsiang-Palais-Terng [HTP] in 4.2.

In fact, the Morse functions concerned in Theorem 3 are all Bott-Samelson type (cf. Theorem 6). The induced cohomology homomorphism of Bott-Samelson cycles enables one to resolve the multiplication in cohomology into the multiplication of symmetric functions of various types (Theorem 7).

4.1. Morse functions on flag manifolds

(Cf. [BS$_1$,BS$_2$].) Let G be a compact connected semi-simple Lie group with the unit $e \in G$ and a fixed maximal torus $T \subset G$. The tangent space $L(G) = T_eG$ (resp. $L(T) = T_eT$) is canonically furnished with the structure of an algebra, known as the *Lie algebra* (resp. the *Cartan subalgebra*) of G. The exponential map induces the commutative diagram

$$\begin{array}{ccc} L(T) & \to & L(G) \\ \exp\downarrow & & \downarrow\exp \\ T & \to & G \end{array}$$

where the horizontal maps are the obvious inclusions. Equip $L(G)$ (hence also $L(T)$) an inner product invariant under the adjoint action of G on $L(G)$.

For a $v \in L(T)$ let $C(v)$ be the centralizer of $\exp(l_v) \subset G$, where $l_v \subset L(T)$ is the line spanned by v. The set of singular points in $L(T)$ is the subspace of the Cartan subalgebra $L(T)$:

$$\Gamma = \{v \in L(T) \mid \dim C(v) > \dim T\}.$$

Lemma 4.1. *Let* $m = \frac{1}{2}(\dim G - \dim T)$. *There are precisely* m *hyperplanes* $L_1, \ldots, L_m \subset L(T)$ *through the origin* $0 \in L(T)$ *so that* $\Gamma = \underset{1\leq i\leq m}{\cup} L_i$.

The planes $L_1, \ldots, L_m$ are known as the singular planes of G, dividing $L(T)$ into finite many convex hulls, known as the *Weyl chambers* of G. Reflections in these planes generate the *Weyl group* W of G.

Fix a regular point $a \in L(T)$. The adjoint representation of G gives rise to a map $G \to L(G)$ by $g \to Ad_g(a)$, which induces an embedding of the *flag manifold* $G/T = \{gT \mid g \in G\}$ of left cosets of T in G into $L(G)$. In this way G/T becomes a submanifold in the Euclidean space $L(G)$.

Consider the function $f_a : G/T \to \mathbb{R}$ by $f_a(x) = \| x - a \|^2$. The following beautiful result of Bott and Samelson [BS$_1$,BS$_2$] tells how to read the critical points information of f_a from the linear geometry of the vector space $L(T)$.

Theorem 4. *f_a is a Morse function on G/T with critical set*

$$\Sigma_a = \{w(a) \in L(T) \mid w \in W\}$$

(the orbit of the W-action on $L(T)$ through the point $a \in L(T)$).

The index function Ind : $\Sigma_a \to \mathbb{Z}$ *is given by*

$$\mathrm{Ind}(w(a)) = 2\#\{L_i \mid L_i \cap [a, w(a)] \neq \emptyset],$$

where $[a, w(a)]$ is the segment in $L(T)$ from a to $w(a)$.

Moreover, Bott and Samelson constructed a set of geometric cycles in G/T that realizes an additive basis of $H_*(G/T;\mathbb{Z})$ as follows.

For a singular plane $L_i \subset L(T)$ let $K_i \subset G$ be the centralizer of $\exp(L_i)$. The Lie subgroup K_i is *very simple* in the sense that $T \subset K_i$ is also a maximal torus with the quotient K_i/T diffeomorphic to the 2-sphere S^2.

For a $w \in W$ assume that the singular planes that meet the directed segments $[a, w(a)]$ are in the order $L_1, \dots, L_r$. Put $\Gamma_w = K_1 \times_T \cdots \times_T K_r$, where $T \times \cdots \times T$ (r-copies) acts on $K_1 \times \cdots \times K_r$ from the left by

$$(k_1, \dots, k_r)(t_1, \dots, t_r) = (k_1 t_1, t_1^{-1} k_2 t_2, \dots, t_{r-1}^{-1} k_r t_r).$$

The map $K_1 \times \cdots \times K_r \to G/T$ by

$$(k_1, \dots, k_r) \to Ad_{k_1 \cdots k_r}(w(a))$$

clearly factors through the quotient manifold Γ_w, hence induces a map

$$g_w : \Gamma_w \to G/T.$$

Theorem 5. *The homology $H_*(G/T;\mathbb{Z})$ is torsion free with the additive basis $\{g_{w*}[\Gamma_w] \in H_*(G/T;\mathbb{Z}) \mid w \in W\}$.*

Proof. Let $e \in K_i (\subset G)$ be the group unit and put $\bar{e} = [e, \dots, e] \in \Gamma_w$. It was actually shown by Bott and Samelson that

(1) $g_w^{-1}(w(a))$ consists of the single point $\bar{e}$;
(2) the composed function $f_a \circ g_w : \Gamma_w \to \mathbb{R}$ attains its maximum only at $\bar{e}$;
(3) the tangent map of g_w at $\bar{e}$ maps the tangent space of Γ_w at $\bar{e}$ isomorphically onto the negative part of $H_{w(a)}(f_a)$.

The proof is completed by Lemma 4.2 in 4.2. □

Remark. It was shown by Chevalley in 1958 [Ch] that the flag manifold G/T admits a cell decomposition $G/T = \bigsqcup_{w \in W} X_w$ indexed by elements in W, with each cell X_w a locally closed subset, known as a Schubert cell on G/T. Hansen [Han] proved in 1971 that $g_w(\Gamma_w) = X_w$, $w \in W$. So the map g_w is currently known as the "*Bott-Samelson resolution* of X_w".

For the description of Bott-Samelson cycles and their applications in algebro-geometric setting, see M. Brion [Br] in this volume.

4.2. Morse function of Bott-Samelson type

In differential geometry, the study of *isoparametric submanifolds* was begun by E. Cartan in 1933. In order to generalize Bott-Samelson's above cited results to these manifolds Hsiang, Palais and Terng introduced the following notation in their work [HPT][1].

Definition 4.1. *A Morse function $f : M \to R$ on a smooth closed manifold is said to be of Bott-Samelson type over $\mathbb{Z}_2$ (resp. $\mathbb{Z}$) if for each $p \in \Sigma_f$ there is a map (called a Bott-Samelson cycle of f at p)*

$$g_p : N_p \to M$$

where N_p is a closed oriented (resp. unoriented) manifold of dimension $\operatorname{Ind}(p)$ *and where*

(1) $g_p^{-1}(p) = \{\overline{p}\}$ *(a single point);*
(2) $f \circ g_p$ *attains absolute maximum only at* $\overline{p}$*;*
(3) *the tangent map $T_{\overline{p}} g_p : T_{\overline{p}} N_p \to T_p M$ is an isomorphism onto the negative space of $H_p(f)$.*

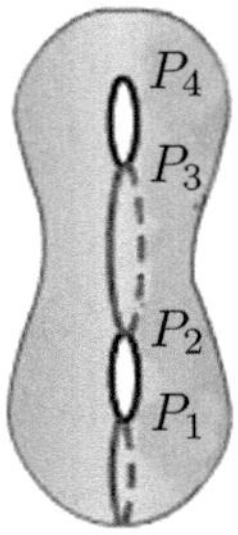

Information that one can get from a Morse function of Bott-Samelson type can be seen from the next result [HPT].

Lemma 4.2. *If $f : M \to R$ is a Morse function of Bott-Samelson type with Bott-Samelson cycles $\{g_p : N_p \to M \mid p \in \Sigma_f\}$, then $H_*(M;\mathbb{Z})$ (resp. $H_*(M;\mathbb{Z}_2)$) has the additive basis*

$$\{g_{p*}[N_p] \in H_*(M;\mathbb{Z}) \mid p \in \Sigma_f\}$$
$$(resp.\ \{g_{p*}[N_p]_2 \in H_*(M;\mathbb{Z}_2) \mid p \in \Sigma_f\}),$$

where $g_{p*} : H_*(N_p;\mathbb{Z}) \to H_*(M;\mathbb{Z})$ is the induced homomorphism and where $[N_p] \in H_*(N_p;\mathbb{Z})$ (resp. $[N_p]_2 \in H_*(N_p;\mathbb{Z}_2)$) is the orientation class (resp. $\mathbb{Z}_2$-orientation class).

Proof. Without loss of generality we may assume (as in the proof of Theorem 2) that $\Sigma_f = \{p_1, \ldots, p_m\}$ and that $f(p_k) < f(p_{k+1})$, $1 \leq k \in m - 1$. Consider the

---

[1] In fact, the embedding $G/T \subset L(G)$ described in 4.1 defines G/T as an isoparametric submanifold in $L(G)$ [HPT].

filtration on M: $M_1 \subset M_2 \subset \cdots \subset M_m = M$ defined by f and Σ_f such that $M_{k+1}\backslash M_k$ contains p_k for every $1 \leq k \leq m-1$.

It suffices to show, that if we put $p = p_{k+1}$, $m = \mathrm{Ind}(p)$, then

$$\text{(D)} \qquad H_r(M_{k+1};\mathbb{Z}) = \begin{cases} H_r(M_k;\mathbb{Z}) & \text{if } r \neq m; \\ H_r(M_k;\mathbb{Z}) \oplus \mathbb{Z} & \text{if } r = m, \end{cases}$$

where the summand $\mathbb{Z}$ is generated by $g_{p*}[N_p]$.

The Bott-Samelson cycle $g_p : N_p \to M$ (cf. Definition 4.1) is clearly a map into M_{k+1}. Let $r : M_{k+1} \to M_k \cup D^m$ be the strong deformation retraction from the proof of Theorem 2, and consider the composed map

$$g : N_p \xrightarrow{g_p} M_{k+1} \xrightarrow{r} M_k \cup D^m.$$

The geometric constraints (1)–(3) on the Bott-Samelson cycle g_p imply that there exists a Euclidean neighborhood $U \subset D^m$ centered at $p = 0 \in D^m$ so that if $V =: g^{-1}(U)$, then g restricts to a diffeomorphism $g \mid V : V \to U$. The proof of (D) (hence of Lemma 4.2) is clearly done by the exact ladder induced by the "*relative homeomorphism*" $g : (N_p, N_p\backslash V) \to (M_k \cup D^m, M_k \cup D^m\backslash U)$

$$\begin{array}{ccccccccc} & & & & \mathbb{Z} & & \mathbb{Z} & & \\ & & & & \| & & \| & & \\ & & 0 \to H_m(N_p) & \overset{\cong}{\to} & H_m(N_p, N_p\backslash V) & \to & H_{d-1}(N_p\backslash V) & \to \cdots & \\ & & & & g_* \downarrow & & g_* \downarrow\cong & & \\ 0 \to H_d(M_k) & \to & H_d((M_k \cup D^m) & \to & H_d((M_k \cup D^m, M_k) & \to & H_{d-1}(M_k) & \to \cdots & \end{array}$$

□

4.3. Bott-Samelson cycles and resolution of Schubert varieties

Le M be one of the following manifolds

$O(n;\mathbb{F})$: orthogonal (or unitary, or symplectic) group of rank n;
$\mathbb{C}S_n$: the Grassmannian of complex structures on $\mathbb{R}^{2n}$;
$G_{n,k}$: the Grassmannian of k-linear subspaces on $\mathbb{C}^n$

and

LG_n: the Grassmannian of Lagrangian subspaces on $\mathbb{C}^n$.

Let $f_a : M \to \mathbb{R}$ be the Morse function considered in Theorem 3 of §3.

Theorem 6. *In each case f_a is a Morse function of Bott-Samelson type which is*
(1) *over $\mathbb{Z}$ for $M = U(n), Sp(n), \mathbb{C}S_n, G_{n,k}$;*
(2) *over $\mathbb{Z}_2$ for $M = O(n)$ and LG_n.*

Instead of giving a proof of this result I would like to show the geometric construction of the Bott-Samelson cycles required to justify the theorem, and to point out the consequences which follow up (cf. Theorem 7).

Let $\mathbb{R}P^{n-1}$ be the real projective space of lines through the origin 0 in $\mathbb{R}^n$; $\mathbb{C}P^{n-1}$ the complex projective space of complex lines through the origin 0 in

$\mathbb{C}^n$, and let $G_2(\mathbb{R}^{2n})$ be the Grassmannian of oriented 2-planes through the origin in $\mathbb{R}^{2n}$.

Construction 1. Resolution $h : \widetilde{M} \to M$ of M.

(1) If $M = SO(n)$ (the special orthogonal group of order n) we let $\widetilde{M} = \mathbb{R}P^{n-1} \times \cdots \times \mathbb{R}P^{n-1}$ (n'-copies, where $n' = 2[\frac{n}{2}]$) and define the map $h : \widetilde{M} \to M$ to be

$$h(l_1, \ldots, l_{n'}) = \Pi_{1 \leq i \leq n'} R(l_i),$$

where $l_i \in \mathbb{R}P^{n-1}$ and where $R(l_i)$ is the reflection on $\mathbb{R}^n$ in the hyperplane $l_i^{\perp}$ orthogonal to l_i.

(2) If $M = G_{n,k}$ we let

$$\widetilde{M} = \{(l_1, \ldots, l_k) \in \mathbb{C}P^{n-1} \times \cdots \times \mathbb{C}P^{n-1} \mid l_i \perp l_j\} \ (k\text{-copies})$$

and define the map $h : \widetilde{M} \to M$ to be $h(l_1, \ldots, l_k) =< l_1, \ldots, l_k >$, where $l_i \in \mathbb{C}P^{n-1}$ and where $< l_1, \ldots, l_k >$ means the k-plane spanned by the $l_1, \ldots, l_k$.

(3) If $M = \mathbb{C}S_n$ we let

$$\widetilde{M} = \{(L_1, \ldots, L_n) \in G_2(\mathbb{R}^{2n}) \times \cdots \times G_2(\mathbb{R}^{2n}) \mid L_i \perp L_j\} \ (n\text{-copies})$$

and define the map $h : \widetilde{M} \to M$ to be $h(L_1, \ldots, L_k) = \Pi_{1 \leq i \leq n} \tau(L_i)$, where $L_i \in G_2(\mathbb{R}^{2n})$ and where $\tau(L_i) : \mathbb{R}^{2n} \to \mathbb{R}^{2n}$ is the isometry which fixes points in the orthogonal complements $L_i^{\perp}$ of L_i and is the $\frac{\pi}{2}$ rotation on L_i in accordance with the orientation.

Construction 2. Bott-Samelson cycles for the Morse function $f_a : M \to \mathbb{R}$ (cf. [Section 3, Theorem 3]).

(1) If $M = SO(n)$ then $\Sigma_a = \{\sigma_0, \sigma_I \in M \mid I \subseteq [1, \ldots, n], \mid I \mid \leq n'\}$. For each $I = (i_1, \ldots, i_r) \subseteq [1, \ldots, n]$ we put

$$\mathbb{R}P[I] = \mathbb{R}P^0 \times \cdots \times \mathbb{R}P^0 \times \mathbb{R}P^{i_1} \times \cdots \times \mathbb{R}P^{i_r} \ (n'\text{-copies}).$$

Since $\mathbb{R}P[I] \subset \widetilde{M}$ we may set $h_I = h \mid \mathbb{R}P[I]$.

The map $h_I : \mathbb{R}P[I] \to SO(n)$ is a Bott-Samelson cycle for f_a at σ_I.

(2) If $M = G_{n,k}$ then $\Sigma_a = \{\sigma_I \in M \mid I = (i_1, \ldots, i_k) \subseteq [1, \ldots, n]\}$. For each $I = (i_1, \ldots, i_k) \subseteq [1, \ldots, n]$ we have

$$\mathbb{C}P^{i_1} \times \cdots \times \mathbb{C}P^{i_k}, \widetilde{M} \subset \mathbb{C}P^{n-1} \times \cdots \times \mathbb{C}P^{n-1} \ (k\text{-copies}).$$

So we may define the intersection $\mathbb{C}P[I] = \mathbb{C}P^{i_1} \times \cdots \times \mathbb{C}P^{i_k} \cap \widetilde{M}$ in $\mathbb{C}P^{n-1} \times \cdots \times \mathbb{C}P^{n-1}$ and set $h_I = h \mid \mathbb{C}P[I]$.

The map $h_I : \mathbb{C}P[I] \to G_{n,k}$ is a Bott-Samelson cycle for f_a at σ_I.

4.4. Multiplication in cohomology: Geometry versus combinatorics

Up to now we have plenty examples of Morse functions of Bott-Samelson type. Let $f : M \to \mathbb{R}$ be such a function with critical set $\Sigma_f = \{p_1, \dots, p_m\}$. From the proof of Lemma 4.2 we see that each descending cell $S(p_i) \subset M$ forms a closed cycle on M and all of them form an additive basis for the homology

$$\{[S(p_i)] \in H_{r_i}(M; \mathbb{Z} \quad \text{or} \quad \mathbb{Z}_2) \mid 1 \leq i \leq m,\ r_i = \mathrm{Ind}(p_i)\},$$

where the coefficients in homology depend on whether the Bott-Samelson cycles are orientable or not.

Many previous work on Morse functions stopped at this stage, for people were content to have found Morse functions on manifolds whose critical points determine an additive basis for homology (such functions are usually called *perfect Morse functions*).

However, the difficult task that one has experienced in topology is not to find an additive basis for homology, but is to understand the multiplicative rule among basis elements in cohomology. More precisely, we let

$$\{[\Omega(p_i)] \in H^{r_i}(M; \mathbb{Z} \quad \text{or} \quad \mathbb{Z}_2) \mid 1 \leq i \leq m,\ r_i = \mathrm{Ind}(p_i)\}$$

be the basis for the cohomology Kronecker dual to the $[S(p_i)]$ as

$$\langle [\Omega(p_i)], [S(p_j)] \rangle = \delta_{ij}.$$

Then we must have the expression

$$[\Omega(p_i)] \cdot [\Omega(p_j)] = \Sigma a_{ij}^k [\Omega(p_k)]$$

in the ring $H^*(M; \mathbb{Z} \text{ or } \mathbb{Z}_2)$, where $a_{ij}^k \in \mathbb{Z}$ or $\mathbb{Z}_2$ depending on whether the Bott-Samelson cycles orientable or not, and where $\cdot$ means *intersection product* in Algebraic Geometry and *cup product* in Topology.

Problem 4. *Find the numbers a_{ij}^k for each triple $1 \leq i, j, k \leq m$.*

To emphasize Problem 4 we quote from N. Steenrod [St, p. 98]:

> "the cup product requires a diagonal approximation $d_\# : M \to M \times M$. Many difficulties experienced with the cup product in the past arose from the great variety of choices of $d_\#$, any particular choice giving rise to artificial looking formulas".

We advise also the reader to consult [La], [K], and [S] for details on multiplicative rules in the intersection ring of $G_{n,k}$ in algebraic geometry, and their history.

Bott Samelson cycles provide a way to study Problem 4. To explain this we turn back to the constructions in 4.3. We observe that

(i) The resolution $\widetilde{M}$ of M are constructed from the most familiar manifolds as
  $\mathbb{R}P^{n-1}$ =the real projective space of lines through the origin in $\mathbb{R}^n$;
  $\mathbb{C}P^{n-1}$ =the real projective space of lines through the origin in $\mathbb{C}^n$;
  $G_2(\mathbb{R}^{2n})$ =the Grassmannian of oriented two-dimensional subspaces in $\mathbb{R}^{2n}$

and whose cohomology are well known as

$$H^*(\mathbb{R}P^{n-1};\mathbb{Z}_2)=\mathbb{Z}_2[t]/t^n; \qquad H^*(\mathbb{C}P^{n-1};\mathbb{Z})=\mathbb{Z}[x]/x^n;$$

$$H^*(G_2(\mathbb{R}^{2n});\mathbb{Z})=\begin{cases}\mathbb{Z}[y,v]/\left\langle x^n-2x\cdot v, v^2\right\rangle & \text{if } n\equiv 1 \bmod 2;\\ \mathbb{Z}[y,v]/\left\langle x^n-2x\cdot v, v^2-x^{n-1}\cdot v\right\rangle & \text{if } n\equiv 0 \bmod 2\end{cases}$$

where

(a) $t(\in H^1(\mathbb{R}P^{n-1};\mathbb{Z}_2))$ is the Euler class for the canonical real line bundle over $\mathbb{R}P^{n-1}$;

(b) $x(\in H^2(\mathbb{C}P^{n-1};\mathbb{Z}))$ is the Euler class of the real reduction for the canonical complex line bundle over $\mathbb{C}P^{n-1}$;

(c) $y(\in H^2(G_2(\mathbb{R}^{2n});\mathbb{Z}))$ is the Euler class of the canonical oriented real 2-bundle γ over $G_2(\mathbb{R}^{2n})$, and where if $s\in H^{2n-2}(G_2(\mathbb{R}^{2n});\mathbb{Z})$ is the Euler class for the orthogonal complement ν of γ in $G_2(\mathbb{R}^{2n})\times\mathbb{R}^{2n}$, then

$$v=\frac{1}{2}(y^{n-1}+s)\in H^{2n-2}(G_2(\mathbb{R}^{2n});\mathbb{Z}).$$ [2]

(ii) the manifolds $\widetilde{M}$ are simpler than M either in terms of their geometric formation or of their cohomology

$$H^*(\widetilde{M};\mathbb{Z})=\mathbb{Z}_2[t_1,\dots,t_{n'}]/\left\langle t_i^n, 1\le i\le n'\right\rangle \quad \text{if } M=SO(n);$$
$$H^*(\widetilde{M};\mathbb{Z})=\mathbb{Z}[x_1,\dots,x_k]/\left\langle p_i, 1\le i\le k\right\rangle \quad \text{if } M=G_{n,k};$$

and

$$H^*(\widetilde{M};\mathbb{Z})=\mathbb{Q}[y_1,\dots,y_n]/\left\langle e_i(y_1^2,\dots,y_n^2), 1\le i\le n-1; y_1\cdots y_n)\right\rangle$$

if $M=\mathbb{C}S_n$, where p_i is the component of the formal polynomial

$$\prod_{1\le s\le i}(1+x_s)^{-1}$$

in degree $2(n-i+1)$ (cf. [D3, Theorem 1]), and where $e_j(y_1^2,\dots,y_n^2)$ is the jth elementary symmetric function in the $y_1^2,\dots,y_n^2$.

(iii) Bott-Samelson cycles on M can be obtained by restricting $h:\widetilde{M}\to M$ to appropriate subspaces of $\widetilde{M}$ (cf. Construction 2).

One can infer from (iii) the following result.

Theorem 7. *The induced ring map $h^*: H^*(M;\mathbb{Z} \text{ or } Z_2)\to H^*(\widetilde{M};\mathbb{Z} \text{ or } \mathbb{Z}_2)$ is injective. Furthermore*

(1) *if $M=SO(n)$, then*

$$h^*(\Omega(I))=m_I(t_1,\dots,t_{n'}),$$

[2]The ring $H^*(G_2(\mathbb{R}^{2n});\mathbb{Z})$ is torsion free. The class $y^{n-1}+s$ is divisible by 2 because of $w_{2n-2}(\nu)\equiv s\equiv y^{n-1} \bmod 2$, where w_i is the ith Stiefel-Whitney class.

where $m_I(t_1, \ldots, t_{n'})$ *is the* monomial symmetric function *in* $t_1, \ldots, t_{n'}$ *associated to the partition* I *([D_2]);*

(2) *if* $M = G_{n,k}$*, then*

$$h^*(\Omega(I)) = S_I(x_1, \ldots, x_k),$$

where $S_I(x_1, \ldots, x_k)$ *is the* Schur symmetric function *in* $x_1, \ldots, x_k$ *associated to the partition* I *([D_1]);*

(3) *if* $M = CS_n$*, then*

$$h^*(\Omega(I)) = P_I(y_1, \ldots, y_n),$$

where $P_I(y_1, \ldots, y_n)$ *is the* Schur P symmetric function *in* $y_1, \ldots, y_n$ *associated to the partition* I*.*

(For definitions of these symmetric functions, see [Ma].)

Indeed, in each case concerned by Theorem 7, it can be shown that the $\Omega(I)$ are the Schubert classes [Ch, BGG].

It was first pointed out by Giambelli [G_1,G_2] in 1902 (see also Lesieur [L]) that multiplicative rule of Schubert classes in $G_{n,k}$ formally coincides with that of Schur functions, and by Pragacz in 1986 that multiplicative rule of Schubert classes in $\mathbb{C}S_n$ formally agree with that of Schur P functions [P, § 6]. Many people asked why such similarities could possibly occur [S]. For instance it was said by C. Lenart [Le] that

> "No good explanation has been found yet for the occurrence of Schur functions in both the cohomology of Grassmannian and representation theory of symmetric groups".

Theorem 7 provides a direct linkage from Schubert classes to symmetric functions. It is for this reason combinatorial rules for multiplying symmetric functions of the indicated types (i.e., the *monomial symmetric functions, Schur symmetric functions* and *Schur P symmetric functions*) correspond to the intersection products of Schubert varieties in the spaces $M = SO(n)$, $G_{n,k}$ and $\mathbb{C}S_n$.

Remark. A link between representations and homogeneous spaces is furnished by Borel [B].

4.5. A concluding remark

Bott is famous for his periodicity theorem, which gives the homotopy groups of the matrix groups $O(n; \mathbb{F})$ with $\mathbb{F} = \mathbb{R}, \mathbb{C}$ or $\mathbb{H}$ in the stable range. However, this part of Bott's work was improved and extended soon after its appearance [Ke], [HM], [AB].

It seems that the idea of Morse functions of Bott-Samelson type appearing nearly half century ago [BS_1, BS_2] deserves further attention. Recently, an analogue of Theorem 7 for the induced homomorphism

$$g_w^* : H^*(G/T) \to H^*(\Gamma_w)$$

of the Bott-Samelson cycle $g_w : \Gamma_w \to G/T$ (cf. Theorem 5) is obtained in [D_4, Lemma 5.1], from which the multiplicative rule of Schubert classes and the Steenrod operations on Schubert classes in a generalized flag manifold G/H [Ch, BGG] have been determined [D_4], [DZ_1], [DZ_2], where G is a compact connected Lie group, and where $H \subset G$ is the centralizer of a one-parameter subgroup in G.

References

[AB] M. Atiyah and R. Bott, On the periodicity theorem for complex vector bundles, Acta Mathematica, 112(1964), 229–247.

[B] A. Borel, Sur la cohomologie des espaces fibré principaux et des espaces homogènes de groupes de Lie compacts, Ann. of Math. (2) 57, (1953). 115–207.

[BGG] I.N. Bernstein, I.M. Gel'fand and S.I. Gel'fand, Schubert cells and cohomology of the spaces G/P, Russian Math. Surveys 28 (1973), 1–26.

[Br] M. Brion, Lectures on the geometry of flag varieties, *this volume.*

[BS_1] R. Bott and H. Samelson, The cohomology ring of G/T, Nat. Acad. Sci. 41 (7) (1955), 490–492.

[BS_2] R. Bott and H. Samelson, Application of the theory of Morse to symmetric spaces, Amer. J. Math., Vol. LXXX, no. 4 (1958), 964–1029.

[Ch] C. Chevalley, Sur les Décompositions Cellulaires des Espaces G/B, in Algebraic groups and their generalizations: Classical methods, W. Haboush ed. Proc. Symp. in Pure Math. 56 (part 1) (1994), 1–26.

[D] J. Dieudonné, A history of Algebraic and Differential Topology, 1900–1960, Boston, Basel, 1989.

[D_1] H. Duan, Morse functions on Grassmannian and Blow-ups of Schubert varieties, Research report 39, Institute of Mathematics and Department of Mathematics, Peking Univ., 1995.

[D_2] H. Duan, Morse functions on Stiefel manifolds Via Euclidean geometry, Research report 20, Institute of Mathematics and Department of Mathematics, Peking Univ., 1996.

[D_3] H. Duan, Some enumerative formulas on flag varieties, Communication in Algebra, 29 (10) (2001), 4395–4419.

[D_4] H. Duan, Multiplicative rule of Schubert classes, to appear in Invent. Math. (cf. arXiv: math. AG/ 0306227).

[DZ_1] H. Duan and Xuezhi Zhao, A program for multiplying Schubert classes, arXiv: math.AG/0309158.

[DZ_2] H. Duan and Xuezhi Zhao, Steenrod operations on Schubert classes, arXiv: math.AT/0306250.

[Eh] C. Ehresmann, Sur la topologie de certains espaces homogènes, Ann. of Math. 35(1934), 396–443.

[G_1] G.Z. Giambelli, Risoluzione del problema degli spazi secanti, Mem. R. Accad. Sci. Torino (2)52(1902), 171–211.

[G_2] G.Z. Giambelli, Alcune proprietà delle funzioni simmetriche caratteristiche, Atti Torino 38(1903), 823–844.

[Han] H.C. Hansen, On cycles in flag manifolds, Math. Scand. 33 (1973), 269–274.

[H] M. Hirsch, Differential Topology, GTM. No.33, Springer-Verlag, New York-Heidelberg, 1976.

[HM] C.S. Hoo and M. Mahowald, Some homotopy groups of Stiefel manifolds, Bull.Amer. Math. Soc., 71(1965), 661–667.

[HPT] W.Y. Hsiang, R. Palais and C.L. Terng, The topology of isoparametric submanifolds, J. Diff. Geom., Vol. 27 (1988), 423–460.

[K] S. Kleiman, Problem 15. Rigorous foundation of the Schubert's enumerative calculus, Proceedings of Symposia in Pure Math., 28 (1976), 445–482.

[Ke] M.A. Kervaire, Some nonstable homotopy groups of Lie groups, Illinois J. Math. 4(1960), 161–169.

[La] A. Lascoux, Polynômes symétriques et coefficients d'intersection de cycles de Schubert. C. R. Acad. Sci. Paris Sér. A 279 (1974), 201–204.

[L] L. Lesieur, Les problèmes d'intersections sur une variété de Grassmann, C. R. Acad. Sci. Paris, 225 (1947), 916–917.

[Le] C. Lenart, The combinatorics of Steenrod operations on the cohomology of Grassmannians, Advances in Math. 136(1998), 251–283.

[Ma] I.G. Macdonald, Symmetric functions and Hall polynomials, Oxford Mathematical Monographs, Oxford University Press, Oxford, second ed., 1995.

[M] C. Miller, The topology of rotation groups, Ann. of Math., 57(1953), 95–110.

[M_1] J. Milnor, Lectures on the h-cobordism theorem, Princeton University Press, 1965.

[M_2] J. Milnor, Morse Theory, Princeton University Press, 1963.

[M_3] J. Milnor, Differentiable structures on spheres, Amer. J. Math., 81(1959), 962–972.

[MS] J. Milnor and J. Stasheff, Characteristic classes, Ann. of Math. Studies 76, Princeton Univ. Press, 1975.

[P] P. Pragacz, Algebro-geometric applications of Schur S- and Q-polynomials, Topics in invariant Theory (M.-P. Malliavin, ed.), Lecture Notes in Math., Vol. 1478, Springer-Verlag, Berlin and New York, 1991, 130–191.

[Sch] H. Schubert, Kalkül der abzählenden Geometrie, Teubner, Leipzig, 1879.

[S] R.P. Stanley, Some combinatorial aspects of Schubert calculus, Springer Lecture Notes in Math. 1353 (1977), 217–251.

[St] N.E. Steenrod and D.B.A. Epstein, Cohomology Operations, Ann. of Math. Stud., Princeton Univ. Press, Princeton, NJ, 1962.

[VD] A.P. Veselov and I.A. Dynnikov, Integrable Gradient flows and Morse Theory, Algebra i Analiz, Vol. 8, no 3.(1996), 78–103; Translated in St. Petersburgh Math. J., Vol. 8, no 3.(1997), 429–446.

[Wh] J.H.C. Whitehead, On the groups $\pi_r(V_{n,m})$, Proc. London Math. Soc., 48(1944), 243–291.

Haibao Duan
Institute of Mathematics
Chinese Academy of Sciences
Beijing 100080
e-mail: dhb@math.ac.cn

Trends in Mathematics:
Topics in Cohomological Studies of Algebraic Varieties, 135–161

Integrable Systems and Gromov-Witten Theory

Ali Ulas Ozgur Kisisel

Abstract. This is an expository paper with an aim of explaining some of the main ideas relating completely integrable systems to Gromov-Witten theory. We give a self-contained introduction to integrable systems and matrix integrals, and their relation to Witten's original conjecture (Kontsevich's theorem). The paper ends with a brief discussion of further developments.

1. Completely integrable systems

1.1. Gelfand-Dikii formalism

Let $\partial = \frac{\partial}{\partial x}$, and $v_i(x)$ be functions of x. The v_i, together with ∂ form an associative, but noncommutative algebra of operators acting on a suitable Hilbert space of functions. By the Leibniz rule, we have $\partial v_i = v_i \partial + v_i'$, considering the effects of these operators on a function. Alternatively, one can take the commutation relation as a definition, and develop the theory algebraically. In this paper we will not be concerned with analytic aspects of the theory of integrable systems, hence we will follow the algebraic point of view. It is easy to check by induction that for every $k \in \mathbb{Z}^+$,

$$\partial^k v_i = v_i \partial^k + \binom{k}{1} v_i' \partial^{k-1} + \cdots + \binom{k}{k} v_i^{(k)}.$$

Now consider a larger algebra of "formal pseudo-differential operators", which by definition consists in expressions of the form $\sum_{i=-\infty}^{n} v_i \partial^i$, such that the commutation relation is extended for all $k \in \mathbb{Z}$ by

$$\partial^k v_i = v_i \partial^k + \binom{k}{1} v_i' \partial^{k-1} + \cdots$$

the sum being infinite whenever $k < 0$. It is straightforward to check that formal pseudo-differential operators form an associative algebra, every nonzero element has an inverse, and every element of the form $\partial^n + \sum_{i=-\infty}^{n-1} v_i \partial^i$ has an nth root.

Actual computations of inverses and roots can be carried out inductively with respect to degree. From now on, denote x derivatives by $(v_i)_x$ rather than v_i'.

Let $L = \partial + v_0 + v_{-1}\partial^{-1} + \cdots$ be the most general formal pseudo-differential operator of degree 1. For any pseudo-differential operator $M = \sum_{i=-\infty}^{n} v_i \partial^i$, let $M_+ = \sum_{i=0}^{n} v_i \partial^i$ denote the differential operator obtained by taking the nonnegative degree part (not positive degree). Let $M_- = M - M_+$. Since for any positive integer m the operator L^m commutes with L, we have,

$$[(L^m)_+, L] = [L^m, L] - [(L^m)_-, L] = -[(L^m)_-, L].$$

The last term above is a pseudo-differential operator of degree at most -1. Therefore the following system of flow equations on v_i makes sense

$$\frac{\partial}{\partial t_m} L = [(L^m)_+, L].$$

Here, equality of the two pseudo-differential operators gives an infinite system of differential equations, one equation for each v_i for a given m. This is a system of PDE's noting that x derivatives of v_i also occur. Alternatively, for every m one has a flow on a fixed infinite-dimensional manifold whose coordinate functions are $v_i, (v_i)_x, (v_i)_{xx}, \ldots$ From now on, let us denote $\frac{\partial}{\partial t_m}$ by ∂_{t_m}, and call the mth system the mth flow.

Notice that $\frac{\partial v_0}{\partial t_m} = 0$ for all m, hence v_0 is constant with respect to every t_m. Without loss of generality, we set $v_0 = 0$.

Theorem 1.1. *Any two flows obtained as above commute.*

Proof. The proof follows beautifully from formal manipulations in the algebra of pseudo-differential operators. First, note that $\partial_{t_m}(M_+) = (\partial_{t_m} M)_+$ for any M. Then, one can check that $\partial_{t_m} L^n = [(L^m)_+, L^n]$ by direct computation. Now,

$$\begin{aligned}\partial_{t_m}(L^n)_+ - \partial_{t_n}(L^m)_+ &= [(L^m)_+, L^n]_+ - [(L^n)_+, L^m]_+ \\ &= [(L^m)_+, (L^n)_+]_+ + [L^m, L^n]_+ - [(L^m)_-, (L^n)_-]_+.\end{aligned} \tag{1.1}$$

But the last two terms are zero, the first since powers of L commute, the second since the inside is of negative degree. Thus one obtains the formula $\partial_{t_m}(L^n)_+ - \partial_{t_n}(L^m)_+ = [(L^m)_+, (L^n)_+]$.

Finally,

$$\begin{aligned}\partial_{t_m}\partial_{t_n} L - \partial_{t_n}\partial_{t_m} L =& [\partial_{t_m}(L^n)_+, L] + [(L^n)_+, \partial_{t_m} L] - [\partial_{t_n}(L^m)_+, L] \\ & - [(L^m)_+, \partial_{t_n} L] \\ =& [[(L^m)_+, (L^n)_+], L] + [(L^n)_+, [(L^m)_+, L]] \\ & - [(L^m)_+, [(L^n)_+, L]] \\ =& 0\end{aligned} \tag{1.2}$$

where the last equality follows from the Jacobi identity. □

1.2. KP and KdV hierarchies

The theorem shows that any two elements of the infinite set of flows above commute. This set of flows is called the Kadomstev-Petviashvili (KP) hierarchy.

There are many interesting ways to impose conditions on L such that the flows on the resulting submanifold make sense. One of these is to require that $(L^n)_- = 0$. Then, since $\partial_{t_m}(L^n)_- = [(L^m)_+, L^n]_- = [(L^m)_+, (L^n)_-]_-$, the right-hand side vanishes if $(L^n)_- = 0$. Hence for any m, an integral curve of the mth flow with its initial point on this submanifold lies completely on the submanifold. Set $L' = L^n$. The resulting hierarchy of flows on the coefficients of L' is called the "nth KdV (Korteweg-deVries) hierarchy". The case $n = 2$ is simply called the KdV hierarchy, and the first nontrivial equation of this hierarchy, obtained for $m = 3$, is the classical KdV equation.

Example. To give a flavor of how the computation of the flows are done, let us find the $m = 3$ flow of the $n = 2$ KdV hierarchy. $L' = L^2 = \partial^2 + u$ is a differential operator of degree 2. Then, setting $L = \partial + v_{-1}\partial^{-1} + v_{-2}\partial^{-2} + \cdots$ we have

$$\begin{aligned} L^2 &= \partial^2 + (\partial v_{-1}\partial + v_{-1}\partial^{-1}\partial) + (\partial v_{-2}\partial^{-2} + v_{-2}\partial^{-2}\partial) \\ &\quad + (2v_{-1}\partial^{-1}v_{-1}\partial^{-1} + \partial v_{-3}\partial^{-3} + v_{-3}\partial^{-3}\partial) + \cdots \\ &= \partial^2 + (2v_{-1} + (v_{-1})_x\partial^{-1}) + (2v_{-2}\partial^{-1} + (v_{-2})_x\partial^{-2}) \\ &\quad + (2(v_{-1})^2\partial^{-2} + 2v_{-3}\partial^{-2} + (v_{-3})_x\partial^{-3} - 2v_{-1}(v_{-1})_x\partial^{-3} \\ &\quad + 2v_{-1}(v_{-1})_{xx}\partial^{-4} - \cdots) + \cdots \\ &= \partial^2 + 2v_{-1} + ((v_{-1})_x + 2v_{-2})\partial^{-1} + ((v_{-2})_x + 2(v_{-1})^2 + 2v_{-3})\partial^{-2} + \cdots \end{aligned} \tag{1.3}$$

where in the final line, like powers of ∂ are collected together. Equating this expression to $\partial^2 + u$ gives

$$\begin{aligned} v_{-1} &= \frac{u}{2} \\ v_{-2} &= -\frac{(v_{-1})_x}{2} = -\frac{u_x}{4} \\ v_{-3} &= \frac{-(v_{-2})_x - 2(v_{-1})^2}{2} = \frac{-u_{xx} - 2u^2}{8} \\ &\cdots \quad \cdots \quad . \end{aligned}$$

Next we wish to find $(L^3)_+$. For degree reasons, only the terms actually displayed below can contribute to nonnegative degree part:

$$\begin{aligned} L^3 =& \partial^3 + \partial^2 v_{-1}\partial^{-1} + \partial v_{-1}\partial^{-1}\partial + v_{-1}\partial^{-1}\partial^{-2} + \partial^2 v_{-2}\partial^{-2} \\ &+ \partial v_{-2}\partial^{-2}\partial + v_{-2}\partial^{-2}\partial^{-2} + \cdots \end{aligned} \tag{1.4}$$

and from here we see that

$$\begin{aligned} (L^3)_+ &= \partial^3 + 3v_{-1}\partial + (3v_{-2} + 3(v_{-1})_x \\ &= \partial^3 + \frac{3u}{2}\partial + \frac{3u_x}{4}. \end{aligned} \tag{1.5}$$

Thus

$$\begin{aligned}\frac{\partial u}{\partial t_3} &= \partial_{t_3} L' \\ &= [(L^3)_+, L'] \\ &= \frac{u_{xxx}}{4} + \frac{3uu_x}{2}.\end{aligned} \tag{1.6}$$

□

1.3. The infinite Grassmannian and τ functions

Consider the KP hierarchy. Since any two of the flows ∂_{t_m} commute, it makes sense to speak about a common solution $L(x, t_1, t_2, \dots)$ of these equations given an initial point $L(x, 0, 0, \dots)$. In what follows, such solutions will be constructed explicitly. It is convenient to introduce an additional invertible formal parameter z. (It also makes sense to take $z \in \mathbb{C}$, or $z \in S^1$ to assure invertibility. For the issues of convergence, and for further details of the following arguments, see [SW].)

The first observation is that, setting $K = 1 + a_1\partial^{-1} + a_2\partial^{-2} + \cdots$ one can recursively find a solution $a_i(x, t_1, \dots)$ for $i = 1, 2, \dots$ of the equation

$$L = K\partial K^{-1}$$

of pseudo-differential operators, with a_i a polynomial in derivatives and antiderivatives of v_i with respect to x. The equation obtained at each step of the recursion is of the form $(a_i)_x = \cdots$ making this solution ambiguous by a constant operator.

Since $\partial e^{xz} = ze^{xz}$, we have $L\Psi = z\Psi$ where $\Psi = Ke^{xz}$. Thus the formal variable z (or any concrete value of z such that all convergence conditions are satisfied) can be visualized as an eigenvalue of L, and Ψ the corresponding eigenfunction.

Now consider the completely general lemma:

Lemma 1.1. *Suppose that $L(t), B(t)$ are operators such that $L(t)$ evolves under the equation $\frac{\partial L}{\partial t} = [B, L]$, and $\Psi(0)$ is a vector such that $L(0)\Psi(0) = z\Psi(0)$. Then the solution $\Psi(t)$ of the differential equation $\frac{\partial \Psi(t)}{\partial t} = B\Psi(t)$ obeys $L(t)\Psi(t) = z\Psi(t)$ for all t.*

Proof. Consider

$$\begin{aligned}\frac{\partial(L(t)\Psi(t))}{\partial t} &= \frac{\partial L(t)}{\partial t}\Psi(t) + L(t)\frac{\partial \Psi(t)}{\partial t} \\ &= B(t)L(t)\Psi(t) - L(t)B(t)\Psi(t) + L(t)B(t)\Psi(t) \\ &= B(t)L(t)\Psi(t).\end{aligned} \tag{1.7}$$

Similarly, notice that

$$\frac{\partial(z\Psi(t))}{\partial t} = B(t)z\Psi(t).$$

Hence the functions $z\Psi(t)$ and $L(t)\Psi(t)$ are solutions of the same differential equation, and they agree at 0. Thus they are equal by the uniqueness theorem for ODE's (from an analysis point of view, we must ensure some regularity of $B(t)$ for the theorem to work). □

A pair of operators L, B as above are called a Lax pair (cf. [Lax]). In our setting, the mth flow is a Lax equation with $B = (L^m)_+$. Let us consider the prototypical case of $L = \partial$. Now, the Lax equation is trivial, however, the evolution equation

$$\frac{\partial \Phi}{\partial t_m} = \partial^m \Phi$$

for an eigenfunction Φ still makes sense. We can immediately write down a solution $\Phi(z, x, t_1, t_2, \dots)$:

$$\Phi = e^{xz + t_1 z + t_2 z^2 + t_3 z^3 + \cdots}.$$

Now, let $\Psi = K\Phi$. Then $L\Psi = z\Psi$ for all $t_1, t_2, \dots$ Motivated by the lemma above, one would like to choose K such that $\frac{\partial \Psi}{\partial t_m} = (L^m)_+ \Psi$ where $L = K\partial K^{-1}$. Assume that there indeed exists some differential operator Q_m of degree m such that $\frac{\partial \Psi}{\partial t_m} = Q_m \Psi$. Then,

$$\begin{aligned} Q_m K \Phi &= \frac{\partial (K\Phi)}{\partial t_m} \\ &= \frac{\partial K}{\partial t_m} \Phi + K z^m \Phi \\ &= \left(\frac{\partial K}{\partial t_m} + K \partial^m\right) \Phi. \end{aligned} \tag{1.8}$$

Since any power series in z can be obtained by choosing suitable x and t_i in Φ, we deduce that

$$Q_m K = \frac{\partial K}{\partial t_m} + K \partial^m. \tag{1.9}$$

Thus $Q_m - K\partial^m K^{-1} = \frac{\partial K}{\partial t_m} K^{-1}$. But the right-hand side has negative degree and Q_m is a differential operator. Thus we deduce that

$$Q_m = (K \partial^m K^{-1})_+ = (L^m)_+$$

hence, $(L^m)_+$ is the only possible differential operator of order m that satisfies this equation. Furthermore if this holds, (using $(L^m)_+$ and Q_m interchangeably) observe that

$$\begin{aligned} \frac{\partial L}{\partial t_m} - [(L^m)_+, L] &= \frac{\partial (K \partial K^{-1})}{\partial t_m} - [Q_m, L] \\ &= \frac{\partial K}{\partial t_m} \partial K^{-1} - K \partial K^{-1} \frac{\partial K}{\partial t_m} K^{-1} - [Q_m, L] \\ &= Q_m L - K \partial^{m+1} K^{-1} - L Q_m + K \partial^{m+1} K^{-1} - [Q_m, L] \\ &= 0. \end{aligned} \tag{1.10}$$

Note that 1.9 is used, and K^{-1} is differentiated noncommutatively. Hence L satisfies the KP hierarchy provided that an operator Q_m of degree m as above exists for every m. Namely we ask when

$$\frac{\partial(1+a_1z^{-1}+a_{-2}z^{-1}+\cdots)(e^{xz+t_1z+t_2z^2+\cdots})}{\partial t_m} = Q_m(1+a_1z^{-1}+a_{-2}z^{-1}+\cdots)(e^{xz+t_1z+t_2z^2+\cdots}) \tag{1.11}$$

has a solution $Q_m = \partial^m + b_{m-2}\partial^{m-2} + \cdots + b_0$ for every $m \in \mathbb{Z}^+$ (Let us disregard the technical problem of multiplying a power series in $\mathbb{C}((z))$ by one in $\mathbb{C}((z^{-1}))$. We may restrict our arguments to the choices of $t_1, t_2, \ldots$ making the second term a polynomial). This is a very strict condition on the a_i, since for an arbitrary choice, Q_m would almost certainly have to have terms of negative degree. The surprising fact, due to Sato, is that this puzzle can be solved very elegantly, and for every m at once (cf. [Sat], [SW]).

Let $H = \mathbb{C}((z))$ be the infinite-dimensional vector space of formal Laurent series with coefficients in $\mathbb{C}$. Let $H_+ = \mathbb{C}[[z]]$ be the subspace consisting on formal power series, and H_- the orthogonal complement with respect to the standard (restricted) inner product. Consider the infinite-dimensional Grassmannian of subspaces of H such that the projection π_+ to H_+ has finite-dimensional kernel and cokernel, and the projection π_- is a compact operator. Say W has index 0 if the the dimensions of $Ker(\pi_+)$ and $Coker(\pi_+)$ are equal. In this paper, we restrict all arguments to subspaces with index 0 (for the general case, see [SW]). As an example, take W consisting in series in the terms $z^{-1}, z, z^2, z^3, \ldots$

For simplicity, consider the set of $x, t_1, t_2, \ldots$ such that $g(z) = e^{xz+\sum t_iz^i}$ is a polynomial. Then $g(0) = 1$. For most such choices, $g^{-1}W \cap H_- = \{0\}$. For instance, for the sample W above, a necessary and sufficient condition is that the linear coefficient of g is nonzero. One constructs an eigenfunction Ψ_W associated to W as follows: For all $g \in W$ that satisfy this condition, demand that

$$\Psi_W(g) = g(z)(1 + a_1(z)z^{-1} + a_2z^{-2} + \cdots) \in W.$$

This puts infinitely many conditions on the a_i. The a_i are thus functions of x, t_i since $g(z)$ is. Complete a_i to a function for non-polynomial $g(z)$ whenever convergence makes sense.

Theorem 1.2. *For every W, and solutions a_i of the equations implied by the constraint above, the function Ψ_W admits operators Q_m for every m as in* 1.11

Proof. Choose the coefficients b_i of $Q_m = \partial^m + b_{m-2}\partial^{m-2} + \cdots + b_0$ by decreasing induction on i using the positive degree terms of the desired equation

$$Q_m g(z)(1 + a_1z^{-1} + \cdots) = \frac{\partial g(z)(1 + a_1z^{-1} + \cdots)}{\partial t_m}.$$

One reduces to

$$Q_m\Psi_W - \frac{\partial \Psi_W}{\partial t_m} \in g(z)H_-.$$

But Q_m does not depend on z. The effect of its constituents on a power series is scaling or derivatives with respect to x, under which W is closed. Also $\frac{\partial \Psi_W}{\partial t_m} \in W$. However $g(z)H_- \cap W = \phi$, hence $Q_m \Psi_W = \frac{\partial \Psi_W}{\partial t_m}$ as desired. □

Therefore, associated to every subspace W in the Grassmannian, one can form an eigenfunction Ψ_W, and consequently a solution L_W of the KP hierarchy. Notice that if W has the property that $z^n W \subset W$, then $z^{nk} W \subset W$ for all $k = 1, 2, \ldots$ as well, and for any $m = nk$ we have

$$\begin{aligned} Q_m \Psi_W &= \frac{\partial \Psi_W}{\partial t_m} \\ &= z^m \Psi_W + g(z)(\frac{\partial a_1}{\partial t_m} + \cdots). \end{aligned} \tag{1.12}$$

So, we again have

$$Q_m \Psi_W - z^m \Psi_W \in g(z)H_-,$$

and the left-hand side belongs to W by assumption. Therefore $Q_m \Psi_W = z^m \Psi_W$ and $\frac{\partial a_i}{\partial t_m} = 0$ for every i. Therefore $\frac{\partial K}{\partial t_m} = 0$, thus also $\frac{\partial L}{\partial t_m} = 0$ for all $m = nk$. Thus L is independent of t_{nk}. Finally, using the formula $Q_m - K\partial^m K^{-1} = \frac{\partial K}{\partial t_m} K^{-1} = 0$, we get $Q_m = K\partial^m K^{-1} = L^m$, thus L^m is a differential operator for $m = nk$. Thus for every W such that $z^n W \subset W$ the associated $(L_W)^n$ is a solution of the nth KdV hierarchy.

Note that every occurrence of x and t_1 above was in the form $x + t_1$, hence one may write just t_1 for $t_1 + x$. We will do this for what follows.

There is a very handy function that can be used to relate the subspace W to the eigenfunction Ψ_W. For simplicity again, assume that π^+ maps W surjectively onto H_+. Then define the "tau function" associated to W

$$\tau_W(g) = \frac{\det(\pi_+(g^{-1}W))}{\det(\pi_+(W))}.$$

The determinants involved are determinants of infinite matrices. For technical issues and a more invariant definition we refer the reader to [SW]. We will contend by giving an example. Since g depends on $x, t_1, \ldots, t_n$, τ is a function of $x, t_1, \ldots, t_n$. To find the explicit dependence, expand $g = e^{xz + t_1 z + t_2 z^2 + \cdots}$. Notice that the coefficients of the resulting power series in z are the ordinary Schur polynomials.

Example. Let W be the set containing power series of the form $c_{-1}(z^{-2} + z^{-1} + 1) + c_1 z + c_2 z^2 + \cdots$. Then $\pi_+(W) = H_+$. Let us write down the matrix of the transformation corresponding to multiplying by $g^{-1} = e^{-xz - t_1 z - t_2 z^2 - \cdots}$ in the basis $\{z^i\}$, and represent the basis elements of W by column vectors. We can write

$g^{-1}W$ conveniently in the form:

$$\begin{bmatrix} \cdots & & & & & & & \\ 0 & 1 & -t_1 & \frac{t_1^2}{2}-t_2 & -\frac{t_1^3}{6}+t_1t_2-t_3 & \cdots & & \\ & 0 & 1 & -t_1 & \frac{t_1^2}{2}-t_2 & -\frac{t_1^3}{6}+t_1t_2-t_3 & \cdots & \\ & & 0 & 1 & -t_1 & \frac{t_1^2}{2}-t_2 & -\frac{t_1^3}{6}+t_1t_2-t_3 & \\ & & & & \ddots & & & \\ & & & & & & & \end{bmatrix}$$

$$\times \begin{bmatrix} \cdots & & & & \\ 0 & 1 & 0 & 0 & \\ 0 & 1 & 1 & 0 & 0 \\ 0 & 1 & 0 & 1 & 0 \\ & & & & \ddots \end{bmatrix}$$

$$= \begin{bmatrix} \cdots & & & & & & & \\ 0 & 1-t_1+\frac{t_1^2}{2}-t_2 & -t_1 & \frac{t_1^2}{2}-t_2 & -\frac{t_1^3}{6}+t_1t_2-t_3 & \cdots & & \\ & 1-t_1 & 1 & -t_1 & \frac{t_1^2}{2}-t_2 & -\frac{t_1^3}{6}+t_1t_2-t_3 & \cdots \\ & 1 & 0 & 1 & -t_1 & \frac{t_1^2}{2}-t_2 & \\ & & & & \ddots & & \\ & & & & & & \end{bmatrix}. \tag{1.13}$$

The projection π_+ has the effect of cropping the right lower quadrant of the resulting matrix, furthermore after the last term displayed, the matrix continues in an upper triangular manner. Hence $\tau_W(t_1, \dots)$ is the following 3×3 determinant:

$$\tau_W = \begin{vmatrix} 1-t_1+\frac{t_1^2}{2}-t_2 & -t_1 & \frac{t_1^2}{2}-t_2 \\ 1-t_1 & 1 & -t_1 \\ 1 & 0 & 1 \end{vmatrix}. \tag{1.14}$$

□

The following theorem enables one to compute Ψ_W, hence L_W in terms of τ_W. We will not give a proof here (cf. [SW] or [vM1]).

Theorem 1.3.

$$\Psi_W(t_1, t_2, \dots) = e^{t_1 z + t_2 z^2 + \dots} \frac{\tau_W(t_1 - z^{-1}, t_2 - \frac{z^{-2}}{2}, t_3 - \frac{z^{-3}}{3}, \dots)}{\tau_W(t_1, t_2, \dots)}. \tag{1.15}$$

One can also find the coefficients of L recursively in terms of τ. In the case of KdV, one obtains $u = 2\partial^2(\log \tau)$ where $L^2 = \partial^2 + u$ (cf. [SW] or [vM1]).

1.4. Discrete counterparts, the Toda hierarchy

A theory parallel to the above, but with x dependence of the operators discrete, can be constructed along almost the same lines. Let us mention some important points of this construction. In the process of discretizing, one replaces each of the functions $v_i(x, \mathbf{t})$ appearing in L by a diagonal matrix obtained from the vector of functions $(\dots, v_{i,-1}(\mathbf{t}), v_{i,0}(\mathbf{t}), v_{i,1}(\mathbf{t}), \dots)$. Denote the diagonal matrix by v_i by abuse of notation. Let Δ be the infinite matrix such that $\Delta_{i,j} = \delta_{ij-1}$. Let

$$L = \Delta + v_0 + v_{-1}\Delta^{-1} + v_{-2}\Delta^{-2} + \cdots .$$

Notice that this time we have the commutation relation

$$\Delta v_i = v_i \Delta + v_{i+}\Delta$$

where $(v_{i+})_{jj} = (v_i)_{j+1j+1} - (v_i)_{jj}$. Note that the only essential difference from the continuous case is that there is an additional factor of Δ remaining with the derivative term v_{i+}. The Gelfand-Dikii formalism applies as before. The flows of the hierarchy are:

$$\frac{\partial L}{\partial t_m} = [(L^m)_+, L]$$

for $m = 1, 2, \dots$ Here, M_+ denotes the upper Borel part of the matrix M (i.e., $(M_+)_{ij} = M_{ij}$ if $j \geq i$ and $(M_+)_{ij} = 0$ otherwise. The flows of the hierarchy commute, explicit solutions can be constructed using τ functions (cf. [AvM2], [vM1]).

The system obtained by the truncation $v_i = 0$ for $i \neq 0, -1$ is called the Toda hierarchy. It can be checked that this submanifold contains all flow curves starting on it. Because of the difference in the commutation relation in the discrete case, one can not discard v_0 unlike in the continuous case. Set $v_0 = a$ and $v_{-1} = b$ for simplicity of notation. Then

$$L = \begin{bmatrix} \cdots & & & & \\ b_1 & a_1 & 1 & 0 & \\ 0 & b_2 & a_2 & 1 & 0 \\ & & & \ddots & \\ & & & & \end{bmatrix} \tag{1.16}$$

is a tridiagonal (Jacobi) matrix. Unlike the case for the KdV hierarchy, the flow for $m = 1$ is not trivial, and it gives the equations:

$$\begin{aligned} \frac{\partial a_i}{\partial t_1} &= b_{i+1} - b_i \\ \frac{\partial b_i}{\partial t_1} &= b_i(a_i - a_{i-1}) \end{aligned} \tag{1.17}$$

which are commonly referred to as the Toda equations. We note that the exponential form of the equations in the physics literature are equivalent to these equations by a change of variables.

An important point is, given a positive integer N, the hierarchy makes sense if we make a further restriction to the submanifold $a_i = 0$ except for $0 < i \leq N$ and $b_i = 0$ except for $1 < i \leq N$. The resulting system is called the finite Toda hierarchy. The periodic counterpart (i.e., $a(i+N) = a(i)$ and $b_{i+N} = b_i$) also makes sense.

As we mentioned, one can develop the formalism of τ functions as in the continuous case. In this case, $\tau(\mathbf{t})$ is a vector whose components are indexed by the discretized variable n; $\tau(\mathbf{t}) = (\dots, \tau_{-1}(\mathbf{t}), \tau_0(\mathbf{t}), \tau_1(\mathbf{t}), \dots)$. For any element of the Grassmannian, the corresponding L can be expressed in terms of τ. For the Toda hierarchy, one has (cf. [AvM1], [vM1]):

Theorem 1.4. $a_i = -\frac{\partial}{\partial_{t_1}} \log(\frac{\tau_{i+1}}{\tau_i}) \qquad b_i = \frac{\tau_{i+1}\tau_{i-1}}{\tau_i^2}.$

We remark that there also exists a 2-Toda hierarchy (cf. [UT1], [UT2]) frequently referred to in the Gromov-Witten theory literature, which is a multi-component system. It is not an example of the setting described above, but can be studied via an analogous extension of the operator theory above (cf. [AvM1]).

2. Random matrices and enumeration of graphs

2.1. Gaussian integrals

Recall that $\int_{-\infty}^{\infty} e^{\frac{-x^2}{2}} dx = \sqrt{2\pi}$. Thus

$$\begin{aligned} \frac{1}{\sqrt{2\pi}} \int_{-\infty}^{\infty} e^{\frac{-x^2}{2}+xy} dx &= \frac{1}{\sqrt{2\pi}} \int_{-\infty}^{\infty} e^{\frac{-(x-y)^2}{2}+\frac{y^2}{2}} dx \\ &= e^{\frac{y^2}{2}}. \end{aligned} \tag{2.1}$$

Consequently, this can be used to compute

$$\begin{aligned} \frac{1}{\sqrt{2\pi}} \int_{-\infty}^{\infty} x^{2k} e^{\frac{-x^2}{2}} dx &= \left(\frac{d^{2k}}{dy^{2k}} \frac{1}{\sqrt{2\pi}} \int_{-\infty}^{\infty} e^{\frac{-x^2}{2}+xy}\right)\Big|_{y=0} \\ &= \left(\frac{d^{2k}}{dy^{2k}} e^{\frac{y^2}{2}}\right)\Big|_{y=0}. \end{aligned} \tag{2.2}$$

(Notice that if the $2k$ above is replaced by an odd number, the answer is 0.) Before evaluation at $y = 0$, the right-hand side expression above is a certain polynomial times $e^{\frac{y^2}{2}}$. After the evaluation, only the constant term of the polynomial remains. Examining the computation, which is performed using the Leibniz rule, one sees that contributions to the constant term arise from different ways of getting k (created) y terms from the exponential, each of which is annihilated by a $\frac{d}{dy}$ acting later in order than the corresponding creator. Therefore, the number above is the number of graphs on $2k$ labelled vertices, such that each vertex is connected to precisely one other vertex (i.e., the number of 1-regular graphs on $2k$ vertices).

The edges of this graph are undirected since the correspondence is as follows: for any given edge, of the two vertices incident with this edge, the vertex with the smaller label corresponds to the creator $\frac{d}{dy}$ and the one with the greater label corresponds to the annihilator $\frac{d}{dy}$. Consequently, the answer of the enumeration problem simply is $(2k-1)!! := (2k-1)(2k-3)\cdots 5\cdot 3\cdot 1$.

One can obtain the same number as the result of a seemingly different enumerative problem. Suppose that $l|2k$, $n = \frac{2k}{l}$, and let H be a 1-regular graph on $2k$ vertices as above. Since H is 1-regular, it seems somewhat redundant to represent it as a graph on $2k$ vertices. So instead, one constructs a new graph G with n vertices from H as follows: For vertices $1, 2, \ldots, l$ in H, place one vertex in G, for $l+1, \ldots, 2l$ place a second vertex and so on. For each vertex v in H, place a half-edge in G incident to the vertex in G representing the set of vertices that v belongs to. Join two half edges into an edge in G if and only if the corresponding vertices in H are connected. The resulting G is an l-regular graph on n vertices. Let us leave G without any labels.

Conversely, given an l-regular graph G on n vertices, one can reconstruct H if additionally labelling on the vertices of G is introduced, and furthermore if every vertex is endowed with an ordering of the half edges incident to that vertex. There are $n!(l!)^n$ ways to place these labels. However, every unlabelled G produces the same H precisely $|Aut(G)|$ times, where this number also includes the involutive symmetries due to loops at a vertex. Let us form a generating function for the number of all l-regular graphs, with a λ^n factor keeping track of the number of vertices. Then, summing over all l-regular G,

$$\begin{aligned}\sum_G \frac{\lambda^n}{|Aut(G)|} &= \sum_n \frac{\lambda^n}{n!(l!)^n}\frac{1}{\sqrt{2\pi}}\int_{-\infty}^{\infty} x^{2k} e^{\frac{-x^2}{2}}\,dx \\ &= \sum_n \frac{1}{\sqrt{2\pi}}\int_{-\infty}^{\infty}\frac{1}{n!}\left(\frac{\lambda x^l}{l!}\right)^n e^{\frac{-x^2}{2}}\,dx \\ &= \frac{1}{\sqrt{2\pi}}\int_{-\infty}^{\infty} e^{\frac{-x^2}{2}+\frac{\lambda x^l}{l!}}\,dx.\end{aligned} \tag{2.3}$$

Denote this generating function by $Z(\lambda)$.

Example. For $l = 3$ one may compute the first few terms of $Z(\lambda)$ as follows: since l is odd, the coefficient of λ^n is 0 unless n is even. For $n = 2$, there are two graphs with automorphism groups of orders 8 and 12 respectively. For $n = 4$, there are 8 graphs with automorphism groups of orders $8, 16, 16, 24, 48, 96, 128, 288$ respectively (see figure 1). Therefore $Z(\lambda) = \frac{5}{24}\lambda^2 + \frac{385}{1152}\lambda^4 + \cdots$ which can also be computed as $Z(\lambda) = \frac{5!!}{3!(2!)^3}\lambda^2 + \frac{7!!}{4!(2!)^4}\lambda^4 + \cdots$

The generating function above enumerates all l-regular graphs, connected or disconnected. Suppose that $F(\lambda) = \sum_{i=0}^{\infty} f_i\lambda^i$ is the generating function for the same enumeration problem, but only over connected l-regular graphs. Also denote $Z(\lambda) = \sum_{i=0}^{\infty} z_i\lambda^i$. Then since a disconnected graph is an assemble of smaller

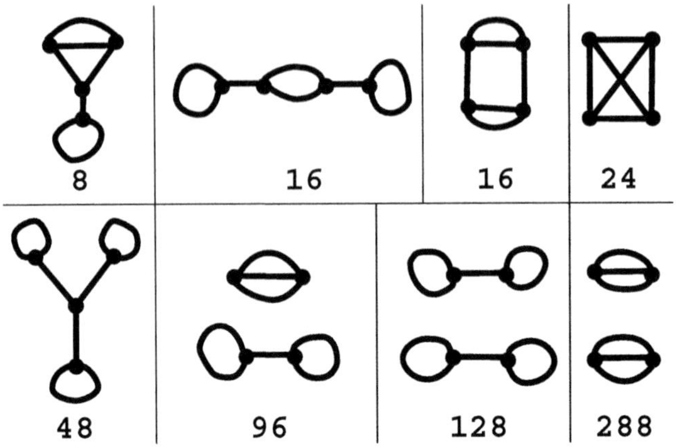

FIGURE 1. 3-valent graphs on 4 vertices, and orders of their automorphism groups

connected graphs, we obtain that

$$z_i = \sum_{i_1+\cdots+i_k=i} \frac{f_{i_1}\cdots f_{i_k}}{k!}$$

and consequently it is easy to check that the following elegant formula holds:

$$Z(\lambda) = e^{F(\lambda)}.$$

For instance, in the example above, $F(\lambda) = \ln(Z(\lambda)) = \frac{5}{24}\lambda^2 + \frac{91}{288}\lambda^4 + \cdots$.

2.2. Matrix integrals and ribbon graphs

A very interesting variant of the construction above gives finer enumerative information about slightly different objects, called ribbon graphs. The generating function above organizes the number of l-regular graphs with respect to the number of vertices. The analysis that follows will end up with an additional parameter producing the number of ribbon graphs for which the surface of minimal genus that they can be embedded on has genus g. The idea is to replace the variable x by a Hermitian matrix.

Say that M denotes an N by N Hermitian matrix with entries M_{ij}. Let dM denote the measure

$$\bigwedge_i dM_{ii} \wedge \bigwedge_{i<j} d(Re(M_{ij}) \wedge \bigwedge_{i<j} d(Im(M_{ij})) = (\frac{\sqrt{-1}}{2})^s \bigwedge_{i,j} M_{i,j}$$

where $s = N(N-1)/2$ and the ordering is chosen so that the signs agree. Identifying the space of Hermitian matrices with $\mathbb{R}^{N^2}$, it is clear that dM is translation invariant. The unitary group $U(N)$ acts on the space of Hermitian matrices by conjugation. It turns out that dM is also invariant under this action. Since M is Hermitian, it is diagonalizable and has real eigenvalues. Therefore M^2 has real nonnegative eigenvalues. So except for the case $M = 0$, the quantity $Tr(M^2)$ is positive. The function $e^{-Tr(M^2)}$ decreases rapidly when the norm of M increases, and it is easy to show that the integral

$$\int f(M)e^{-Tr(M^2)}dM$$

converges for any polynomial f in the entries of M.

To avoid computation of constants which are not central to our discussion, scale dM such that $\int e^{-Tr(M^2)}dM = 1$.

Consider the following matrix integral which is analogous to 2.3:

$$Z(N,\lambda) = \int e^{-Tr(\frac{M^2}{2} - \lambda M^l/((l!)N^{\frac{l}{2}-1}))}dM. \tag{2.4}$$

This integral may not converge for any definite value of λ, therefore the above expression is just a formal series in λ each coefficient of which converges. Notice that an additional factor of $\frac{1}{N^{\frac{l}{2}-1}}$ is inserted in the second term of the exponential, and this will be explained later. Expanding this second term, the computation reduces to evaluation of integrals of the form $\int \frac{(Tr(M^l))^n}{(l!)^n N^{n(l-2)/2}} e^{-Tr(\frac{M^2}{2})}dM$. Since $Tr(M^l) = \sum M_{i_1 i_2} M_{i_2 i_3} \cdots M_{i_l i_1}$, it certainly suffices to compute any integral of the form $\int M_{i_1 j_1} \cdots M_{i_r j_r} e^{-Tr(\frac{M^2}{2})}dM$. Essentially the same computation as in 2.2 shows:

$$\int M_{i_1 j_1} \cdots M_{i_r j_r} e^{-Tr(\frac{M^2}{2})}dM = \frac{\partial}{\partial Y_{i_r j_r}} \cdots \frac{\partial}{\partial Y_{i_1 j_1}} e^{Tr(\frac{Y^2}{2})}|_{Y=0}. \tag{2.5}$$

Since $Tr(M^2) = \sum_{i,j} M_{ij}M_{ji}$, each contribution to the right-hand side in this case results from a creator $\frac{\partial}{\partial Y_{ij}}$ paired with an annihilator $\frac{\partial}{\partial Y_{ji}}$ acting later. Thus, we observe again that r must be even (say $r = 2k$) in order for the integral not to vanish. Again, we have a bijective correspondence between contributing actions and labelled 1-regular graphs on $2k$ vertices, however, this time on the graph H we must put additional labelling specifying which partial differential operator each vertex corresponds to. Given two vertices connected by an edge, if the first vertex corresponds to $\frac{\partial}{\partial Y_{ij}}$, then by the discussion above the second should correspond to $\frac{\partial}{\partial Y_{ji}}$.

As before, assume that $2k = nl$. One constructs an l-regular graph G from H. The n vertices are placed as before. Also a half-edge in G for each vertex in H is placed as before, however this time every half edge carries an additional labelling (i,j). Replace this information by the doubling of each half-edge into one outgoing half-edge and one incoming half-edge, labelled by i and j respectively. Two half edges in G are joined to an edge if and only if the corresponding vertices in H are joined, thus one sees that the final labelling on edges is compatible (see figure 2). Conversely, given an l-regular graph on n vertices and labelled directed double edges, one can label the vertices and order the half-edges at every vertex in $n!(l!)^n$

ways to retrieve H, and the same H is obtained $|Aut(G)|$ times where now $Aut(G)$ denotes the group of automorphisms of G that preserve the edge labelling.

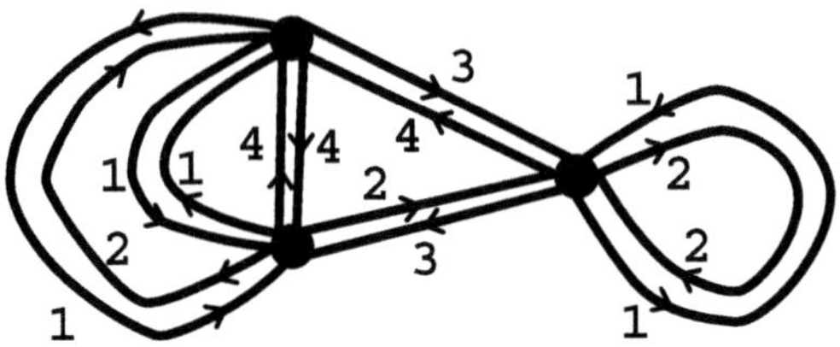

FIGURE 2. A non-ribbon graph representing $(M_{12}M_{21}M_{34}M_{23})(M_{43}M_{44}M_{11}M_{21})(M_{32}M_{44}M_{11}M_{12})$

Let us reconsider $\int (Tr(M^l))^n e^{-Tr(\frac{M^2}{2})} dM$. The upshot of having traces is that terms in the expansion of such an integral are in one to one correspondence with a subset of the graphs above which satisfy an additional compatibility condition at the vertices. Since $Tr(M^l) = \sum M_{i_1 i_2} M_{i_2 i_3} \cdots M_{i_l i_1}$, in the expansion of $(Tr(M^l))^n$ it is possible to partition each monomial of degree ln in groups of l such that each group contains a monomial of the form $M_{i_1 i_2} M_{i_2 i_3} \cdots M_{i_l i_1}$. But now at each vertex of G the half edges can be ordered cyclically in a way compatible with the indexing, and thus each outgoing half edge has the same labelling as the neighboring incoming half edge. For better visualization, one may remove the vertex with an l-way crossroad (see figure 3). The complete figure is composed of closed paths each of which has a coherent label. It is easy to see that conversely every such figure, henceforth called a ribbon graph, corresponds to a contributing term in the integral above.

It is possible to obtain a closed surface by attaching 2-disks to each of the closed paths mentioned above. It is clear that the surface obtained has genus minimal among the genera of the surfaces that the ribbon graph can be embedded in. A ribbon graph in the expansion above has n vertices and $\frac{nl}{2}$ (thickened) edges. Suppose also that it has f closed paths. Then the Euler characteristic formula gives $\chi(G) = 2 - 2g = f - \frac{nl}{2} + n$. Now remove the edge labels of the ribbon graphs. For a given ribbon graph, there are N^f ways to label the edges coherently using integers $i \in \{1, \ldots, N\}$. Considering also the factor $N^{n(1-\frac{l}{2})}$ in 2.4, we deduce that the unlabelled ribbon graphs of genus g all contribute to the coefficient of N^{2-2g} in the expansion. Therefore one has

$$Z(N, \lambda) = \sum_{n,g} N^{2-2g} \lambda^n \sum_G \frac{1}{|Aut(G)|} \tag{2.6}$$

where the inner sum is over all unlabelled l-regular ribbon graphs of n vertices and of genus g.

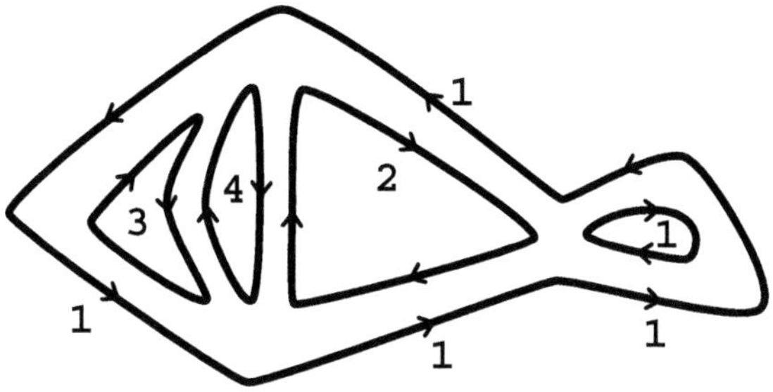

FIGURE 3. A ribbon graph representing the $(M_{11}M_{11}M_{12}M_{21})(M_{12}M_{24}M_{43}M_{31})(M_{13}M_{34}M_{42}M_{21})$ term in $(Tr(M^4))^3$

2.3. Matrix integrals and the Toda lattice

In this section consider a matrix integral of the form

$$Z(N, t_1, t_2, \dots) = \int e^{-Tr(\frac{M^2}{2} - \sum t_i M^i)} dM. \tag{2.7}$$

(The symbol λ will be reserved for another purpose in this section). As a first step, following [BIZ] and [vM2], reduce the matrix integral to an ordinary integral. The idea is the following. Every Hermitian matrix M with eigenvalues $\lambda_1, \dots, \lambda_N$ can be written in the form $M = U\Lambda U^*$ with U unitary, where $\Lambda = (\lambda_1, \dots, \lambda_N)$. We mentioned without proof that the volume form dM is invariant under the action of the unitary group. Thus it might be a good idea to integrate over the unitary group first, since both the integrand and the form posses this symmetry. Notice that every unitary U can be expressed as e^A for A a skew-Hermitian matrix. Let E_{ij} denote the elementary matrix with $(E_{ij})_{kl} = \delta_{ik}\delta_{jl}$. Then $\exists a_{ij}, b_{ij} \in \mathbb{R}$ for $i < j$, uniquely determined by A such that

$$A = \sum_{1 \le i < j \le N} a_{ij}(E_{ij} - E_{ji}) + \sqrt{-1} b_{ij}(E_{ij} + E_{ji}).$$

An easy computation shows that

$$[A, \Lambda] = (\lambda_j - \lambda_i) \sum_{1 \le i < j < N} a_{ij}(E_{ij} + E_{ji}) + \sqrt{-1} b_{ij}(E_{ij} - E_{ji}).$$

Notice

$$M = e^A \Lambda e^{-A} = \Lambda + [A, \Lambda] + \cdots$$

where the omitted terms have more commutators. It is possible to show by a stationary phase argument that the higher commutator terms do not contribute to the integral, hence we have

$$
\begin{aligned}
dM =& \bigwedge_i d\lambda_i \wedge \bigwedge_{1\le i<j\le N} d((\lambda_j-\lambda_i)a_{ij}) \wedge d((\lambda_j-\lambda_i)b_{ij})\\
=& \prod_{i<j}(\lambda_i-\lambda_j)^2 \bigwedge_i d\lambda_i \wedge \bigwedge_{1\le i<j\le N} da_{ij}\wedge db_{ij}\\
=&\Delta^2 \bigwedge_i d\lambda_i \wedge dU
\end{aligned}
$$

where $\Delta^2 = \prod_{i<j}(\lambda_i - \lambda_j)$, which is also equal to the determinant of the Vandermonde matrix. Integration over the unitary group then produces a constant independent of Λ, the volume of the unitary group. We disregard this constant since it does not intervene the dynamics of $Z(N, t_1, \dots)$. Also notice that $Tr(M^k) = \sum_{i=1}^N \lambda_i^k$. Thus the problem is reduced to studying the integrals

$$\int_{\mathbb{R}^N} \Delta^2 e^{\sum_{i=1}^N V(\lambda_i)} \bigwedge_{i=1}^N d\lambda_i \tag{2.8}$$

where V is a power series with coefficients $t_1, t_2 + \frac{1}{2}, t_3, \dots$.

The next observation, following [Wit1], makes it possible to decouple this integral into a product of N 1-variable integrals. This uses the theory of orthogonal polynomials. The measure $d\mu_i = e^{V(\lambda_i)}d\lambda_i$ defines an inner product on real-valued polynomials in λ by

$$\langle p_1, p_2\rangle = \int p_1 p_2 d\mu_i.$$

It is easily seen by induction that there exists a unique orthogonal basis $\{p_n\}$ for polynomials in λ of degree $\le N$ such that $p_n = \lambda^n + \cdots$, here insisting that the first coefficient of p_n is 1. Define $h_i = \langle p_i, p_i\rangle$. Now a key observation is that the matrix P with entries $P'_{ij} = p_{i-1}^{j-1}$ can be obtained from the Vandermonde matrix by elementary column operations not changing the determinant. Hence we can replace Δ by $\det(P)$. Expanding, one gets

$$
\begin{aligned}
&\int_{\mathbb{R}^N} \Delta^2 e^{\sum_{i=1}^N V(\lambda_i)} \bigwedge_{i=1}^N d\lambda_i\\
&\quad = \int_{\mathbb{R}^N} \sum_{\pi,\pi'\in S_N} (-1)^{\sigma(\pi)+\sigma(\pi')} p_{\pi_i}(\lambda_i) p_{\pi'_i}(\lambda_i) \bigwedge_{i=1}^N d\lambda_i\\
&\quad = \sum_{\pi,\pi'\in S_N} (-1)^{\sigma(\pi)+\sigma(\pi')} \prod_i \Big(\int_{\mathbb{R}} p_{\pi_i}(\lambda_i) p_{\pi'_i}(\lambda_i) d\lambda_i\Big)\\
&\quad = N! h_0 h_1 \cdots h_N
\end{aligned} \tag{2.9}
$$

since for a fixed π, the only way that the term can survive for every i is that $\pi' = \pi$. (Here, S_N is the symmetric group on N letters.) Thus the problem reduces to determining $h_n = \int_{\mathbb{R}} p_n(\lambda)p_n(\lambda)d\lambda$ as a function of $t_1, t_2, \dots$.

Since $\lambda p_n(\lambda)$ is a polynomial of degree $n+1$, it should be uniquely expressible in the form:

$$\lambda p_n(\lambda) = p_{n+1}(\lambda) + a_n p_n(\lambda) + b_n p_{n-1}(\lambda) + q_{n-2}(\lambda)$$

where q_{n-2} is a polynomial of degree at most $n-2$, and a_n, b_n depending on t_i but not λ. Notice that because of the inductive construction of the basis, the inner product of p_n with any polynomial of degree less then n is 0. Now, here is a beautiful classical observation: Take the inner product of both sides with q_{n-2}. The right-hand side is the norm squared of q_{n-2}. The left-hand side can be written as $\int_{\mathbb{R}}(\lambda q_{n-2})p_n d\lambda$. But λq_{n-2} has degree less than n. Hence the left-hand side is 0. Thus $q_{n-2}=0$. This gives

$$\lambda p_n(\lambda) = p_{n+1}(\lambda) + a_n p_n(\lambda) + b_n p_{n-1}(\lambda).$$

Hence the operator of multiplication by λ can be represented by the tridiagonal matrix L of 1.16, with respect to the basis consisting in the orthogonal polynomials p_i. Note that both the basis and L depend on t_i, whereas λ (hence the operator) does not. Taking inner products of both sides with p_{n-1} we get

$$\begin{aligned} b_n &= \langle \lambda p_n, p_{n-1}\rangle \\ &= \langle p_n, \lambda p_{n-1}\rangle \\ &= \langle p_n, p_n + a_{n-1}p_{n-1} + b_{n-1}p_{n-2}\rangle \\ &= \langle p_n, p_n\rangle \\ &= h_n. \end{aligned} \tag{2.10}$$

Let us investigate how L depends on t_i. It is important to notice that $L(t_1,\dots)$ is always tridiagonal, but with respect to a basis that depends on $(t_1,\dots)$. Say that $L = KL_0K^{-1}$ where L_0 is the value of L at $t_i = 0 \forall i$, and K is the change of basis matrix from $\mathbf{t} = (0,\dots)$ to $\mathbf{t} = (t_1,\dots)$. Then

$$\begin{aligned} \frac{\partial L}{\partial t_i} &= \frac{\partial (KL_0K^{-1})}{\partial t_i} \\ &= \frac{\partial K}{\partial t_i}L_0K^{-1} - KL_0K^{-1}\frac{\partial K}{\partial t_i}K^{-1} \\ &= \frac{\partial K}{\partial t_i}K^{-1}L - L\frac{\partial K}{\partial t_i}K^{-1} \\ &= [B_i, L] \end{aligned} \tag{2.11}$$

where $B_i = \frac{\partial K}{\partial t_i}K^{-1}$. Thus the flow equations for L are in Lax form, and they commute since we know that there exists a well-defined solution L valid for all t_i.

This shows that L is subject to a set of N commuting flows. That these have to be the Toda flows, hence $\frac{\partial L}{\partial t_i}$ can again be shown using a purely algebraic argument (cf. [vM1]).

3. Witten's conjecture and Kontsevich's solution

3.1. $\overline{M}_{g,n}$ and ψ classes

Let $M_{g,n}$ denote the moduli space of smooth irreducible, complete algebraic curves over $\mathbb{C}$ of genus g with n distinct marked points. Assume that $2 - 2g - n < 0$, so that each one of the curves parameterized by this space, if punctured at its marked points, becomes hyperbolic. This is assumed in order to ensure that the automorphism group of each of the marked curves is a finite group. Notice, this requirement implies only that if $g = 0$ then $n \geq 3$ and if $g = 1$ then $n \geq 1$. It can be shown that $M_{g,n}$ is algebraic, but it is certainly not a complete variety (i.e., not compact in the usual topology). Among various possibilities, one way to compactify $M_{g,n}$ is to allow degenerations of smooth curves into the so called "stable curves" only. Call C a stable curve with data (g, n), if C is a connected but possibly reducible curve of arithmetic genus g with at most nodal singularities, C has n distinct marked points which are not singular points, and every irreducible component of C punctured at its special points is hyperbolic (this time we are regarding nodes, as well as marked points, as special points). See Figure 4 for examples.

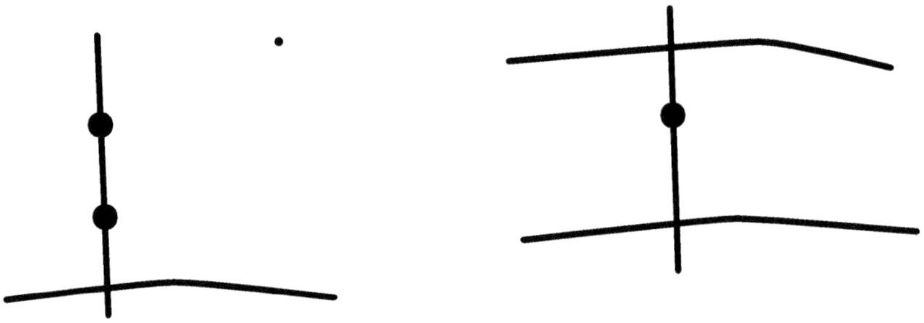

FIGURE 4. Examples of stable curves. Unmarked components have genus > 0.

We wish to explain intuitively why compactification is established despite the fact that further degenerations are not allowed. As a first example, consider a 1-parameter family of curves in which the complex structure and $n - 2$ of the marked points stay fixed, but two of the marked points are approaching each other at the point $x \in C$. Recall that one tries to avoid allowing repeated marked points in the compactification. Instead of this, take the limit object to be the curve obtained by attaching C a new genus 0 component at x, and two marked points on this component. Notice that up to isomorphism, there is only one genus 0 curve with 3 marked points, so this limit curve is well defined. As a second example, take a one parameter family of nodal curves such that a third branch is approaching a node in a way that the tangent directions of the three branches are distinct. Define the limit curve to be the blow up of the ordinary limit curve at its singularity. The exceptional divisor is a new component of genus 0, and it has 3 marked points, so the limit is a stable curve. These two examples generalize, and it is a theorem that $M_{g,n}$ can be compactified into a projective variety allowing only stable degenerations. The resulting object is called the "Deligne-Mumford compactification" of $M_{g,n}$, and is denoted by $\overline{M}_{g,n}$. This variety is irreducible, connected and of dimension $3g - 3 + n$ (cf. [DM]).

Denote the ith marked point on a curve by x_i. Consider the map $\pi : \overline{M}_{g,n+1} \to \overline{M}_{g,n}$ obtained by forgetting the $(n+1)$st marked point x_{n+1}, and if x_{n+1} is lying on a genus 0 component with only 2 other special points, contracting this component to a single point. It can be shown that this map is an algebraic morphism. Let us look at the fiber of π over a point $C \in \overline{M}_{g,n}$. Then for every point $x \in C$, there is a corresponding point in $\pi^{-1}(C)$: If x is nonsingular and not among $x_1, \dots, x_n$, then the corresponding point in the fiber is the $n+1$ pointed curve with $x_{n+1} = x$. If x is a node, then the corresponding point is the blow up of C at x with x_{n+1} on the exceptional component, and if $x \in \{x_1, \dots, x_n\}$, then the corresponding point is the unique stable curve with an additional genus 0 component at x, and x_i, x_{n+1} on this component. Now, suppose that C has an automorphism φ such that $\varphi(x_i) = x_i$ for every $i = 1, \dots, n$. Then for every x in C, the curves represented in $\pi^{-1}(C)$ corresponding to x and $\varphi(x)$ are isomorphic $n + 1$ pointed curves, therefore are the same point in $\overline{M}_{g,n+1}$. Hence the fiber $\pi^{-1}(C)$ is $C/Aut(C)$ (where $Aut(C)$ is the group of automorphisms of C fixing each of the n markings). So $\pi : \overline{M}_{g,n+1} \to \overline{M}_{g,n}$ is not actually a universal curve over $\overline{M}_{g,n}$; for a family $\pi' : S \to B$ of pointed curves such that at least one ${\pi'}^{-1}(x)$ has automorphisms, there does not exist any map $f : B \to \overline{M}_{g,n}$ such that $S \to B$ is the pullback by f of $\overline{M}_{g,n+1} \to \overline{M}_{g,n}$.

The discussion above suggests that it is natural to keep the information of the automorphism group (which is finite) along with the curve, hence the natural category to view $\overline{M}_{g,n}$ in is the category of orbifolds. For cases such as $g = 1, n = 1$ or $g = 2$, every curve has a hyperelliptic involution, thus one may wish to work algebraically (e.g., if one wishes to do intersection theory) to capture nonreduced structure as well. The same remark applies if one is interested in subloci in $\overline{M}_{g,n}$, such that each member of the sublocus represents a curve with an automorphism (example: the hyperelliptic locus). Then the correct context is to work in is the (2-)category of Deligne-Mumford stacks.

For $i \in \{1, 2, \dots, n\}$, let $\sigma_i : \overline{M}_{g,n} \to \overline{M}_{g,n+1}$ be the map taking C to the point in the fiber of π representing the $n+1$ pointed curve for $x = x_i$ as discussed above. Therefore the image consists in points of $\overline{M}_{g,n+1}$ representing stable curves which are the unions of at least two components where one of the components is a genus 0 curve with x_i, x_{n+1} as the only marked points and joined to the rest of the curve at one node situated at the location of x_i in C. Since x_i is not a singular point in C, such points in $\overline{M}_{g,n+1}$ are not singular points of the fiber. Therefore the cotangent space along the fiber to $\sigma_i(\overline{M}_{g,n})$ is one-dimensional at every point. Denote the pullback of the resulting line bundle to $\overline{M}_{g,n}$ by $\mathcal{L}_i$, and set $\psi_i = c_1(\mathcal{L}_i)$. Notice that we are glossing over curves with automorphisms in the argument above. In order for everything to make sense, one has to develop and work with a theory of line bundles, cohomology and the Chow ring in the framework of Deligne-Mumford stacks. In this context, the intersection numbers are not always integers, but are rather in $\mathbb{Q}$. The classes ψ_i are analogous to the Chern classes of the universal k-plane bundle of the Grassmannian $\mathrm{Gr}(k,n)$, which are known to generate the Chow ring of $\mathrm{Gr}(k,n)$. There is a simplified model of the Chow ring of $\overline{M}_{g,n}$, called the tautological ring of $\overline{M}_{g,n}$, consisting in many geometrically interesting classes. Given two stable curves of genera g_1, g_2 and having n_1+1 and n_2+1 marked points respectively, one can attach the two curves at their last marked points to obtain a new stable curve of genus g_1+g_2 and having n_1+n_2 marked points. This construction can be done in families and one obtains a morphism:

$$\Delta : \overline{M}_{g_1,n_1+1} \times \overline{M}_{g_2,n_2+1} \to \overline{M}_{g_1+g_2,n_1+n_2}.$$

The tautological ring $R^*(\overline{M}_{g,n})$ of $\overline{M}_{g,n}$ is defined to be the subring of the Chow ring generated by the classes ψ_i, and their pullbacks and push-forwards under π and Δ for different values of n,n_i,g_i. This ring has been of much interest, for an exposition see [Vak]. As an example of the geometric uses of the tautological ring, the hyperelliptic locus can be represented by an element in $R^*(\overline{M}_{g,n})$ (cf. [Mum]).

The conjectures and theorems that will be described concern evaluation of top-dimensional products of the ψ classes against the fundamental class of $\overline{M}_{g,n}$. Such an evaluation will be denoted by $\int_{\overline{M}_{g,n}} \psi_1^{k_1} \cdots \psi_n^{k_n}$, and vanishes for dimensional reasons unless $k_1+\cdots+k_n = 3g-3+n$. There are two alternative notations for this quantity. The first is $\langle \tau_1^{k_1} \cdots \tau_n^{k_n} \rangle$. Since the expression is symmetric under a permutation of marked points, it only depends on the unordered partition $k_1, \ldots, k_n$ of $3g-3+n$, and thus the following notation unambiguously represents it: $\langle \tau_{k_1}\tau_{k_2}\cdots\tau_{k_n}\rangle$. Both notations are used, sometimes simultaneously (beginning with [Wit1]) in the literature.

3.2. Witten's conjecture

In [Wit1], Witten made a remarkable conjecture about the intersection numbers discussed above.

Theorem 3.1. *(Witten's conjecture, Kontsevich's theorem) Let*

$$F(t_1,t_2,\ldots) = \langle e^{\sum_{i=1}^{\infty} t_i\tau_i}\rangle := \sum_j \sum_{k_1,\ldots,k_j} \langle \tau_{k_1}\cdots\tau_{k_j}\rangle \frac{t_1^{k_1}}{k_1!}\cdots\frac{t_j^{k_j}}{k_j!}$$

and let $Z(t_1,t_2,\ldots) = e^F$ (Notice that both notations for the intersection number is used in the expression above). Then Z is a tau function of the KdV hierarchy.

This conjecture was proven in [Ko] by Kontsevich. One important corollary of this result is that, the KdV equations provide enough recursions among the numbers $\langle \tau_{k_1}\tau_{k_2}\cdots\tau_{k_n}\rangle$, allowing the computation of all of them.

The motivation for the conjecture, together with some evidence in low genus, came from physics. The reader is referred to the original article [Wit1], and [Dij]. Here, we will attempt to summarize the idea, but please be warned that the next two paragraphs should only be considered as a naive impression of the author about profound theories of physics: In quantum mechanics, an integral Z of the quantity $e^{\frac{i}{\hbar}\int Ldt}$ over all possible paths between two points, where L is the relevant Lagrangian, denotes an amplitude (which in turn determines the infinitesimal probability of the particle to move from the first point to the second in the physical system that the Lagrangian describes). Such integrals over infinite-dimensional spaces are often not completely understood mathematically, however computations that physicists perform using them can be in excellent agreement with experiment. The formalism of the path integral can be applied in more recent and more general theories of fundamental forces of the universe, such as relativistic quantum mechanics or string theory. The context relevant for the conjecture of Witten is that of plausible two-dimensional gravity theories compatible with quantum theory, which are conceptual models built in analogy to Einstein's theory of general relativity rather than being theories aiming at describing a physical situation. The analog of the path integral in this case is first of all a sum over all g, furthermore it is an integral over all metrics on a 2-manifold of genus g rather than an integral over all paths. The main problem comes from what to write in place of $e^{\frac{i}{\hbar}\int Ldt}$. This term is written in the light of physical principles, which dictate certain symmetries, and conditions such that infinity terms which would flood the computation otherwise do not come up.

We remark that, the presence of terms of different genera in the sum is an essential feature for the mathematical applications of the theorem: the result enables computations in high genus to be done recursively in terms of lower genus terms.

Witten considers two different previously proposed models obeying these principles. The first of these models, topological gravity, reduces the integration over all metrics to an integration over all conformal, hence complex structures, thus the quantity becomes a sum of integrals over moduli spaces of each genus. The second model, named the 1-matrix model, approximates an arbitrary metric on the 2-manifold M by a discrete metric, in other words a metric such that for each approximation the curvature is supported as δ functions at finitely many points. One way to do this to consider all tessellations of M by a given number n of polygons. Assume that the tessellation is simplicial, then a vertex of one polygon can meet another polygon only at a vertex of the second. Now, if we declare that each of the polygons is flat, regular, and of a fixed volume, then the condition of fitting these together at the vertices produces a metric on M supported only at the vertices. When $n \to \infty$, one expects that every continuous metric on M is approximated by a sequence of tessellations. Next, replace the simplicial complex obtained from the polygons by its dual graph. For instance, if the tessellation consists in l-sided polygons only, one obtains an l-regular graph on n vertices. Finally, the integral in question is approximated by a sum over all such graphs. The graph sum is equal to a matrix integral in the manner described above. Therefore one replaces the path integral problem with a study of the asymptotic behavior of a matrix integral as $n \to \infty$.

3.3. Kontsevich's proof

Now we will outline one part of Kontsevich's proof of this conjecture. This important step involves a combinatorial description of $\overline{M}_{g,n}$. This will reveal the connection of $\overline{M}_{g,n}$ to ribbon graphs having one extra structure. The graphs arise in a way quite different from what Witten suggests in [Wit1]. The rest of the proof will not be explained, and we refer the reader to [Ko] and [Lo].

It is a classical theorem that given an orientable 2-manifold C of genus g equipped with a Riemannian metric ds^2, one can always solve the Beltrami equation and find a new metric locally of the form $h(dx^2+dy^2)$ (isothermal coordinates) conformally equivalent to the original metric. This in turn, allows one to define a complex structure on C, thus making C a Riemann surface. Let us identify an n pointed compact Riemann surface C of genus g, with the punctured Riemann surface $C' = C - \{x_1, \ldots, x_n\}$, where we keep the punctures labelled. Assume that $2-2g-n<0$ as before. By the uniformization theorem for Riemann surfaces, every such C' has the open unit disk as its universal cover. Endowing the unit disk with the Poincaré metric and letting it induce a metric on every quotient would amount to choosing a constant curvature $= -1$ metric from the conformal class of every Riemann surface in $\overline{M}_{g,n}$.

The idea that leads to a combinatorial description of $\overline{M}_{g,n}$ is to make a different choice of (singular) metric from each conformal class. Notice that given a holomorphic quadratic differential on C', that is a $(0,2)$ tensor locally of the form $h(z)dz \otimes dz$, one obtains a metric $|h(z)||dz|^2$ away from the zero locus of the differential, in the same conformal class with the metric on C'. Before proceeding further, we will discuss some examples.

Example 1. Let D' be the punctured open unit disk, $\{z| \ |z|<1\} - \{0\}$ in $\mathbb{C}$. Consider the following holomorphic quadratic differential on D':

$$\omega = \frac{p^2}{(2\pi)^2 z^2} dz \otimes dz.$$

Then by a simple residue computation, every circle centered at the origin has length p. Thus the Riemann surface D' with respect to the metric induced by ω is a cylinder.

Example 2. Next, consider $P = \mathbb{C} - \{-1,1\}$. This domain is biholomorphic to the Riemann sphere punctured at 3 points, ∞ being the third puncture. Let $h(z)dz \otimes dz$ be the quadratic form

$$h(z)dz \otimes dz = \left(\frac{a}{(z-1)^2} + \frac{b}{(z+1)^2} + \frac{c}{(z-1)^2(z+1)^2}\right) dz \otimes dz.$$

Then $h(z)dz \otimes dz$ has quadratic poles at $-1, 1, \infty$, and it is an easy exercise that any triple of positive quadratic residues at these points can be obtained by selecting suitable (a, b, c). Denote the three residues by $(\frac{p_{-1}}{2\pi})^2, (\frac{p_1}{2\pi})^2, (\frac{p_\infty}{2\pi})^2$. Take a small closed path γ around -1. It has length

$$\oint_\gamma \sqrt{|h(z)|}|dz| = \oint_\gamma |\frac{p_{-1}}{2\pi(z-1)} + \cdots||dz|.$$

If γ is chosen such that $h(z)dz \otimes dz$ is real on γ, then we can remove the absolute values, so by Cauchy's theorem the answer is p_{-1}. Moreover, moving outwards starting from this curve, every member of the set of trajectories on which $h(z)dz \otimes dz$ is real has to have constant length p_{-1} up to some bifurcation locus, after which real trajectories begin to wind around a different set of poles. Thus we see that the real trajectories around each pole are geodesic circles of a cylinder in the new metric. We associate the Riemann surface P together with $h(z)dz \otimes dz$ the combinatorial data consisting in the graph of the critical locus of real trajectories, an orientation at each vertex of the graph (which remembers the order of gluing of the three cylinders to the graph), and lengths of each edge of the graph. Some of the different possibilities for the combinatorics of the graph are demonstrated in figure 5. Notice that the graph will be trivalent unless p_1, p_{-1}, p_∞ are subject to a nontrivial constraint.

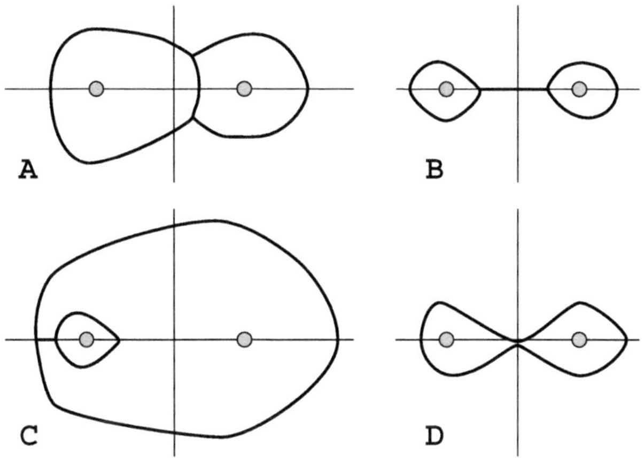

FIGURE 5. Some of the possibilities for the ribbon graph: A. p_{-1}, p_1, p_∞ satisfy the triangle inequality. B. $p_\infty > p_1 + p_{-1}$. C. $p_1 > p_\infty + p_{-1}$ (analogous to B). D. $p_\infty = p_{-1} + p_1$

Notice that giving an orientation to every vertex is equivalent to thickening the given graph into a ribbon graph. The additional information one has here, compared to the ribbon graphs discussed previously, is that every edge is assigned a length by the metric.

Given any Riemann surface with punctures, say that a quadratic differential on C is a Jenkins-Strebel quadratic differential if it shares the following feature with the differentials described in the examples: the requirement is that it has at most quadratic poles at the punctures (higher-order poles can be ruled out by insisting that the real trajectories except for the measure 0 critical trajectories are closed). Then the nearby real trajectories foliate a cylinder by essentially the same argument. The observations in the two examples above generalize beautifully to the following theorem.

Theorem 3.2. *(Jenkins, Strebel* [Jen], [Str]*) Let C be a Riemann surface with n punctures, $n > 0$ is an integer, $2g-2-n<0$, and $p_1,\dots,p_n$ positive real numbers. Then there exists a unique Jenkins-Strebel quadratic differential on C such that the complement of the critical locus consists in n cylinders of circumferences $p_1,\dots,p_n$ in the metric induced.*

Thus, for any C with additional data $p_1,\dots,p_n$, the critical locus can be given the structure of a ribbon graph with length assignments (henceforth called a metric ribbon graph) uniquely by the theorem. Conversely, given any metric ribbon graph, one can retrieve C and $p_1,\dots,p_n$ uniquely by attaching n cylinders to it.

This gives $M_{g,n}\times\mathbb{R}^n_+$ a combinatorial description. The orbifold parameterizing metric ribbon graphs is denoted by $M^{comb}_{g,n}$. The compactification $\overline{M}^{comb}_{g,n}$ of $M^{comb}_{g,n}$ used in Kontsevich's paper is slightly different from what would bijectively correspond to the Deligne-Mumford compactification. Some details about this point are discussed in [Zvo].

The next step of the proof is to express $c_1(\psi_i)$ in terms of the combinatorial data. Kontsevich expresses the 2-form representing this class explicitly in terms of the lengths l_i of the edges of the graph and shows that Witten's generating function Z can be expressed as a matrix integral which additionally involves the data l_i. This matrix integral looks like (cf. [Ko], [Lo]):

$$\int e^{TrM^2\Lambda+TrM^3}dM. \tag{3.1}$$

Afterwards, the proof still requires the final step of showing that this matrix integral is a tau function (cf. [Ko], [Lo]).

4. Some of the further developments

The outline in this paper explains how integrable systems come up in the intersection theory of $\overline{M}_{g,n}$. This corresponds to the Gromov-Witten theory of a point. One may ask whether or not extensions to the case where the target space X is not a point is possible. In this case, $\overline{M}_{g,n}$ is replaced by the space of stable maps to X, $\overline{M}_{g,n}(X,\beta)$, where $\beta\in A_2(X)$ (for definitions, see [KM], [FP], [KV]). For the case of X not a point, there are geometrically interesting classes besides the ψ classes; namely the enumerative classes arising from constraints imposing that the image of each marked point of the curve lands on a certain cycle of X.

In [EY], Eguchi and Yang make a conjectural generalization the matrix model discussed above to the case $X=\mathbb{P}^1$. Based on computations on $\mathbb{P}^1$ and more generally on $\mathbb{P}^N$ and certain Fano varieties, Eguchi, Hori and Xiong propose a conjecture, called the Virasoro conjecture in [EHX]. The presence of the Virasoro algebra is natural because of the following:

Consider the Grassmannian formulation discussed above. Functions on S^1 (or formal Laurent series in z) act naturally on the Grassmannian by multiplication. Also note that taking a restricted choice of functions, one can produce natural actions on the solutions of each of the n-KdV hierarchies: Since W such that $z^nW \subset W$ correspond to the solutions of the n-KdV equation, multiplication by a Laurent series in z^n will act on the set of W corresponding to these solutions. The natural question then is, whether or not this action lifts to an action on the tau function, and henceforth on the actual solutions of the KP or n-KdV hierarchies. However, the presence of the projection π_+ in the equation for τ prevents the action of the Abelian group of functions on S^1 to lift to an action on the tau functions. However, the upshot is, a central extension of this group can be arranged to act on the tau functions appropriately (cf. [SW]).

Considering the action at the Lie algebra level, one obtains many representations of the Virasoro algebra (cf. [KR]) acting on the solutions of the KP or n-KdV hierarchies, and their tau functions. In [DVV], Witten's conjecture was expressed in terms of the tau function being annihilated by a half of the Virasoro operators L_i, precisely, the ones with $i \in \{-1, 0, \dots\}$. The conjecture of [EHX], generalizing this observation, was later extended to a conjecture for an arbitrary variety X. For a survey of the Virasoro conjecture, see [Get]. The point is that, even though the relevant matrix model or corresponding integrable hierarchy for a target variety X may be missing, it is conceivable that one can prove the Virasoro relations.

The Virasoro conjecture has been proven by Givental for $X = \mathbb{P}^n$ (cf. [Giv1], [Giv2]), and by Okounkov and Pandharipande for curves (cf. [OP1], [OP2], [OP3]). The general conjecture remains open.

In another direction, in [Wit2], Witten made a conjecture corresponding to the n-KdV hierarchy, instead of the usual KdV hierarchy in Witten's original conjecture. For developments in this direction, see [JKV] and [Lee].

References

[AvM1] Adler, M., Van Moerbeke, P. *String Orthogonal Polynomials, String Equations and 2-Toda Symmetries* Comm. Pure and Appl. Math. **50**(1997), 241–290

[AvM2] Adler, M., Van Moerbeke, P. *Vertex Operator Solutions of the Discrete KP Hierarchy* Comm. Math. Phys. **203**(1999), 185–210

[BIZ] Bessis, D., Itzykson, C., Zuber, J.B. *Quantum Field Theory Techniques in Graphical Enumeration* Advances in Applied Math. **1**(1980), 109–157

[DM] Deligne, P., Mumford, D. *The Irreducibility of the Space of Curves of Given Genus* I.H.E.S. Publ. Math. **45** (1969), 75–109

[Dic] Dickey, L.A. *Lectures on Classical $\mathcal{W}$ Algebras* Acta Applicandae Math. **47** (1997), 243–321

[Dij] Dijkgraaf, R. *Intersection Theory, Integrable Hierarchies and Topological Field Theory* "New Symmetry Principles in Quantum Field Theory" (Cargese, 1991) NATO Adv. Sci. Inst. Ser. B. Phys., **295**, 95–158

[DVV] Dijkgraaf, R., Verlinde, H., Verlinde, E. *Loop equations and Virasoro constraints in nonperturbative two-dimensional quantum gravity.* Nuclear Phys. B **348**, 435–456, (1991)

[EHX] Eguchi, T., Hori, K., Xiong, C.S. *Quantum cohomology and Virasoro algebra.* Phys. Lett. B **402** (1997), 71–80

[EY] Eguchi, T., Yang, S.K. *The topological CP^1 model and the large-N matrix integral.* Modern Phys. Lett. A **9** (1994), 2893–2902

[FP] Fulton, W., Pandharipande, R. *Notes on stable maps and quantum cohomology* Proc. Sympos. Pure Math. **62** Part 2,(1997), 45–96

[Get] Getzler, E. *The Virasoro conjecture for Gromov-Witten invariants.* Algebraic geometry: Hirzebruch 70 (Warsaw, 1998), Contemp. Math. **241** (1999), 147–176

[Giv1] Givental, A.B. *Semisimple Frobenius structures at higher genus.* Internat. Math. Res. Notices **23** (2001), 1265–1286

[Giv2] Givental, A.B. *Gromov-Witten invariants and quantization of quadratic Hamiltonians.* Dedicated to the memory of I. G. Petrovskii on the occasion of his 100th anniversary. Mosc. Math. J. **1** (2001), 551–568

[JKV] Jarvis, T.J., Kimura, T., Vaintrob, A. *Gravitational descendants and the moduli space of higher spin curves.* Advances in algebraic geometry motivated by physics (Lowell, MA, 2000) Contemp. Math. **276** 167–177

[Jen] Jenkins, J.A. *On the Existence of Certain General Extremal Metrics.* Ann. of Math. (2) **66** (1957), 440–453

[KR] Kac, V., Raina, *Bombay lectures on highest weight representations of infinite-dimensional Lie algebras.* Advanced Series in Mathematical Physics, 2. World Scientific Publishing Co., Inc., Teaneck, NJ, (1987)

[KV] Kock, J., Vainsencher, I. *A formula de Kontsevich para curvas racionais planas.* 22° Colóquio Brasileiro de Matemática. [22nd Brazilian Mathematics Colloquium] Instituto de Matemática Pura e Aplicada (IMPA), Rio de Janeiro, (1999)

[KM] Kontsevich, M., Manin, Yu. *Gromov-Witten classes, quantum cohomology, and enumerative geometry.* Commun. in Math. Phys. **164** (1994), 525–562

[Ko] Kontsevich, M. *Intersection Theory on the Moduli Space of Curves and the Matrix Airy Function* Commun. in Math. Phys. **147** (1992), 1–23

[Lax] Lax, P.D. *Periodic Solutions of Korteweg-de Vries Equation* Comm. in Pure and Appl. Math. **28** (1975), 141–188

[Lee] Lee, Y.P. *Witten's conjecture, Virasoro conjecture, and invariance of tautological equations* (preprint) math. AG/0311100

[Lo] Looijenga, E. *Intersection theory on Deligne-Mumford compactifications (after Witten and Kontsevich).* Séminaire Bourbaki, Vol. 1992/93. Astérisque **216** (1992), 187–212.

[vM1] Van Moerbeke, P. *Integrable Foundations of String Theory* Lectures on Integrable Systems (Sophia-Antipolis, 1991), 163–267

[vM2] Van Moerbeke, P. *Integrable Lattices: Random Matrices and Random Permutations* Random Matrix Models and Their Applications, M.S.R.I. Publ. **40** , Cambridge Univ. Press (2001), 321–406

[Mul] Mulase, M. *Algebraic Theory of the KP Equations* Perspectives in Mathematical Physics, Conf. Proc. Lecture Notes Math. Phys. III (1994), 151–217

[Mum] Mumford, D. *Towards an Enumerative Geometry of the Moduli Space of Curves* Arithmetic and Geometry, Vol II, Progr. Math. **36**, Birkhäuser Boston, Boston, MA (1983), 271–328

[OP1] Okounkov, A., Pandharipande, R. *Gromov-Witten theory, Hurwitz theory, and completed cycles* (preprint) math. AG/0204305

[OP2] Okounkov, A., Pandharipande, R. *The equivariant Gromov-Witten theory of* P^1. (preprint) math. AG/0207233

[OP3] Okounkov, A., Pandharipande, R. *Virasoro constraints for target curves.* (preprint) math. AG/0308097

[Sat] Sato, M., Sato, Y. *Soliton equations as dynamical systems on infinite-dimensional Grassmann manifold.* Nonlinear partial differential equations in applied science (Tokyo, 1982) North-Holland Math. Stud.**81**(1983), 259–271

[SW] Segal, G., Wilson, G. *Loop Groups and Equations of KdV Type* Publ. Math. I.H.E.S. **61**(1985), 5–65

[Str] Strebel, K. *Quadratic Differentials* Ergebnisse der Mathematik und ihrer Grenzgebiete (3), Springer-Verlag, Berlin (1984), 5–65

[UT1] Ueno, K., Takasaki, K. *Toda Lattice Hierarchy I* Proc. Japan Acad. Ser. A Math. Sci. **59** no. 5 (1983), 167–170

[UT2] Ueno, K., Takasaki, K. *Toda Lattice Hierarchy II* Proc. Japan Acad. Ser. A Math. Sci. **59** no. 5 (1983), 215–218

[Vak] Vakil, R. *The Moduli Space of Curves and Its Tautological Ring* Notices of the AMS **50** No. 6 (1991), 647–658

[Var] Varadarajan, V.S. *Lie Groups, Lie Algebras, and their Representations* GTM **102** Springer Verlag, (1984)

[Wit1] Witten, E. *Two-Dimensional Gravity and Intersection Theory on Moduli Space* Surveys in Diff. Geo. **1**(1991), 243–310

[Wit2] Witten, E. *Algebraic geometry associated with matrix models of two-dimensional gravity.* Topological methods in modern mathematics (Stony Brook, NY, 1991), Publish or Perish, Houston, TX, (1993), 235–269

[Zvo] Zvonkine, D. *Strebel Differentials on Stable Curves and Kontsevich's Proof of Witten's Conjecture* preprint, arXiv:math. AG/0209071, (2004)

Ali Ulas Ozgur Kisisel
Dept. of Mathematics
Middle East Technical University
Ankara 06531, Turkey
e-mail: `akisisel@metu.edu.tr`

Trends in Mathematics:
Topics in Cohomological Studies of Algebraic Varieties, 163–174

Multiplying Schubert Classes

Piotr Pragacz

Abstract. We show how to compute the structure constants for cohomological multiplication of Schubert classes by exploiting the action of the Weyl group and that of BGG-operators, on the cohomology ring of a flag variety. We illustrate this method with simple proofs of the Chevalley and Pieri formulas.

1. Introduction

One of the main problems of Schubert calculus on flag varieties (or generalized flag varieties G/B) is to give expressions for the structure constants for the cohomological multiplication of Schubert classes. The *main problem* is to describe these structure constants as *the cardinalities of some sets*, but also "closed" formulas for the structure constants are of some interest. This is a classical topic starting with the work of Schubert, Giambelli, Pieri, Lesieur, Hodge-Pedoe, Littlewood-Richardson[1], Borel, Chevalley, Monk, and Horrocks, and continuing in recent years with the work of Bernstein-Gelfand-Gelfand, Demazure, Koch,Lascoux-Schützenberger, Kleiman-Laksov, Stoll, Carrell, Kostant-Kumar, Hiller-Boe, Stembridge, Akyildiz, Sertöz, Fulton, Pragacz-Ratajski, Bergeron-Sottile, Knutson, Vakil, Buch-Kresch-Tamvakis, Duan, and Gatto – to mention a few. We do not attempt to survey this activity here, but in the bibliographical references the reader may find a vast discussion of the structure constants.

The purpose of this note is to give a closed formula (Theorem 1) for these constants. This formula evolved from a sequence of papers [24], [25], and [26] (see also [23] and [9]). The first main tool that we use is the action of the Weyl group on $H^\bullet(G/B, \mathbf{Q})$ expressed in terms of Schubert classes. The second main tool is the theory of BGG-operators acting as skew derivations on $H^\bullet(G/B, \mathbf{Q})$. These

Research supported by KBN grant 2P03A 024 23.

[1] The famous combinatorial Littlewood-Richardson rule governing the multiplication of Schubert classes on Grassmannians, was found, in fact, in a parallel context of representation theory. The same remark applies to the contribution of Stembridge.

fundamental tools were developed mainly by Bernstein-Gelfand-Gelfand [1] and Demazure [6], [7] in the 70's, as a continuation of the work of Borel [2].

We illustrate our method with short proof the Chevalley formula (Theorem 2), and transparent, purely algebro-combinatorial proof of the classical Pieri formula (Theorem 3). (These proofs were mentioned in [11], p. 122 and [26], p. 50, respectively.)

Background. The content of this note was obtained in the 90's and has not been written up until now.[2] Following an encouragement of Michel Brion, we have decided to publish it now because of an increasing interest in the structure constants. The story told here is closely related to the lecture notes by Brion [3], Buch [4], Duan [8], and Tamvakis [27] in the present volume. This is, in fact, the main reason for the appearance of this note here.

2. Characteristic map and BGG-operators

A general reference for group-theoretic notions used in this note is [17].

Let G be a semisimple algebraic group and $B \subset G$ a Borel subgroup.

Let X be a variety on which B acts freely (from the right). Suppose that the quotient X/B exists so that $p : X \to X/B$ is a principal B-bundle. On the other hand, let $\mu : B \to GL(V)$ be a linear representation. We denote by $\mathcal{L}_\mu$ the vector bundle $X \times^B V$ that is the quotient of $X \times V$ by the equivalence relation

$$(x, v) \sim \big(xb, \mu(b)^{-1}v\big),$$

where $x \in X$, $b \in B$, and $v \in V$. Equivalently, if U is an open subset of X/B, then $\Gamma(U, \mathcal{L}_\mu)$ is the set of morphisms $\varphi : p^{-1}(U) \to V$ such that $\varphi(xb) = \mu(b)^{-1}\varphi(x)$.

In particular, with any character χ of B (that is, a homomorphism of B into the multiplicative group) there is associated a line bundle $\mathcal{L}_\chi$; this induces a homomorphism of groups $X^*(B) \to \mathrm{Pic}(X/B)$, where $X^*(B)$ denotes the group of characters of B.

Composing this homomorphism with the homomorphism of the first Chern class from $\mathrm{Pic}(X/B)$ to $H^2(X/B, \mathbf{Z})$, one gets a homomorphism from $X^*(B)$ to $H^2(X/B, \mathbf{Z})$, which extends to a homomorphism of graded rings

$$c : S^\bullet\big(X^*(B)\big) \to H^\bullet(X/B, \mathbf{Z})$$

from the symmetric algebra of the $\mathbf{Z}$-module $X^*(B)$ to the cohomology ring of X/B; this homomorphism is called the *characteristic map* of the fiber bundle $p : X \to X/B$. In this note, $S = \oplus S^k$ will denote the symmetric algebra $S^\bullet\big(X^*(B)\big) = \oplus S^k\big(X^*(B)\big)$.

[2]The material of this note was presented at various Impanga seminars, and, e.g., at the Littelmann-Mathieu seminar in Strasbourg (December, 1994), at the Summer School "Schubert Varieties" in Thurnau (June, 1995), and at the topology seminar at CAS in Beijing (June, 2002).

Choose a maximal torus $T \subset B$ with the Weyl group $W = N_G(T)/T$ of (G,T). Then W acts on the group of characters $X^*(T)$ of T, and since $X^*(B) = X^*(T)$, this induces an action of W on S.

The root system of (G,T) is denoted by R; the set R^+ of positive roots consists in the opposites of roots of (B,T). Let $\Delta \subset R^+$ be the associated basis of R. The Weyl group W is generated by simple reflections, i.e., by the reflections associated with the elements of Δ. For any root $\alpha \in R$, we denote by s_α the reflection associated with α. The reflection s_α can be realized as a linear endomorphism of the Euclidean space $X^*(T) \otimes \mathbf{R}$, equipped with a W-invariant inner product $(\ ,\ )$. We have $s_\alpha(\lambda) = \lambda - (\alpha^\vee, \lambda)\alpha$, where $\alpha^\vee = 2\alpha/(\alpha,\alpha)$.

By a *reduced decomposition* of an element $w \in W$ we understand a presentation $w = s_{\alpha_1} \cdots s_{\alpha_l}$ where all $\alpha_p \in \Delta$, and l is the smallest number occurring in such a presentation, called the *length* of w and denoted $l(w)$.

By w_0 we denote the longest element of W, the unique element of W with length equal to the cardinality of R^+.

We shall need the following "BGG-operators" A_w, $w \in W$, acting on the ring S (cf. [1], [6], and [7]).

Definition 1. *Given a root α and $f \in S$, we set*

$$A_\alpha(f) := \frac{f - s_\alpha(f)}{\alpha}.$$

The operator A_α is a well-defined (group) endomorphism on S lowering the degree by 1. Note that $A_\alpha(f) = (\alpha^\vee, f)$ for $f \in S^1$; this will be used in the proof of Theorem 2.

We now record (cf. [1], Theorem 3.4 and [6], Théorème 1):

Lemma 1. *If $\alpha_1, \ldots, \alpha_k$ and $\beta_1, \ldots, \beta_k$ are simple roots such that*

$$s_{\alpha_1} \cdots s_{\alpha_k} = s_{\beta_1} \cdots s_{\beta_k}$$

are two reduced decompositions, then

$$A_{\alpha_1} \cdots A_{\alpha_k} = A_{\beta_1} \cdots A_{\beta_k}.$$

Thus for $w \in W$, given its reduced decomposition $w = s_{\alpha_1} \cdots s_{\alpha_k}$, the operator

$$A_w := A_{\alpha_1} \cdots A_{\alpha_k}$$

is well defined (i.e., does not depend on a reduced decomposition of w).

The following result says how the BGG-operators act on products (cf., e.g., [6], Eq. (6), p. 289):

Lemma 2. *We have for $f, g \in S$ and a simple root α,*

$$A_\alpha(fg) = A_\alpha(f)g + s_\alpha(f)A_\alpha(g). \tag{1}$$

Geometric interpretations of BGG-operators are related to correspondences in flag bundles (cf., e.g., [11], Chap. 2 and 6), and Gysin maps for *Bott-Samelson schemes.* These schemes are described in the notes by Brion [3] and Duan [8] in the present volume. This last aspect of BGG-operators is discussed in [11], Appendix C.

The reader may also consult [15] for a detailed treatment of the so-called Schubert calculus of the coinvariant algebra, that is based on BGG-operators.

3. Structure constants for Schubert classes

In the geometry of flag manifolds G/B a large role is played by the *Schubert cells* BwB/B and their closures called *Schubert varieties.* We set $X^w := \overline{Bw_0wB/B}$. The cohomology class $[X^w]$ of X^w lies in $H^{2l(w)}(G/B, \mathbf{Z})$. The Schubert cells form a cellular decomposition of G/B, so the classes $[X^w]$ form an additive basis for the cohomology.

Our goal, in this section, is to give a closed formula for the constants c^u_{wv}, appearing in the decomposition of the product

$$[X^w] \cup [X^v] = \sum_u c^u_{wv} [X^u] \tag{2}$$

of Schubert classes.

We shall need a couple of tools that we describe now.

The characteristic map $c : S \to H^\bullet(G/B, \mathbf{Z})$ of the fibration $G \to G/B$ is usually called the *Borel characteristic map.* Its kernel is generated by positive degree W-invariants, and $c \otimes \mathbf{Q}$ is surjective (cf. [2]), so that the cohomology ring $H^\bullet(G/B, \mathbf{Q})$ is identified with the quotient of $S \otimes \mathbf{Q}$ modulo the ideal generated by positive degree W-invariants. By combining this last property with Lemma 2, we infer that the BGG-operators induce – via the characteristic map – operators A_w on $H^\bullet(G/B, \mathbf{Q})$ lowering the degree by $2l(w)$.

In particular, for $a, b \in H^\bullet(G/B, \mathbf{Q})$ and a simple root α, we have

$$A_\alpha(a \cup b) = A_\alpha(a) \cup b + s_\alpha(a) \cup A_\alpha(b)\,. \tag{3}$$

Iterations of this equation will play an important role in the present section and the next one.

Note also that the action of W on S induces – via the characteristic map – an action of W on $H^\bullet(G/B, \mathbf{Q})$. (This action will be described below in terms of Schubert classes – cf. Lemma 4.)

We record the following equation relating three "heroes" of the present note: the characteristic map, BGG-operators, and Schubert classes (cf. [7, Section 4] and [1, Section 4]): for $f \in S^k$, in $H^\bullet(G/B, \mathbf{Z})$ we have

$$c(f) = \sum_{l(w)=k} A_w(f)[X^w]\,. \tag{4}$$

This equation is closely related to the question of finding polynomial representatives of Schubert classes – a problem that we do not address in the present note (cf. [11] for a discussion of this issue).

The next result says how the operators A_w act on Schubert classes (cf. [1], Theorem 3.14 (i)):

Lemma 3. *For $l(vw^{-1}) = l(v) - l(w)$, we have*

$$A_w([X^v]) = [X^{vw^{-1}}], \tag{5}$$

and in the opposite case, $A_w([X^v]) = 0$.

We have also the following formula for the action of a simple reflection on a Schubert class (cf. [1], Theorem 3.12 (iv) and [7], Proposition 3):

Lemma 4. *For a simple root α and $w \in W$,*

$$s_\alpha([X^w]) = [X^w] \qquad \text{if } l(ws_\alpha) = l(w) + 1; \tag{6}$$

$$s_\alpha([X^w]) = -[X^w] - \sum (\beta^\vee, \alpha)[X^{ws_\alpha s_\beta}] \qquad \text{if } l(ws_\alpha) = l(w) - 1, \tag{7}$$

where the sum is over all positive roots $\beta \neq \alpha$ such that $l(ws_\alpha s_\beta) = l(w)$.

We now proceed towards computing the structure constants c^u_{wv}. By combining Equations (2) and (5), we can express the coefficient c^v_{wv} as follows:

$$c^u_{wv} = A_u([X^w] \cup [X^v]). \tag{8}$$

Suppose that $l(w) = k$ and $l(v) = l$. Take a reduced decomposition of u:

$$u = s_{\alpha_1} \cdots s_{\alpha_{k+l}}.$$

Iterating (3) we obtain

$$c^u_{wv} = A_{\alpha_1} \cdots A_{\alpha_{k+l}}([X^w] \cup [X^v]) = \sum A_I([X^w]) \cup A^I_\alpha([X^v]),$$

where the sum is over all subsequences $I = (i_1 < \cdots < i_k) \subset \{1, 2, \ldots, k+l\}$, $A_I := A_{\alpha_{i_1}} \cdots A_{\alpha_{i_k}}$, and A^I_α is obtained by replacing in $A_{\alpha_1} \cdots A_{\alpha_{k+l}}$ each A_{α_i} by s_{α_i} for $i \in I$. By Lemma 3 we infer the following result.

Theorem 1. *With the above notation,*

$$c^u_{wv} = \sum A^I_\alpha([X^v]), \tag{9}$$

where the sum runs over all I such that $s_{\alpha_{i_1}} \cdots s_{\alpha_{i_k}}$ is a reduced decomposition of w.

Applying successively to the summands in (9) the formulas (5), (6) and (7), we get an expression for the constants c^u_{wv}.

Recall the following formula for multiplication by the classes of Schubert divisors in $H^\bullet(G/B, \mathbf{Z})$:

Theorem 2. (*Chevalley,* [5]) *For $w \in W$, and a simple root α,*

$$[X^w] \cup [X^{s_\alpha}] = \sum (\beta^\vee, \omega_\alpha)[X^{ws_\beta}], \tag{10}$$

where β runs over positive roots such that $l(ws_\beta) = l(w) + 1$ and ω_α denotes the fundamental weight associated with α.

Proof. We prove Equation (10) using Theorem 1. By the definition of a fundamental weight, we have for $\gamma \in \Delta$, $(\omega_\alpha, \gamma^\vee) = \delta_{\alpha\gamma}$, the Kronecker delta. This implies that $A_\gamma(\omega_\alpha) = \delta_{\alpha\gamma}$, and using Equation (4) we get $c(\omega_\alpha) = [X^{s_\alpha}]$. Fix $w \in W$ and pick $f \in S \otimes \mathbf{Q}$ such that $(c \otimes \mathbf{Q})(f) = [X^w]$. Then in $H^\bullet(G/B, \mathbf{Q})$,

$$[X^{s_\alpha}] \cup [X^w] = (c \otimes \mathbf{Q})(\omega_\alpha \cdot f), \tag{11}$$

and by Theorem 1 we obtain that the coefficient of the Schubert class $[X^u]$ in the expansion of (11) can be evaluated as the sum (9) with $[X^v]$ replaced by ω_α.

Take a reduced decomposition $u = s_{\alpha_1} \cdots s_{\alpha_h}$. By the "Exchange Condition" (cf. [18], pp.14–15), a reduced decomposition for w can be gotten from the one for u by omitting one simple reflection if $u = ws_\beta$ for some (positive) root β. Conversely, if $w = s_{\alpha_1} \cdots s_{\alpha_{p-1}} s_{\alpha_{p+1}} \cdots s_{\alpha_h}$, then

$$w^{-1}u = s_{\alpha_h} \cdots s_{\alpha_p} \cdots s_{\alpha_h} = s_\beta$$

for $\beta = s_{\alpha_h} \cdots s_{\alpha_{p+1}}(\alpha_p)$. The root β is positive by, e.g., [15, Proposition 3.6] because $s_{\alpha_h} \cdots s_{\alpha_1}$ is reduced. Since the omitted simple reflection is unique, the looked at sum (9) has exactly one summand

$$s_{\alpha_1} \cdots s_{\alpha_{p-1}} A_{\alpha_p} s_{\alpha_{p+1}} \cdots s_{\alpha_h}(\omega_\alpha) = A_{\alpha_p} s_{\alpha_{p+1}} \cdots s_{\alpha_h}(\omega_\alpha).$$

The latter expression equals $(\beta^\vee, \omega_\alpha)$ because $A_{\alpha_p}(g) = (\alpha_p^\vee, g)$ for $g \in S^1$, the inner product $(\ ,\ )$ is W-invariant, and $s_{\alpha_h} \cdots s_{\alpha_{p+1}}(\alpha_p^\vee) = \beta^\vee$. This proves the theorem. □

For an algebraic proof in the SL_n-case[3] along these lines, see [20]. A geometric proof in the SL_n-case is given in the notes by Brion [3].

The same method works for all spaces G/P, where P is a parabolic subgroup of G. Let θ be a subset of Δ and let W_θ be the subgroup of W generated by $\{s_\alpha\}_{\alpha\in\theta}$. We set $P_\theta := BW_\theta B$. Denote by W^θ the set

$$W^\theta := \{w \in W : \ l(ws_\alpha) = l(w) + 1 \quad \forall \alpha \in \theta\}.$$

This last set is the set of *minimal length left coset representatives* of W_θ in W.

The projection $G/B \to G/P_\theta$ induces an injection

$$H^\bullet(G/P_\theta, \mathbf{Z}) \hookrightarrow H^\bullet(G/B, \mathbf{Z})$$

which additively identifies $H^\bullet(G/P_\theta, \mathbf{Z})$ with $\bigoplus_{w\in W^\theta} \mathbf{Z}[X^w]$. Multiplicatively, $H^\bullet(G/P_\theta, \mathbf{Q})$ is identified with the ring of invariants $H^\bullet(G/B, \mathbf{Q})^{W_\theta}$. We refer for details to [1], Sect. 5.

[3]This case was obtained by Monk [21] using different methods.

The restriction $c : S^{W_\theta} \to H^\bullet(G/P_\theta, \mathbf{Z})$ of the Borel characteristic map satisfies, for any W_θ-invariant f from S^k, the following equation in $H^\bullet(G/P_\theta, \mathbf{Z})$:

$$c(f) = \sum_{\substack{w \in W^\theta \\ l(w)=k}} A_w(f)[X^w]\,. \tag{12}$$

For maximal parabolic subgroups P of the symplectic and orthogonal groups, this method led to combinatorial expressions for the structure constants in the products of arbitrary Schubert classes by some "special Schubert classes" in $H^\bullet(G/P, \mathbf{Z})$ (cf. [24], [25], [26], [9], and [23]).

Remark 1. *Equation* (3) *is often called the "Leibniz-type formula". Kostant and Kumar* [19] *discovered independently, in the context of the "nil Hecke ring", that the structure constants can be computed via the iteration of the Leibniz-type formula.*

4. A combinatorial proof of the Pieri formula

In this section, we give a proof of the classical Pieri formula for the Grassmannian $\mathrm{Gr}(n, m)$ of n-dimensional subspaces in $\mathbf{C}^m$ via the above method. In fact, there are two Pieri formulas: for multiplication by the Chern classes[14] of the tautological subbundle on $\mathrm{Gr}(n, m)$, and for multiplication by the Chern classes of the tautological quotient bundle on $\mathrm{Gr}(n, m)$. The latter version appears more often mainly because the Chern classes of the tautological quotient bundle enjoy a simple interpretation in terms of the classical "Schubert conditions": the kth Chern class is represented by the locus of all n-planes in $\mathbf{C}^m$ which have positive-dimensional intersection with a fixed $(m - n - k + 1)$-plane in $\mathbf{C}^m$. By passing to the dual Grassmannian, we see that both formulas are, in fact, equivalent. We shall treat in detail the latter case. We also make a link with the ring of symmetric functions, known since Giambelli (cf. [12] and [13]).

For the remainder of this note, we set $q := m - n$.

In the following, I, J will denote *strict* partitions contained in the partition $(m, m-1, \ldots, q+1)$ with exactly n parts[4]. (We identify partitions with their Young diagrams, as is customary.) Note that such partitions contain the "upper-left triangle"

$$\delta = (n, n-1, \ldots, 1)\,.$$

On the other hand, λ, μ will denote "ordinary" partitions contained in (q^n). In fact, there is a bijection between these two sets: with I, we associate λ defined by $\lambda_p = i_p - n + p - 1$ for $p = 1, \ldots, n$.

Also, we associate with I the following permutation w_I in the symmetric group S_m:

$$w_I = \cdots (s_{q-\lambda_3+3} \cdots s_{q+1} s_{q+2})(s_{q-\lambda_2+2} \cdots s_q s_{q+1})(s_{q-\lambda_1+1} \cdots s_{q-1} s_q)\,. \tag{13}$$

[4]In other words, $I = (i_1, \ldots, i_n)$ where $m \geq i_1 > \cdots > i_n \geq 1$.

It is easy to see, that the right-hand side of (13) gives a reduced decomposition of w_I.

Take for example $m = 7$, $n = 3$, and $I = (6,4,3)$. Then $\lambda = (3,2,2)$ and $w_I = s_5s_6s_4s_5s_2s_3s_4$ which is the permutation $[1,3,6,7,2,4,5]$ (we display a permutation as the sequence of its consecutive values).

In general, for $I = (m \geq i_1 > \cdots > i_n \geq 1)$, we have in S_m,

$$w_I = [j_1 < \cdots < j_q \, , m+1-i_n < \cdots < m+1-i_1] \, ,$$

where $j_1, \ldots, j_q$ are uniquely determined by I.

Let $B \subset SL_m(\mathbf{C})$ be the Borel group of lower triangular matrices. Using the notation of the previous section, we set $P = P_\theta$, where θ is obtained by omitting the simple root $\varepsilon_n - \varepsilon_{n+1}$ in the basis $\varepsilon_1 - \varepsilon_2, \varepsilon_2 - \varepsilon_3, \ldots, \varepsilon_{m-1} - \varepsilon_m$ of the root system of type (A_{m-1}):

$$\{\varepsilon_i - \varepsilon_j \mid i \neq j\} \subset \oplus_{i=1}^m \mathbf{R}\varepsilon_i \, .$$

We have an identification $SL_m(\mathbf{C})/P = Gr(n,m)$. We set $X^I := \overline{Bw_0w_IP/P}$, where $w_0 = [m, m-1, \ldots, 1]$, and $X^\lambda := X^I$ for λ associated with I as above. Note that $[X^\lambda] \in H^{2|\lambda|}(\mathrm{Gr}(n,m), \mathbf{Z})$, where $|\lambda|$ denotes the sum of the parts of λ.

Denote by $(k+)$ the strict partition $(k+n, n-1, \ldots, 1)$, so that its associated λ is a one-row partition (k).

We want to compute the coefficients c_J in the expansion:

$$[X^I] \cup [X^{(k+)}] = \sum_J c_J [X^J] \, .$$

Set $x_i := -\varepsilon_{m+1-i}$ for $i = 1, \ldots, m$, so that $c(x_1), \ldots, c(x_q)$ are the Chern roots of the tautological quotient bundle on $\mathrm{Gr}(n,m)$. The Borel characteristic map allows us to treat $H^\bullet(\mathrm{Gr}(n,m), \mathbf{Z})$ as a quotient of the ring S' of polynomials symmetric in $x_1, \ldots, x_q$ and in $x_{q+1}, \ldots, x_m$. (Recall that for type (A_{m-1}), the characteristic map is surjective without tensoring by $\mathbf{Q}$.) The operators s_α and A_α indexed by the simple roots corresponding to P are induced by the following operators s_i and A_i, $i = 1, \ldots, q-1, q+1, \ldots, m-1$, on S'. The operator s_i interchanges x_i with x_{i+1}, leaving other variables invariant, and A_i is the ith simple (Newton's) divided difference ∂_i: for $f \in S'$,

$$\partial_i(f) = \frac{f - s_i(f)}{x_i - x_{i+1}} \, .$$

The operator A_w on S', in this case ($w \in S_m$), will be denoted by ∂_w, as is customary.

Let $e_k = e_k(x_1, \ldots, x_q)$ be the kth elementary symmetric polynomial in $x_1, \ldots, x_q$. We now record:

Lemma 5. *For any $k = 1, \ldots, q$, the following equation holds in $H^\bullet(\mathrm{Gr}(n,m), \mathbf{Z})$:*

$$c(e_k) = [X^{(k)}] \, .$$

Proof. By virtue of Equation (12), it suffices to show that

$$\partial_w(e_k) = 0 \quad \text{unless} \quad w = w_{(k+)}\,, \quad \text{and} \quad \partial_{w_{(k+)}}(e_k) = 1\,.$$

Note that $w_{(k+)} = s_{q-k+1} \cdots s_{q-1}s_q$. The displayed assertion follows by induction on the number of variables, by invoking the following properties of ∂_i:

$$\partial_i(e_r(x_1, \dots, x_j)) \neq 0 \quad \text{only if} \quad j = i\,,$$
$$\partial_i(e_r(x_1, \dots, x_i)) = e_{r-1}(x_1, \dots, x_{i-1})\,.$$

The lemma is proved. □

This lemma says that $X^{(k)}$ represents the kth Chern class of the tautological rank q quotient bundle on $\mathrm{Gr}(n, m)$.

Number the successive columns of J from left to right with $m, m-1, \dots, 1$, the successive rows from top to bottom with $1, \dots, n$, and use the matrix coordinates for boxes in J.

Let J^* be the effect of subtracting the triangle δ from J. In the following, D will denote a subset of J^*.

Definition 2. *Read J^* row by row from left to right and from top to bottom. Every box from D (resp. from $J^* \setminus D$) in column i gives us s_i (resp. ∂_i). Then ∂_J^D is the composition of the resulting s_i's and ∂_i's (the composition written from right to left), and r_D is the word obtained by erasing all the ∂_i's from ∂_J^D.*

In particular, r_{J^*} is the reduced decomposition (13) of w_J, and $\partial_J^{\emptyset} = \partial_{w_J}$.

Take for example $m = 8$, $n = 3$, and $J = (8, 6, 5)$. In the following picture, "•" depicts a box in D and "∘" stands for a box in $J^* \setminus D$. Moreover, the row-numbers and column-numbers are displayed.

	8	7	6	5	4	3	2	1
1	×	×	×	•	•	•	•	•
2	×	×	•	∘	∘	∘		
3	×	•	∘	•	•			

Then we have

$$\partial_J^D = s_4 s_5 \partial_6 s_7 \partial_3 \partial_4 \partial_5 s_6 s_1 s_2 s_3 s_4 s_5 \qquad \text{and} \qquad r_D = s_4 s_5 s_7 s_6 s_1 s_2 s_3 s_4 s_5\,.$$

If r_D is a reduced decomposition of w_I, then D is a disjoint union of the following "p-ribbons". For fixed $p = 1, \dots, n$, the *p-ribbon* consists of all boxes of D giving rise to those s_i (in r_D) which "transport" the item "$m + 1 - i_p$" from its position in $[1, 2, \dots, m]$ to its position in the sequence w_I.

In the above example, for $I = (7, 5, 2)$, the 1-ribbon consists of the dots in the first row, the 2-ribbon is $\{(3, 4), (3, 5), (2, 6)\}$, and the 3-ribbon is $\{(3, 7)\}$.

It can happen that some p-ribbon is empty. Suppose that p is such that the p-ribbon is not empty (this is equivalent to the fact that the box $(p, n + p - 1)$ belongs to the p-ribbon). Then the column-numbers of boxes in the p-ribbon are $m + 1 - i_p, \dots, n + p - 2, n + p - 1$, and their row-numbers weakly increase while reading D from left to right and from top to bottom.

By Theorem 1 and Lemma 5, we have

$$c_J = \sum \partial_J^D(e_k)\,, \tag{14}$$

where the sum is over all subsets $D \subset J^*$ such that r_D is a reduced decomposition of w_I.

We need the following lemma.

Lemma 6. *If there are boxes (i,j) and $(i-1,j-1)$ in $J^* \setminus D$, then $\partial_J^D(e_k)=0$.*

Proof. We set $E := \prod_{i=1}^n (1+x_i)$ and we shall prove that $\partial_J^D(E)=0$. To compute with compositions of the s_i's and ∂_i's in ∂_J^D, it is handy to introduce the following more general functions. For $\mathbf{a}=(a_1,a_2,\ldots,a_m)\in\{0,1\}^m$, we set $E_{\mathbf{a}} := \prod_{i=1}^m (1+a_i x_i)$, so that $E = E_{(1,\ldots,1,0,\ldots,0)}$ with q 1's. We have:

$$s_i(E_{\mathbf{a}}) = E_{\mathbf{a}'} \quad \text{where } \mathbf{a}' = (a_1,\ldots,a_{i-1},a_{i+1},a_i,a_{i+2},\ldots,a_m)\,; \tag{15}$$

$$\partial_i(E_{\mathbf{a}}) = d\cdot E_{\mathbf{a}'} \quad \text{if } a_{i+1} = a_i + d\,, \tag{16}$$

where $\mathbf{a}'=(a_1,\ldots,0,0,\ldots,a_n)$ is $\mathbf{a}$ with a_i,a_{i+1} replaced by zeros. Using (15) and (16), we see that the operator ∂_j in ∂_J^D, corresponding to the box (i,j) "kills" the function $E_{\mathbf{a}}$ that has been obtained by applying the previous operators s_r and ∂_r (in ∂_J^D) to E. This proves the lemma. □

It follows from this lemma that there is *at most one* $D\subset J^*$ such that r_D is a reduced decomposition of w_I and $\partial_I^D(e_k)\neq 0$, namely $D=I^*$. (Indeed, the p-ribbon must exactly coincide with the pth row of I^*.) In other words, the sum in (14) has at most one summand.

Second, applying Lemma 6 again, we see that $D=I^*$ gives a non-zero contribution to the sum in (14) iff $J\setminus I$ is a *horizontal strip with pairwise separated rows*[5]. In this case, using (15) and (16), we obtain $\partial_J^{I^*}(e_k)=1$.

We rewrite the outcome of the above considerations in terms of Schubert classes $[X^\lambda]\in H^\bullet(\mathrm{Gr}(n,m),\mathbf{Z})$ in part (i) of the following theorem. Part (ii) follows from part (i) by passing to the dual Grassmannian.

Theorem 3. *(Pieri, [22]) (i) For any partition $\lambda\subset(q^n)$ and $k=1,\ldots,q$,*

$$[X^\lambda]\cup[X^{(k)}] = \sum_\mu [X^\mu]\,, \tag{17}$$

where $|\mu|=|\lambda|+k$ and $\mu\setminus\lambda$ is a horizontal strip.

(ii) For any partition $\lambda\subset(q^n)$ and $p=1,\ldots,n$,

$$[X^\lambda]\cup[X^{(1,\ldots,1)}] = \sum_\mu [X^\mu]\,, \tag{18}$$

where 1 appears p times, $|\mu|=|\lambda|+p$ and $\mu\setminus\lambda$ is a vertical strip.

---

[5] Recall that a *horizontal strip* is a skew diagram with at most one box in each column, and a *vertical strip* is a skew diagram with at most one box in each row.

For example, we have in $H^\bullet(\mathrm{Gr}(3,8),\mathbf{Z})$:

$$[X^{(4,2)}] \cup [X^{(3)}] = [X^{(5,4)}] + [X^{(5,3,1)}] + [X^{(5,2,2)}] + [X^{(4,4,1)}] + [X^{(4,3,2)}].$$

Remark 2. *There are several (really) different proofs of the Pieri formula. We do not attempt to make a survey here. The proof that appears most often in monographs is based on studying the triple intersection of general translates of Schubert varieties. This proof appeared originally in Hodge's paper* [16]. *Cf. also* [10], § 9.4, *and the notes of Brion* [3], *where this proof is discussed in the context of Richardson varieties.*

Remark 3. *The Schubert classes $X^{(k)}$ and $X^{(1,\dots,1)}$ are often called "special". These classes and the "special Schubert classes" in* [24], [25], [26] *and* [9] *have the following property: the corresponding $w \in W$ has a unique reduced decomposition. This seems to be a proper group-theoretic characterization of a "special Schubert class", and was also remarked by Kirillov and Maeno.*

Acknowledgments. I wish to thank Michel Brion for his encouragement to write up this material, and for some valuable comments.

References

[1] I.N. Bernstein, I.M. Gelfand and S.I. Gelfand, *Schubert cells and cohomology of the spaces G/P,* Russian Math. Surveys, **28:3** (1973), 1–26.

[2] A. Borel, *Sur la cohomologie des espaces fibrés principaux et des espaces homogènes de groupes de Lie compacts,* Ann. of Math. **57** (1953), 115–207.

[3] M. Brion, *Lectures on geometry of flag varieties,* this volume.

[4] A.S. Buch, *Combinatorial K-theory,* this volume.

[5] C. Chevalley, *Sur les décompositions cellulaires des espaces G/B,* Algebraic Groups and their generalizations, (W.S. Haboush and B.J. Parshall, eds.), Proc. Symp. Pure Math. **56**, Part I (1994), AMS, 1–23.

[6] M. Demazure, *Invariants symétriques entiers des groupes de Weyl et torsion,* Invent. Math., **21** (1973), 287–301.

[7] M. Demazure, *Désingularisation des variétés de Schubert généralisées,* Ann. Sci. École Norm. Sup. **7** (1974), 53–88.

[8] H. Duan, *Morse functions and cohomology of homogeneous spaces,* this volume.

[9] H. Duan, P. Pragacz, *Divided differences of type D and the Grassmannian of complex structures,* in: "Algebraic Combinatorics and Quantum Groups", N. Jing ed., World Scientific (2003), 31–60.

[10] W. Fulton, *Young tableaux,* Cambridge University Press, 1997.

[11] W. Fulton, P. Pragacz, *Schubert varieties and degeneracy loci,* LNM **1689**, Springer, 1998.

[12] G.Z. Giambelli, *Risoluzione del problema degli spazi secanti,* Mem. R. Accad. Sci. Torino (2) **52** (1902), 171–211.

[13] G.Z. Giambelli, *Alcune proprietà delle funzioni simmetriche caratteristiche,* Atti Torino **38** (1903), 823–844.

[14] A. Grothendieck, *La théorie des classes de Chern,* Bull. Soc. Math. France **86** (1958), 137–154.

[15] H. Hiller, *Geometry of Coxeter groups,* Pitman, 1982.

[16] W.V.D. Hodge, *The intersection formulae for a Grassmannian variety,* J. London Math. Soc. **17** (1942), 48–64.

[17] J.E. Humphreys, *Linear Algebraic Groups,* Springer, 1975.

[18] J.E. Humphreys, *Reflection groups and Coxeter groups,* Cambridge University Press, 1990.

[19] B. Kostant, S. Kumar, *The nil Hecke ring and cohomology of G/P for a Kac-Moody group G,* Adv. in Math. **62** (1986), 187–237.

[20] I.G. Macdonald, *Notes on Schubert polynomials,* Publ. LACIM **6**, UQUAM, Publ. Montréal, 1991.

[21] D. Monk, *The geometry of flag manifolds,* Proc. London Math. Soc. **9** (1959), 253–286.

[22] M. Pieri, *Sul problema degli spazi secanti,* Rend. Inst. Lombardo **2** (1893), 534–556 and **27** (1894), 258–273.

[23] P. Pragacz, *Symmetric polynomials and divided differences in formulas of intersection theory,* in: Parameter Spaces, Banach Center Publications **36** (1996), 125–177.

[24] P. Pragacz, J. Ratajski, *Pieri type formula for isotropic Grassmannians; the operator approach,* Manuscripta Math. **79** (1993), 127–151.

[25] P. Pragacz, J. Ratajski, *A Pieri-type theorem for Lagrangian and odd orthogonal Grassmannians,* J. Reine Angew. Math. **476** (1996), 143–189.

[26] P. Pragacz, J. Ratajski, *A Pieri-type formula for even orthogonal Grassmannians,* Fund. Math. **178** (2003), 49–96.

[27] H. Tamvakis, *Gromov-Witten invariants and quantum cohomology of Grassmannians,* this volume.

Piotr Pragacz
Polish Academy of Sciences
Institute of Mathematics
Śniadeckich 8
PL-00-956 Warszawa, Poland
e-mail: P.Pragacz@impan.gov.pl

Trends in Mathematics:
Topics in Cohomological Studies of Algebraic Varieties, 175–201

Lectures on Characteristic Classes of Constructible Functions

Jörg Schürmann

Notes by Piotr Pragacz and Andrzej Weber

Abstract. The following lectures were delivered at the Mini-School "Characteristic classes of singular varieties" in Banach Center, 23–27 April 2002, by Jörg Schürmann. These lectures discuss the calculus of characteristic classes associated with constructible functions on possibly singular varieties, and focus on the specialization properties. The point of view of characteristic classes of Lagrangian cycles is emphasized. A Verdier-type Riemann-Roch theorem is discussed.[1]

1. History

1.1. Characteristic classes of vector bundles

The notion of a characteristic class of a vector bundle has appeared in several contexts.

1.1.1. Chern class of a complex vector bundle. Consider two functors on the category of topological spaces:

- the set of isomorphism classes of complex vector bundles over X,
- the cohomology group of X.

The Chern class is a contravariant functorial class in integral cohomology, i.e., a natural transformation

$$c^* : \mathrm{Vect}_{\mathbf{C}}(-) \to H^{2*}(-;\mathbf{Z}),$$

For a complex vector bundle V over X

$$c^*(V) = 1 + c^1(V) + c^2(V) + \cdots + c^{rk(V)}(V), \qquad c^i(V) \in H^{2i}(X;\mathbf{Z}).$$

[1]The contribution of Piotr Pragacz (partially supported by KBN grant No. 2P03A 024 23) and Andrzej Weber is – apart of some aspects of proofs, mentioned in the text – of purely redactional nature. The text was authorized by J. Schürmann.

For a map $f : X' \to X$ one can compute the Chern class of the induced bundle f^*V by the formula:

$$c^*(f^*V) = f^*(c^*(V)) \in H^{2*}(X'; \mathbf{Z}) .$$

(Cf. [Ch].)

1.1.2. Stiefel-Whitney class of a real vector bundle. It is a contravariant functorial class in mod 2 cohomology, i.e., a natural transformation

$$w^* : \mathrm{Vect}_{\mathbf{R}}(-) \to H^*(-; \mathbf{Z}/2) .$$

(Cf. [MiSt].)

1.1.3. Euler class of a real oriented vector bundle. It is a contravariant functorial class in integral cohomology, i.e., a natural transformation

$$e : \text{Oriented } \mathrm{Vect}_{\mathbf{R}}(-) \to H^*(-; \mathbf{Z}) .$$

For a bundle of rank r over X the Euler class lies in $H^r(X; \mathbf{Z})$.
(Cf. [MiSt].)

1.1.4. Characteristic class of an algebraic vector bundle over an algebraic variety. The Chern classes can be defined as *operators* acting on the Chow groups (Fulton, [Fl, Ch.3]). The action is denoted by $\cap$

$$c^i(V)\cap : A_k(X) \to A_{k-i}(X) ,$$

$$[\alpha] \mapsto c^i(V) \cap [\alpha] .$$

1.2. Properties of characteristic classes of vector bundles

We will concentrate on the case of complex bundles but the formulas hold for real bundles with Chern classes replaced by Stiefel-Whitney classes.

1.2.1. Whitney formula. For a short exact sequence of vector bundles:

$$0 \to V' \to V \to V'' \to 0$$

we have

$$c^*(V) = c^*(V') \cup c^*(V'') .$$

Also for a pair of bundles $V \in Vect_{\mathbf{C}}(X)$ and $W \in Vect_{\mathbf{C}}(Y)$

$$c^*(V \times W) = c^*(V) \times c^*(W) \in H^{2*}(X \times Y) .$$

1.2.2. Relation with fundamental class. For an algebraic subvariety in a complex variety one can define its fundamental class $Cl(X) = [X]$ in homology with closed supports (so-called Borel-Moore homology). This construction was extended by Borel-Heafliger to a map

$$Cl : A_*(X) \to H_{2*}^{BM}(X; \mathbf{Z}) .$$

The Chern classes in $A^*(X)$ and in $H^{2*}(X)$ are related by the formula:

$$Cl(c^i(V) \cap a) = c^i(V) \cap Cl(a) \in H_{2(k-i)}^{BM}(X; \mathbf{Z})$$

for any class of an algebraic cycle $a \in A_k(X)$.

In the real case (by a result of Sullivan, [Su]) one can define a cycle map

$$Cl_{\mathbf{R}} : A_*(X) \to H_*^{BM}(X(\mathbf{R}); \mathbf{Z}/2) .$$

Here $X(\mathbf{R})$ denotes the real points of an algebraic variety over $\mathbf{R}$. Then

$$Cl_{\mathbf{R}}(c^i(V) \cap a) = w^i(V(\mathbf{R})) \cap Cl_{\mathbf{R}}(a) \in H_{k-i}^{BM}(X(\mathbf{R}); \mathbf{Z}/2)$$

for any class of a real algebraic cycle $a \in A_k(X)$.

1.2.3. Projective bundle theorem. The following construction shows that the first Chern class is the most important.

Let $\pi : V \to X$ be a vector bundle of rank r. Let $\hat{\pi} : \mathbf{P}(V) \to X$ be its projectivization. As usual $\mathcal{O}(-1) \subset \hat{\pi}^* V$ denotes the universal tautological bundle and $\mathcal{O}(1) = \mathcal{O}(-1)^*$ its dual. Then the cohomology of $\mathbf{P}(V)$ is a free module over $H^*(X)$:

$$H^*(\mathbf{P}(V) = H^*(X)[t]/t^r ,$$

with $t = c^1(\mathcal{O}(1))$. The algebra structure of $H^*(\mathbf{P}(V))$ determines the Chern classes of V due to the unique relation

$$t^r + c^1(V)t^{r-1} + \cdots + c^r(V) = 0 .$$

1.2.4. Grothendieck-Riemann-Roch theorem. The Chern character

$$ch^* : K^0(X) \to H^{2*}(X; \mathbf{Q})$$

is a homomorphism of rings. Here $K^0(X)$ is the Grothendieck group of vector bundles equipped with a ring structure coming from tensor product. We have

$$ch^*([V \otimes W]) = ch^*([V]) \cup ch^*([W]) , \qquad ch^*([\mathcal{O}_X]) = 1 \in H^0(X; \mathbf{Q}) .$$

The Chern character commutes with pull-backs. Grothendieck-Riemann-Roch theorem describes how the Chern character behaves with respect to a proper push-forward. Let $f : X \to Y$ be a proper map of smooth algebraic varieties. Due to smoothness we can identify K-theory of vector bundles with K-theory of coherent sheaves. Every sheaf has a finite locally free resolution. This is a form of Poincaré duality. We set $K_0(-) = K^0(-)$. We have the induced map

$$f_* : K_0(X) \to K_0(Y)$$

derived from the push-forward of sheaves. On the level of cohomology we have usual Poincaré duality $H^*(-) = H_*^{BM}(-)$ with a switch of degrees. We also have the induced map f_* of homology. Nevertheless the Chern character does not commute with f_* in general. It has to be corrected by the Todd class. The following diagram does commute:

$$\begin{array}{ccccccc} K_0(X) & = & K^0(X) & \xrightarrow{ch^* \cup Td(TX)} & H^*(X;\mathbf{Q}) & = & H_*^{BM}(X;\mathbf{Q}) \\ \downarrow f_* & & & & & & \downarrow f_* \\ K_0(Y) & = & K^0(Y) & \xrightarrow{ch^* \cup Td(TY)} & H^*(Y;\mathbf{Q}) & = & H_*^{BM}(Y;\mathbf{Q}) \end{array} .$$

If we take Y to be a point then we obtain the Hirzebruch-Riemann-Roch theorem:

$$\chi(H^*(X,V)) = \int_X ch^*(V) \cup Td(TX) .$$

Let us set $Td(X) = Td(TX)$. The Todd class is multiplicative with respect to the Cartesian product:

$$Td(X \times Y) = Td(X) \times Td(Y) .$$

This formula can be generalized for singular varieties, see 1.3.5.

1.3. Characteristic classes of singular varieties

The following classes are studied:

Characteristic classes of vector bundles c^*
in cohomology, contravariant functoriality

$\bigcup$

Smooth spaces

$c_*(X) = c^*(TX) \cap [X]$

$\bigcap$

Characteristic classes of singular spaces c_*
in homology, covariant functoriality for proper maps

This picture is contained in a wider context of bivariant theory (Fulton-MacPherson, [FlMa]).

1.3.1. The systematic study of singular spaces started in the sixties. One should mention the work of Thom, Whitney, Hironaka, Łojasiewicz and others on triangulations, stratifications, resolution of singularities (in characteristic zero).

1.3.2. In 1965 M.-H. Schwartz defined certain characteristic classes using obstruction theory of so-called *radial fields*, [Sw]. Her classes were defined for a singular variety X embedded in a smooth manifold M. One obtains Chern-Schwartz classes in $H_X^{2*}(M;\mathbf{Z})$. By Alexander duality this group can be identified with $H_{2*}^{BM}(X;\mathbf{Z})$.

1.3.3. Around 1969 D. Sullivan, [Su], proved that real analytic spaces are mod 2 *Euler spaces* (i.e., the Euler characteristic of the link of any point is even). It follows that the sum of i-simplices in the first barycentric subdivision of any triangulation is mod 2 cycle. This allowed him to define Stiefel-Whitney classes $w_i(X) \in H^{BM}_{\dim X - i}(X;\mathbf{Z}/2)$. For a smooth X these classes are Poincaré dual to the usual ones.

1.3.4. In 1974 R. MacPherson, [Ma], solved the problem of existence of covariantly functorial classes, answering thus a "Deligne-Grothendieck conjecture" (see [Gr, Deuxième Partie, pp.361–368] for an interesting discussion of this conjecture and its general context). MacPherson classes are natural transformations from constructible functions to homology with the normalization condition $c_0(1_X) = [X]$. There are two ingredients used by MacPherson: *Euler obstruction* and *Chern-Mather classes*. To show that the constructed classes are indeed functorial MacPherson uses his famous *graph construction*. Originally MacPherson classes were constructed for complex algebraic varieties. Later his construction was carried for varieties over arbitrary (algebraically closed) fields of characteristic zero as well as for analytic spaces.

1.3.5. A generalization of Todd classes was studied by Baum-Fulton-MacPherson [BFM]. For a singular variety embedded in a smooth ambient space M one constructs a *localized Chern character* ch_X^M. Then

$$\tau_* = Td(TM) \cap ch_X^M : K_0(X) \to H_{2*}^{BM}(X;\mathbf{Z})$$

is functorial with respect to push-forward and it is independent from the embedding. For a precise construction of ch_X^M as well as for its properties, see [BFM] (compare also with Fulton, [Fl, Ch.18]).

The Todd class of a variety is defined by $Td_*(X) = \tau_*([\mathcal{O}_X])$. It satisfies

$$Td(X \times Y) = Td(X) \times Td(Y) .$$

1.3.6. Gonzalez-Sprinberg and Verdier gave an algebraic formula for the Euler obstruction, [G-S].

1.3.7. J.-P. Brasselet and M.-H. Schwartz have shown that MacPherson classes "coincide" with Schwartz classes, [BrSw].

1.3.8. Bivariant theory was developed by Fulton and MacPherson, [FlMa]. Bivariant versions of Stiefel-Whitney classes were given in piecewise linear context as well as a bivariant theory of Todd homology class. Chern classes of *cellular maps* were studied by Brasselet, [Br].

1.3.9. Lagrangian or microlocal approach to Chern and Stiefel-Whitney classes was initiated by M. Kashiwara and his work on the local index theorem for holonomic $\mathcal{D}$-modules, [Ka1] and [Ka2, thm.6.3.1]. A certain topological invariant studied by Kashiwara appeared to be the Euler obstruction (it's a result of Dubson, [Du], compare [BDK, thm.1]). Moreover one should mention the following references and techniques:

- V. Ginzburg: holonomic $\mathcal{D}$-modules for complex algebraic varieties, [Gi];
- C. Sabbah: approach for analytic varieties using geometric methods, [Sa];
- G. Kennedy generalized the construction of Sabbah to algebraic varieties over an arbitrary (algebraically closed) field of characteristic zero, [Ke];
- Kashiwara-Schapira: a microlocal approach to characteristic cycles of constructible sheaves in their book *Sheaves on manifolds*, [KaSp]

- J. Fu: geometric measure theory, [Fu];
- Fu-McCrory: Lagrangian approach to Stiefel-Whitney classes for real algebraic varieties, [FuMc].

1.4.

In the remaining lectures we will use the language of the book of Goresky and MacPherson "Stratified Morse theory", [GoMa], to develop a Morse theory for constructible functions (following [GrMa]). For a more sophisticated Morse theory of constructible sheaves one should compare with [Sch1] and [KaSp].

2. Calculus of constructible functions

2.1. Motivation

The motivation for this lecture is the singular Grothendieck-Riemann-Roch theorem [BFM], which states that for a singular subvariety X embedded in a smooth ambient space the map

$$\tau_* = Td(TM) \cap ch^M_X : K_0(X) \to H^{BM}_*(X; \mathbf{Q})$$

is well defined and it is functorial with respect to push-forward. Moreover it is independent from the embedding. The topological counterpart would be a localized class $?^M_X$ making the following diagram commutative

$$\begin{array}{ccccc} K_0(X) & & \xrightarrow{c^*(TM)\cap ?^M_X} & & H^{BM}_*(X;\mathbf{Z}) \\ & \chi_X \searrow & & \nearrow c_* & \\ & & CF(X)\,. & & \end{array}$$

Here $CF(X)$ is the group of constructible functions on X. The map χ_X from the Grothendieck group of constructible sheaves is taking the Euler characteristic stalkwise (it is surjective) and c_* is Chern-MacPherson transformation.

2.2. Category

We will work in the following categories

algebraic varieties over a field of characteristic zero

$\cup$

algebraic complex varieties

$\cap$

complex analytic varieties

$\cup$

complex analytic varieties with a fixed stratification

We will use techniques developed for real sub- and semi-analytic varieties.

2.3. Definitions of constructible functions

2.3.1. Let X be an algebraic variety defined over an algebraically closed field k with $char(k) = 0$. A *constructible set* is a subset defined by finitely many unions, intersections, and complements of (closed) subvarieties. A *constructible function* $\alpha : X \to \mathbf{Z}$ is a function such that $\alpha^{-1}(m)$ is constructible for any $m \in \mathbf{Z}$ and non-empty only for finitely many $m \in \mathbf{Z}$. The set of constructible functions is a group denoted by $CF(X)$. The assumption $char(k) = 0$ will only be used later on (in the definition of or) for the properties of some transformations, e.g., for Theorem 2.7 and Proposition 2.8 in the algebraic context (where one uses l-adique cohomology for the definition of the Euler characteristic).

Each function can be written as a finite sum $\alpha = \sum 1_{Z_i}$, where Z_i are (closed) subvarieties of X.

2.3.2. For a field k of characteristic zero which is not algebraically closed we define

$$CF(X) = CF(X(\bar{k}))^{Gal(\bar{k}/k)}\,.$$

2.3.3. For complex analytic varieties we consider functions which are locally finite sum of the form $\alpha = \sum 1_{Z_i}$ with Z_i being closed analytic subvarieties. Similarly constructible functions are defined for real analytic, subanalytic etc. varieties.

2.4. Transformation of constructible functions

2.4.1. Contravariant functoriality. Let $f : X \to Y$ be a map of varieties. Contravariant transformation

$$f^* : CF(Y) \to CF(X)$$

is just the composition

$$f^*(\alpha) = \alpha \circ f\,.$$

2.4.2. Multiplication. We can multiply functions pointwise

$$(\alpha \cdot \beta)(x) = \alpha(x)\beta(x)\,.$$

2.4.3. Exterior product. We have an operation

$$\times : CF(X) \times CF(Y) \to CF(X \times Y)\,,$$
$$(\alpha \times \beta)(x, y) = \alpha(x)\beta(y)\,.$$

2.4.4. Euler characteristic. We work here in the real world. Denote by $CF_c(X)$ the group of constructible functions with compact support. (By the definition the support is always closed.) Euler characteristic is the unique linear map

$$\chi : CF_c(X) \to \mathbf{Z}\,,$$
$$\chi(1_Z) = \chi(H^*(Z))$$

for Z compact subanalytic. We introduce the following notation for $\alpha \in CF(X)$ and subsets $A \subset B \subset X$, with B compact

$$\chi(B, A; \alpha) := \chi(\alpha \cdot 1_{B \setminus A})\,.$$

We can think about $\chi(B,A;\alpha)$ as a relative homology of the pair (B,A) with coefficients in α. We set $\chi(B;\alpha)=\chi(B,\emptyset;\alpha)$.

Example 2.1. Let $\alpha = 1_Z$. Then

$$\begin{aligned}\chi(B,A;1_Z) &= \chi(1_Z(1_B-1_A))\\ &= \chi(1_{Z\cap B})-\chi(1_{Z\cap A})\\ &= \chi(H^*(Z\cap B))-\chi(H^*(Z\cap A))\\ &= \chi(H_c^*(Z\cap(B\setminus A))\\ &=: \chi_c(1_{Z\cap(B\setminus A)})\end{aligned}$$

The last term is called the *Euler characteristic with compact support.*

Example 2.2. Let Z be an open ball in $\mathbf{R}^n$. Then $\chi_c(1_Z)=(-1)^n$.

Euler characteristic with compact support is additive. Suppose a compact set X is a finite disjoint union of locally closed subsets X_i. Then

$$\chi(X;\alpha)=\sum\chi_c(X_i;\alpha)\,.$$

In the complex case we have[1]:

Proposition 2.3. *Let X be a compact (or compactifiable) complex variety decomposed into a disjoint union of locally closed subvarieties X_i. Then for every constructible function α,*

$$\chi(X;\alpha)=\sum\chi(X_i;\alpha)\,.$$

One can propose the following reasoning to prove the above assertion. It is enough to take $\alpha=1_X$ and $\{X_i\}$ to be the strata of a Whitney stratification of X with $\chi(Z):=\chi(H^*(Z))$ for $Z\subset X$ locally closed.

We record the following result of Sullivan: if L is a compact real Whitney stratified space with strata of even codimension, and dimension of L is odd, then $\chi(L)=0$. (Cf. [Su].)

We will now show that $\chi(X)=\sum\chi(X_i)$ by induction on the depth $d(X)$. We recall that *depth* is the maximal (real) codimension of strata.

If $d(X)=0$, then X has a unique stratum, and the assertion is obvious.

Suppose that $d(X)=2d$. Let X_g be the deepest stratum. This stratum has a neighborhood U, which is a fibered over X_g with the fiber being the cone over the link L_g (cf. [Mat]). Since L_g is compact of dimension $2d-1$, and the strata are of even codimension (as in X), we have $\chi(L_g)=0$. Finally,

$$\begin{aligned}\chi(X)&=\chi(X\setminus X_g)+\chi(U)-\chi(U\setminus X_g)\\ &=\sum_{i\neq g}\chi(X_i)+\chi(X_g)-\chi(L_g)\chi(X_g)=\sum_i\chi(X_i)\,.\end{aligned}$$

This proves Proposition 2.3. Of course, if the deepest stratum is not connected, the argument should be slightly modified.

[1]This paragraph is added by PP-AW.

Corollary 2.4. *If X is a complex algebraic variety (or just admitting a compactification stratified by even codimension strata), then $\chi_c(X) = \chi(X)$.*

Corollary 2.4 for a complex algebraic variety is a special case of [Sch1, formula (6.42), p.413] or [La], and compare with 2.4.6.

The following example will be invoked several times.

Example 2.5. Let $a < b < c \in \mathbf{R}$ and α_-, α_0 and α_+ be integers. Define a constructible function on $\mathbf{R}$

$$\alpha = \alpha_- 1_{[a,b[} + \alpha_0 1_{\{b\}} + \alpha_+ 1_{]b,c]} \,.$$

Then

$$\chi([a,c];\alpha) = (\alpha_- - \alpha_-) + \alpha_0 + (\alpha_+ - \alpha_+) = \alpha_0 \,.$$

We call this property the *conic structure* (of the interval).

2.4.5. Proper push-forward. Let $f : X \to Y$ be a proper map. We define a transformation of constructible functions

$$f_* : CF(X) \to CF(Y) \,,$$
$$f_*(\alpha)(y) = \chi(f^{-1}(y);\alpha) \,.$$

(It is enough to assume that f is proper on the support of α.) In particular

$$f_*(1_Z)(y) = \chi(f^{-1}(y) \cap Z) \,.$$

Example 2.6. If Y is a point, then $f_*(\alpha) = \chi(\alpha) 1_{pt}$.

Theorem 2.7. *The proper push-forward of a constructible function is constructible.*

Proposition 2.8. *For a sequence of proper maps $X \xrightarrow{f} Y \xrightarrow{g} Z$ we have*

$$(g \circ f)_* = g_* \circ f_* \,.$$

The proofs of these results rely on the stratification properties of maps.

Consider a real-valued function $f : X \to \mathbf{R}^{\geq 0}$ which is proper and a constructible function $\alpha \in CF(X)$. Let $X_0 = f^{-1}(0)$. By 2.7 the function $f_*(\alpha)$ is constructible. It follows that there exists $\epsilon > 0$, such that $f_*(\alpha)_{|]0,\epsilon]}$ is constant. By the *conic structure* of the interval (see Example 2.5)

$$\chi([0,\epsilon]; f_*\alpha) = f_*(\alpha)(0) = \chi(f^{-1}(0);\alpha) = \chi(X_0;\alpha) \,.$$

2.4.6. Verdier duality for constructible functions. We work in the real world. The duality map

$$D : CF(X) \to CF(X)$$

is defined by

$$(D\alpha)(x) = \alpha(x) - \chi(S_\epsilon(x);\alpha) \,,$$

where $S_\epsilon(x)$ is a small sphere centered at x. One can show that $D \circ D = Id$. If we deal with complex varieties then due to Sullivan, [Su], we know that $\chi(S_\epsilon(x);\alpha) = 0$. Thus $D = Id$. (For real varieties we only have $\chi(S_\epsilon(x);\alpha) = 0 \bmod 2$.) In particular, in the complex algebraic context one gets $\chi = \chi_c$, since in general $\chi \circ D = \chi_c$ in the semi-algebraic context.

2.4.7. Nearby cycles. Let $k = \mathbf{R}$ or $\mathbf{C}$. Let $f : X \to k$ be a function. Set $X_0 = f^{-1}(0)$. In the complex case $k = \mathbf{C}$ we have a Milnor fibration for appropriate choice of $0 < \epsilon \ll \delta \ll 1$ with the fiber

$$M_{f,x} = B_\delta(x) \cap \{f = \epsilon\} \cap X .$$

The nearby cycle operation

$$\psi_f : CF(X) \to CF(X_0)$$

is defined by

$$\psi_f(\alpha)(x) = \chi(M_{f,x}; \alpha) .$$

The value of $\psi_f(\alpha)$ depends only on the values of α outside of X_0.

In the real world we have two nearby cycle operations (studied by McCrory and Parusiński, [McPa1]): the lower ψ_f^- and the upper ψ_f^+, due to the possibility of taking negative values of ϵ. Negative and positive Milnor fibers might differ.

2.4.8. Vanishing cycles. Let us concentrate on the complex case. The vanishing cycle operation

$$\phi_f : CF(X) \to CF(X_0)$$

is defined by

$$\phi_f(\alpha) = \psi_f(\alpha) - \alpha_{|X_0} .$$

The motivation comes from the world of derived category of sheaves. The analogous operations form a distinguished triangle

$$\begin{array}{ccc} i^*F & \longrightarrow & \psi_f F . \\ {\scriptstyle +1} \nwarrow & & \swarrow \\ & \phi_f F & \end{array}$$

where i denotes the inclusion of X_0 into X.

Remark 2.9. On the $\mathcal{D}$-module level $\phi_f[-1]$ corresponds to vanishing cycles.

2.4.9. Specialization. The specialization of Verdier [Ve2], has also its counterpart for constructible functions. Let $X \subset Y$ be a closed subset. The specialization is an operation

$$sp : CF(Y) \to CF_{mon}(C_X Y) .$$

Here $C_X Y$ denotes the normal cone of X in Y. The subscript *mon* stands for *monodromic*, i.e., conic function. Let A^1 be the affine line and $A^* = A^1 \setminus \{0\}$. The normal cone is contained in the deformation space M (cf. [Fl, Ch.5]).

$$M = Bl_{X \times \{0\}}(Y \times A^1) \setminus Bl_{X \times \{0\}}(Y \times \{0\}) ,$$

$$\begin{array}{ccccc} C_X Y & \subset & M & \supset & Y \times A^* \\ \downarrow & & \downarrow g & & \downarrow \\ \{0\} & \subset & A^1 & \supset & A^* . \end{array}$$

Specialization is defined by the formula

$$sp(\alpha) = \psi_g(\tilde{\pi}^*(\alpha)),$$

where $\tilde{\pi}$ is the composition of the blow up $M \to Y \times A^1$ and the projection to Y.

The value of $sp(\alpha)$ depends only on the values of $\tilde{\pi}^*(\alpha)$ outside of $C_X Y$ and $sp(\alpha)_{|X} = \alpha_{|X}$. Here we consider X as embedded in $C_X Y$ as the vertex-section.

For some aspects of operations on constructible functions, cf. [Sp].

3. Stratified Morse theory for constructible functions and Lagrangian cycles

In this lecture, we work in the real world (following [GrMa]). By *dimension* we mean the real dimension.

3.1. Stratification

Let X be a closed subset of a smooth manifold M. We can work over $\mathbf{C}$ as well as over $\mathbf{R}$. A stratification of X is a filtration

$$X_\bullet : \emptyset = X_{-1} \subset X_0 \subset \cdots \subset X_n = X,$$

where each inclusion is assumed to be a closed embedding in the appropriate category. Moreover each difference $X^i := X_i \setminus X_{i-1}$ should be a smooth manifold of dimension i or empty. A stratum of $X_\bullet$ is a connected component of some X^i.

The group of constructible functions with respect to a stratification $CF(X_\bullet)$ consists of functions constant on strata. The group of all constructible functions is the direct limit of the groups $CF(X_\bullet)$.

3.2. Whitney conditions

Whitney conditions are traditionally called 'a' and 'b' (although 'b' is stronger then 'a'). Consider the following situation for two strata S_α and S_β. Suppose that there are given sequences of points $x_i \in S_\alpha$ and $y_i \in S_\beta$, both converging to $x \in S_\alpha \cap \bar{S}_\beta$. Assume that the secant line $\overline{x_i y_i}$ converges in the projective space to ℓ and that the tangent space $T_{y_i} S_\beta$ converges to a space V in an appropriate Grassmannian. Then

a): $T_x S_\alpha \subset V$,
b): $\ell \subset V$.

Example 3.1. There are three strata: circle S^1, open disk D^2 and $X \setminus \overline{D}^2$. The Whitney condition b) is not satisfied.

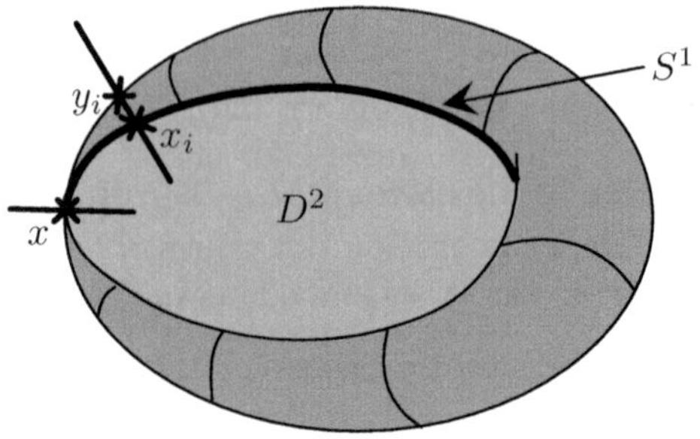

Perhaps, the simplest counterexample for the Whitney condition b) is the "Whitney cusp" $X := \{x^2 + z^2 \cdot (z - y^2)\} \subset \mathbf{R}^3$, with $X_1 := \{x = 0 = z\}$ and $X_2 := X$. It is a-regular, but not b-regular in the point 0. Consider a sequence $(c_n \neq 0)$ converging to 0, and look at $x_n := (0, c_n, 0) \in X^1 = X_1$ and $y_n := (0, c_n, c_n^2) \in X^2 = X \backslash X_1$. Then $T_{y_n} X^2 = (0, -2c_n, 1)^\perp \to (0, 0, 1)^\perp =: V$, but the corresponding secant line converges to $l = \{x = y = 0\}$, which is orthogonal to $V = \{z = 0\}$.

Let

$$\Lambda := \bigsqcup T_S^* M \subset T^* M$$

be the disjoint union of the conormal spaces to the strata. The Whitney condition a) is equivalent to the statement that Λ is closed in T^*M.

The following is an exercise for using the condition b). Let $S_\alpha = \{x_0\}$. In some local coordinates $x_0 = 0$. Let $\nu = \sum x_i^2$ be the distance function.

Proposition 3.2. *If the Whitney condition b) is satisfied at x_0, then the restriction of ν to each stratum has no critical points near x_0.*

Proof. Assume the converse. Then there exists a sequence $y_i \in S_\beta$, converging to x_0, such that $T_{y_i} S_\beta \subset ker\, d\nu$. The secant $\overline{x_0 y_i} \perp ker\, d\nu$. We can assume that $\overline{x_0 y_i}$ converges to a line ℓ and $T_{y_i} S_\beta$ converges to a space V. Then $\ell \perp V$. This contradicts b).

Definition 3.3. In the conormal space Λ we distinguish the subspace of *good covectors*

$$\Lambda_S^o := T_S^* M \setminus \bigcup_{S' \neq S} \overline{T_{S'}^* M}$$

$$\Lambda^o := \bigsqcup_S \Lambda_S^o .$$

Example 3.4. Let $X = [a, b] \subset \mathbf{R}$ with the stratification $X_0 = \{a, b\} \subset X_1 = X$.

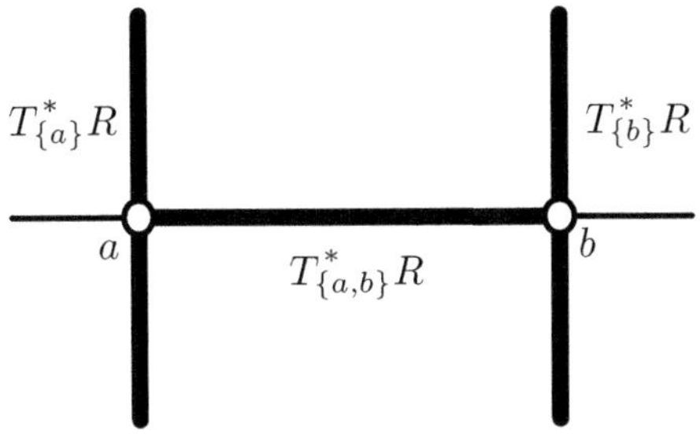

3.3. Morse theory

We will formulate results coming form Morse theory which we will treat as a "black box" here.

By the *normal slice* of a stratum S we mean a germ of a submanifold $N \subset M$ which is transverse to S and $\dim N = \operatorname{codim} S$. Suppose $N \cap S = \{x\}$. Fix a real function with a good differential df_x at x. Define the upper/lower half-link of f by

$$L_f^\pm = N \cap X \cap B_\delta(x) \cap \{f = f(x) \pm \epsilon\}\,,$$

for $0 < \epsilon \ll \delta \ll 1$. Fix a constructible function $\alpha \in CF(X_\bullet)$. For each $x \in S$ we will define a *normal index*

$$i(df_x;\alpha) = \alpha(x) - \chi(L_f^-;\alpha)\,.$$

In the complex case we can rewrite this quantity as

$$\alpha(x) - \psi_{g_{|N}}(x) = -\phi_{g_{|N}}(x)\,.$$

Here g is a holomorphic function with $g(x)=0$ and the real part $f:=re(g)$ as before. Moreover it is related to the Euler characteristic of the complex link, [GoMa].

The following facts follow from the Morse theory:

- $i(df_x;\alpha)$ depends only on $df_x \in \Lambda^o$ and it is locally constant on Λ^o,
- Duality: $i(df_x;D\alpha) = (-1)^{\dim S} i(-df_x;\alpha)$ where D is Verdier duality 2.4.6. In particular for complex varieties $i(df_x;D\alpha) = i(df_x;\alpha) = i(-df_x,\alpha)$,
- $i(df_x \times dg_y;\alpha\times\beta) = i(df_x;\alpha)\cdot i(dg_y;\beta)$,
- Change of Euler characteristic is equal to the index. Suppose $f: M \to \mathbf{R}$ is a function such that $f^{-1}[a,b)\cap X$ is compact and f has at most one critical point in $f^{-1}[a,b]$ at x. Assume $a < f(x) < b$. Then for any constructible function $\alpha \in CF(X_\bullet)$

$$\chi(\{a \le f \le b\},\{f=a\};\alpha) = \\ = \begin{cases} 0 & \text{if } x \text{ is not a critical point,} \\ (-1)^\lambda i(df_x;\alpha) & \text{if } x \text{ is a stratified Morse critical point.}\end{cases}$$

Here λ is the Morse index of $f_{|S}$.

Remark 3.5. We say that x is a stratified Morse critical point if df_x is good and $f_{|S}$ is a classical Morse critical point. Then $df(M) \cap \Lambda \subset \Lambda^o$ near df_x and $df(M)$ intersects transversally $T^*_S M$ at df_x. The right choice of orientation for $T^*_S M$ allows one to express $(-1)^\lambda$ as the intersection number $\sharp[df(M)] \cap [T^*_S M]$.

3.4. Lagrangian cycles

The references for this subsection are [GrMa], [ScVi], and [Sa]. For simplicity let us assume that the ambient space M is oriented. The space of good covectors decomposes into the connected components

$$\Lambda^o = \bigsqcup \Lambda^o_j \,.$$

Each component has its index $i(\Lambda^o_j; \alpha)$.

Example 3.6. The indices in Example 3.4 are

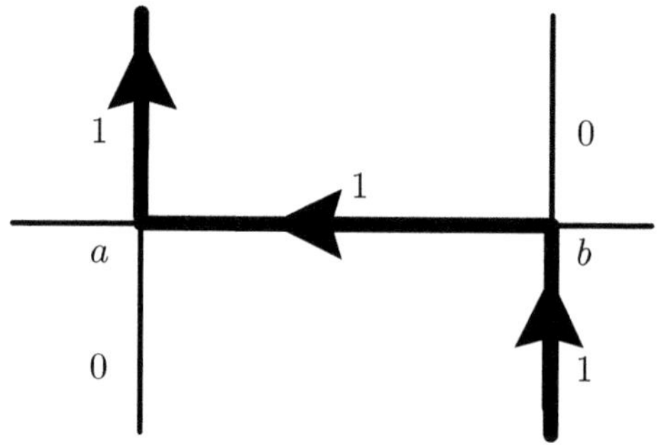

We claim:

Proposition 3.7. *The sum* $\sum i(\Lambda^o_j; \alpha)\Lambda^o_j$ *is a homology cycle.*

Remark 3.8. The conormal space $T^*_S M$ can always be oriented. Unfortunately there are several conventions to do this. In the complex case our orientation differs from the complex one by $(-1)^{\dim_{\mathbb{C}} S}$. Moreover, in the complex case, the fundamental class $[\overline{T^*_S M}]$ is already a homology cycle. So Proposition 3.7 is non-trivial only in the real context. Here it goes back originally to Kashiwara (and also to Fu).

Our construction gives us an assignment called *the characteristic cycle* or *microlocal Euler class*

$$CC = \mu eu : CF(X_\bullet) \longrightarrow H^{BM}_{top}(\Lambda) = L(X_\bullet, M) \,.$$

In our notation $L(X_\bullet, M)$ denotes the group of Lagrangian cycles supported on the covectors of the conormal spaces to the strata of $X_\bullet$. One can directly show by induction on $n = dim(X)$, that the map CC is an isomorphism. Here one only uses $i(df_x; \alpha) = \alpha(x)$ for x in a top-dimensional stratum (so that $L^-_f = \emptyset$).

Theorem 3.9. *There exists an inverse transformation to CC*

$$Eu^\vee : L(X_\bullet, M) \to CF(X_\bullet) \,.$$

It is given by

$$L \mapsto \alpha \,,$$

where the value of α *at* $x_0 \in X$ *is given by the intersection number*

$$\alpha(x_0) = \sharp_{d\nu_0}[d\nu(M)] \cap [L] \,.$$

Here, as before, $\nu(x) = \sum x_i^2$ *in some local coordinates in which* $x_0 = 0$.

The map $Eu^\vee$ is in the complex context induced by the local Euler obstruction – cf. Remark 4.5; this motivates the notation.

The theorem follows from the formula (compare Example 4.4):

$$\sharp_{d\nu_0}[d\nu(M)] \cap [CC(\alpha)] = \alpha(x_0)\,.$$

Consider now the complex context. If $CC(\alpha) = \overline{T_S^* M}$, that is $\alpha = Eu^\vee(\overline{T_S^* M})$, then one can recover the values of the Euler obstruction of $\bar{S}$:

$$Eu_{\bar{S}}(x_0) = (-1)^{\dim_{\mathbf{C}} S} Eu_{\bar{S}}^\vee(x_0) = (-1)^{\dim_{\mathbf{C}} S} \alpha(x_0)\,.$$

The sign comes from a different choice of orientation when comparing with [Ma].

4. Characteristic classes of Lagrangian cycles

4.1. Recollection of the previous lecture

Let $X_\bullet$ be a Whitney stratified space contained in an oriented smooth manifold M. We use stratified Morse theory for constructible functions as a "black box". Let

$$CC : CF(X_\bullet) \to H^{BM}_{\dim M}(\Lambda) = L(X_\bullet, M)$$

be the characteristic cycle map. (As before $\Lambda = \bigcup_S T_S^* M$ is a closed subspace of T^*M.) Recall that df_x has a stratified Morse critical point on S if and only if in the neighborhood of df_x

$$df(M) \pitchfork (T_S^* M)^o = \{df_x\}$$

and df_x is good, see Def. 3.3. Then

$$\sharp_{df_x}[df(M)] \cap [T_S^* M] = (-1)^\lambda\,,$$

where λ is the Morse index of $f|S$ in x. If M and S are complex manifolds, then the canonical orientation of $T_S^* M$ differs from the complex orientation and

$$[T_S^* M] = (-1)^{\dim_{\mathbf{C}} S}[(T_S^* M)_{\mathbf{C}\text{-orientation}}]\,.$$

Remark 4.1. The Lagrangian cycle $CC(\alpha)$ does not depend on the choice of a stratification in which α is constructible. If one subdivides a stratification, then the indices of the additional strata vanish.

Now consider the group $L(T^*M_{|X})$ of all Lagrangian cycles in $T^*M_{|X}$, which by definition is the direct limit of the groups $L(X_\bullet, M)$. In the complex case it is generated by $[\overline{T_{S_{\mathrm{reg}}}^* M}]$, where S is a subvariety of X. These cycles are conic.

Warning. If we work in the real subanalytic context, then *conic* means invariant with respect to the multiplicative action of $\mathbf{R}^+$, not $\mathbf{R}^*$.

Exercise: Show that $CC(\alpha) \subset T_M^* M$ if and only if α is locally constant. Recall that $T_M^* M$ is the zero section of T^*M.

4.2. Intersection formula

Theorem 4.2. *Suppose $f : M \to [a, \infty[$ is a real analytic function, and $\alpha \in CF(M)$ is a constructible function, such that $f_{|\,\mathrm{supp}(\alpha)}$ is proper and*

$$\pi(\mathrm{supp}\ CC(\alpha) \cap df(M)) \subset \{a \leq f \leq b\}\,.$$

Here $\pi : T^*M \to M$ is the projection. Then

$$\sharp[df(M)] \cap [CC(\alpha)] = \chi(\{f \leq c\}; \alpha)$$

for any $c > b$.

For the proof of the theorem one deforms f to a Morse function. By additivity of the Euler characteristic with compact support, one reduces to the case of exactly one Morse critical point $x \in \mathrm{supp}(\alpha)$. Then the claim follows from the last result of our "black box" about Morse theory.

Example 4.3. If f is a constant function and $\pi(\mathrm{supp}(CC(\alpha)) \cap T^*_M M)$ is compact (this is so when supp is compact), then

$$\chi(M; \alpha) = \sharp[T^*_M M] \cap [CC(\alpha)]\,.$$

If $\alpha = 1_M$ for a compact oriented manifold M, the Theorem 4.2 reduces to the well-known Poincaré-Hopf theorem. We can also write

$$\chi(X) = \sharp(s^! CC(1_X))\,,$$

where $s : M \to T^*M$ is the zero section. The last formula can be decomposed into steps:

- $CC(1_X)$ is the microlocal Euler class,
- $s^! CC(1_X)$ is the Euler class,
- $\sharp s^! CC(1_X)$ is the degree of the Euler class, i.e., the Euler characteristic.

Example 4.4. Let ν be the function as in Proposition 3.2

$$\nu = \sum x_i^2 : \{\nu < \epsilon\} \to [0, \epsilon[\,.$$

Then

$$\sharp_{d\nu_0}[d\nu(M)] \cap [CC(\alpha)] = \chi(\{\nu \leq \epsilon\}; \alpha)\,.$$

Due to conic structure the latter is equal to

$$\chi(\{\nu = 0\}; \alpha) = \alpha(0)$$

for $0 < \epsilon \ll 1$. This formula gives us an inverse $Eu^\vee$ to the characteristic cycle map $CC : CF(X) \to L(T^*M_{|X})$.

Remark 4.5. In the complex category the value of $Eu^\vee[\overline{T^*_{S_{reg}} M}] \in CF(X)$ at a point x is just the Euler obstruction $Eu_S(x)$ corrected by the sign $(-1)^{\dim_{\mathbb{C}} S}$, see 3.9.

4.3. Variety of categories

In other categories our approach works as well. We obtain a map

$$Eu^\vee : L(T^*M_{|X}) \to \mathcal{H}_*(X)\,,$$

where $\mathcal{H}_*(-)$ may denote:

- $A_*(-/k)$ – group of algebraic cycles if we work in the algebraic category of varieties over a field k with $char(k) = 0$. Here one associates to $[\overline{T^*_{S_{reg}}M}]$ (for a subvariety S) the dual Euler obstruction of S using the algebraic definition of [G-S] as in 1.3.6.
- $H^{BM}_{2*}(-)$ – in the analytic or algebraic category over $\mathbf{C}$,
- $H^{BM}_*(X(\mathbf{R}); \mathbf{Z}/2)$ – in the algebraic category over $\mathbf{R}$.

In the last case we obtain Stiefel-Whitney classes. With a suitable modification an Euler class can be obtained in the category of real oriented varieties.

4.4. Segre classes

In the complex category we work with conic cycles which are invariant with respect to $\mathbf{C}^*$. It is convenient to pass to the projectivization of T^*M. With a conic cycle ξ one can associate its Segre class

$$s_*(\xi) = \hat{\pi}_*(c(\mathcal{O}(-1))^{-1} \cap [\hat{\xi}]) \in \mathcal{H}_*(X)\,.$$

Here $\hat{\pi} : \mathbf{P}(T^*M) \to X$ is the projection and $\hat{\xi}$ denotes the projectivization of ξ. If ξ has components in the zero section, then one should use this definition for the projective completion of ξ in $P(T^*M \oplus 1)$.

Remark 4.6. For real algebraic varieties the above construction works since the conic cycles are invariant with respect to $\mathbf{R}^*$. On the level of constructible functions it means that $D(\alpha) = \alpha \bmod 2$. This is just the condition for being an Euler space. See [FuMc].

Remark 4.7. In [Fl], the Segre class of a subscheme of M is defined by a formula similar to the one given above, involving the normal cone of the subscheme in M.

If X is a hypersurface in M, then $s_*CC(1_X)$ can be expressed in terms of the Segre class of X in M and of a correction term, which amounts essentially to the Segre class of the Jacobian subscheme of X in M, cf., e.g., [Al1] or Aluffi's lectures in the present volume.

4.5. Definition of Chern-MacPherson classes

We will introduce dual Chern classes of a constructible function. They are related to the usual Chern-MacPherson classes via the formula

$$c_i^\vee = (-1)^i c_i\,.$$

We remark that the involution

$$\vee : A_*(-) \to A_*(-)\,,$$
$$\vee : H_{2*}(-) \to H_{2*}(-)$$

is useful in various contexts. The classes $c_*^\vee$ are defined by the commuting diagram

$$\begin{array}{ccc} CF(X) & \overset{CC}{\underset{Eu^\vee}{\rightleftarrows}} & L(T^*M_{|X}) \,. \\ {\scriptstyle c_*^\vee} \searrow & & \swarrow {\scriptstyle c^*(T^*M)\cap s_*} \\ & \mathcal{H}_*(X)\,. & \end{array}$$

The class $?_X^M := s_*(CC(\cdot))$ (cf. 2.1) resembles the localized Chern character in singular Riemann-Roch theorem of Baum-Fulton-MacPherson, cf. 1.3.5.

4.6. Proper push-forward

In order to prove functoriality of c_* we will define *the proper push-forward* of Lagrangian cycles. Let $f : X \to Y$ be a proper map which extends to a map of ambient spaces, also denoted by $f : M \to N$. (We can even assume that f is submersion. Just replace M by $M \times N$ which contains M as the graph of f.) We have a diagram of bundles

$$\begin{array}{ccccc} T^*M & \overset{df}{\longleftarrow} & f^*T^*N & \overset{\tau}{\longrightarrow} & T^*N \\ & \searrow \swarrow & & & \downarrow \\ & M & \overset{f}{\longrightarrow} & & N\,. \end{array}$$

Definition 4.8. A Lagrangian cycle ξ supported over X is transformed by f to

$$f_*([\xi]) := \tau_*(df^*[\xi])\,.$$

Here one uses Poincaré duality for the definition of df^*. That $f_*([\xi])$ really defines a Lagrangian cycle follows from a suitable Whitney stratification of the map f : $X \to Y$ (or from generic smoothness in the algebraic context over a field over characteristic zero).

We will show that the proper push-forward of Lagrangian cycles agrees with the proper push-forward of constructible functions. It is enough to check that

$$Eu^\vee f_* CC(\alpha) = f_*(\alpha)\,.$$

We compute

$$Eu^\vee f_* CC(\alpha)(y) = Eu^\vee(\tau_* df^*[CC(\alpha)])(y)$$

for each $y \in Y \subset N$. By the definition (in Theorem 3.9) of $Eu^\vee$ it is equal to

$$\sharp_{d\nu_y}([d\nu(N)] \cap \tau_* df^*[CC(\alpha)])\,.$$

Let $a_\nu : M \to f^*T^*N$ be the section induced by $d\nu$. We rewrite the last expression as:

$$\begin{aligned} &\sharp_{d\nu_y} \tau_*(a_{\nu *} f^*[N]) \cap \tau_* df^*[CC(\alpha)] \\ &\quad = \sharp df_*(a_{\nu *}[M] \cap df^*[CC(\alpha)]) \\ &\quad = \sharp df_* a_{\nu *}[M] \cap [CC(\alpha)] \\ &\quad = \sharp d(\nu \circ f)[M] \cap [CC(\alpha)]\,, \end{aligned}$$

where we have used twice the projection formula and $\sharp$ stands for the intersection number. By the Intersection Formula §4.2 combined with the conic structure (as in Example 2.5) the last expression is equal to

$$\chi(\{\nu \circ f = 0\}; \alpha) = f_*(\alpha)(y)\,.$$

This completes the proof of the functoriality of CC. The functoriality of $c^*(T^*M) \cap s_*(\cdot)$ under this proper push-forward of Lagrangian cycles is much simpler. The main point then is the fact, that $c^*(T^*M) \cap s_*(\xi)$ is just the sum of all components of $[\hat{\xi}] \in H_*(P(T^*M))$ in the decomposition of $H_*(P(T^*M))$ as in the projective bundle theorem 1.2.1 (compare [Al2, lem. 4.2, 4.3]).

5. Verdier-Riemann-Roch theorem and Milnor classes

We will assume, that the considered spaces are algebraic or complex analytic.

5.1. Multiplicativity

Chern-MacPherson classes were invented as the unique classes which are functorial with respect to proper push-forwards. It turns out that they behave well with respect to the exterior products. By [Kw] we have

$$c_*(\alpha \times \beta) = c_*(\alpha) \times c_*(\beta)\,.$$

5.2. Verdier-Riemann-Roch theorem

Let $i : X \subset Z$ be a regular embedding and $p : Z \longrightarrow Y$ be a smooth submersion. All spaces are allowed to have singularities. The normal cone of X in Z is the normal bundle $C_X Z = N_X Z$. Let T_p be the tangent bundle of the fibers.

Question. Does the following diagram commute?

$$\begin{array}{ccc} CF(Y) & \xrightarrow{c_*} & \mathcal{H}_*(Y) \\ \downarrow{\scriptstyle p^*} & & \downarrow{\scriptstyle p^* \cap c^*(T_p)} \\ CF(Z) & \xrightarrow{c_*} & \mathcal{H}_*(Z) \\ \downarrow{\scriptstyle i^*} & & \downarrow{\scriptstyle i^* \cap c^*(N_X Z)^{-1}} \\ CF(X) & \xrightarrow{c_*} & \mathcal{H}_*(X)\,. \end{array}$$

According to Yokura [Yo] the answer is as follows:

Claim 5.1. *The upper square commutes.*

The lower square does not commute in general. The proof of the Yokura result is the following: we have to show that

$$c_*(p^*\alpha) = c^*(T_p) \cap p^* c_*(\alpha)\,.$$

We can assume that $\alpha = q_* 1_M$ with smooth M (we use a resolution of singularities). The map p is covered by a map p'

$$\begin{array}{ccccc} Z \times_Y M & = & M' & \xrightarrow{p'} & M \\ & & \downarrow{\scriptstyle q'} & & \downarrow{\scriptstyle q} \\ & & Z & \xrightarrow{p} & Y\,. \end{array}$$

Note that M' is smooth and $c^*(TM') = c^*(T_{p'}) \cup p'^* c^*(TM)$. Then

$$c_*(1_{M'}) = c^*(TM') \cap [M'] = (c^*(T_{p'}) \cup p'^* c^*(TM)) \cap [M']$$
$$= c^*(T_{p'}) \cap (p'^* c^*(TM) \cap [M']) = c^*(T_{p'}) \cap (p'^* c_*(1_M)).$$

Now we apply q'_* to both sides. The left-hand side can be rewritten as

$$q'_* c_*(1_{M'}) = c_*(q'_* 1_{M'}) = c_*(p^*\alpha)\,.$$

The right-hand side is equal to

$$q'_*(c^*(q'^* T_p) \cap (p'^* c_*(1_M))) = q'_*(q'^* c^*(T_p) \cap (p'^* c_*(1_M)))$$

and by the projection formula it is equal to

$$c^*(T_p) \cap (q'_* p'^* c_*(1_M)) = c^*(T_p) \cap (p^* q_* c_*(1_M))$$
$$= c^*(T_p) \cap (p^* c_* q_*(1_M)) = c^*(T_p) \cap (p^* c_* \alpha)\,.$$

To see that the lower diagram does not commute in general, consider a regular embedding of X into a smooth $Z = M$. On one hand

$$c_*(i^*(1_M)) = c_*(1_X)$$

is the Chern-MacPherson class. On the other hand

$$c^*(N_X M)^{-1} \cap i^*(c_*(1_M)) = c^*(N_X M)^{-1} \cap i^*(c^*(TM) \cap [M])$$
$$= (c^*(N_X M)^{-1} \cup i^*(c^*(TM))) \cap [X] = c_*^{FJ}(X)$$

is the Fulton-Johnson class, which in general differs from the MacPherson class.

5.3. Definition of Milnor class

The difference between MacPherson class and Fulton-Johnson class is usually called Milnor class. In our approach we do not have to assume that $Z = M$ is smooth.

Definition 5.2. For a regular embedding $X \subset Z$, and a constructible function $\alpha \in CF(Z)$, the difference

$$\mathcal{M}(X \subset Z; \alpha) = c^*(N_X Z)^{-1} \cap i^* c_*(\alpha) - c_*(i^*\alpha) \in \mathcal{H}_*(X)$$

is called *Milnor class of the pair $X \subset Z$ relative* to α.

Let k be the zero-section and π be the projection in $N_X Z$. By 5.1 for π and $i^*\alpha$

$$c_*(\pi^* i^* \alpha) = c^*(T_\pi) \cap \pi^* c_*(i^*\alpha)\,.$$

We apply k^* to both sides:

$$k^* c_*(\pi^* i^* \alpha) = k^*(c^*(T_\pi) \cap \pi^* c_*(i^*\alpha))$$
$$= c^*(N_X Z) \cap k^* \pi^* c_*(i^*\alpha) = c^*(N_X Z) \cap c_*(i^*\alpha)\,.$$

Then $c_*(i^*(\alpha))$ can be rewritten as

$$c^*(N_X Z)^{-1} \cap k^*(c_*(\pi^* i^*(\alpha)))\,.$$

Thus

$$\mathcal{M}(X \subset Z; \alpha) = c^*(N_X Z)^{-1} \cap \left(i^* c_*(\alpha) - k^*(c_*(\pi^* i^*(\alpha)))\right).$$

5.4. Milnor class can be computed from vanishing cycles

Note that $i^* = k^* \circ sp$, where $sp : H_*^{BM}(Z) \to H_*^{BM}(C_X Z)$ is the homology specialization map (cf. [Ve1] and [Fl, Ch. 5]). Hence

$$i^* c_*(\alpha) = k^* sp\, c_*(\alpha)\,.$$

We state without a proof the following crucial property:

Theorem 5.3. *Let $X \subset Z$ be a closed embedding and let $\alpha \in CF(Z)$. Then*

$$sp(c_*(\alpha)) = c_*(sp(\alpha)) \in H_*^{BM}(C_X Z)\,.$$

(For a sketch of the proof and discussion of a general context of this theorem, due essentially to Verdier [Ve1], we refer to [Sch2, Thm. 1.1].)

According to 5.3 we rewrite the preceding formula

$$i^* c_*(\alpha) = k^* c_* sp(\alpha)\,.$$

Corollary 5.4. *The Milnor class can be expressed as*

$$\mathcal{M}(X \subset Z; \alpha) = c^*(N_X Z)^{-1} \cap k^* c_*(\Phi_i(\alpha))\,,$$

where

$$\Phi_i : CF(Z) \to CF_{mon}(C_X Z)\,,$$
$$\Phi_i(\alpha) = sp(\alpha) - \pi^* \circ i^*(\alpha) \in CF_{mon}(C_X Z)\,.$$

Note that $k^*\Phi_i(\alpha) = 0$.

We will now show that $\Phi_i(\alpha)$ is a generalization of the vanishing cycle operation ϕ_f as defined in 2.4.8 [2]. Assume that $X \subset Z$ is a regular embedding of codimension 1. Then X is locally described by a function $f : Z \to A^1$, i.e., $X = f^{-1}(0)$. Let us consider the deformation space M and the map $g : M \to A^1$ such that $sp(\alpha) = \psi_g(\tilde{\pi}^*\alpha)$, see 2.4.9. The function f determines a section s of $\tilde{\pi}$. The image of s does not intersect the zero-section. We have $gs = f$.

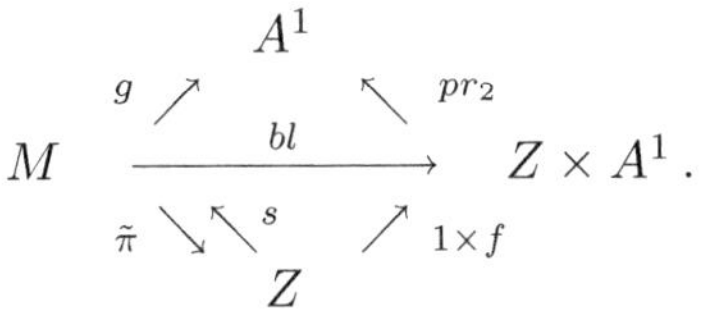

The Milnor fiber of g at points away from the zero-section of $N_X Z$ is the product of the Milnor fiber of f and the disk. Thus

$$s^* sp(\alpha) = s^* \psi_g(\pi^*(\alpha)) = \psi_{gs}(\alpha) = \psi_f(\alpha)\,,$$

and

$$s^*(\Phi_i(\alpha)) = \psi_f(\alpha) - i^*(\alpha) = \phi_f(\alpha)\,.$$

[2] This paragraph is added by PP-AW.

In particular, the formula from Corollary 5.4 gives a generalization of Parusiński-Pragacz's formula, [PaPr2], for the Milnor class of a hypersurface, to the higher codimension case. In fact, in [PaPr2], the formula is stated equivalently in terms of a Whitney stratification of the hypersurface, i.e., in the form conjectured by Yokura, generalizing an earlier work of Parusiński-Pragacz [PaPr1] on the Euler characteristic of a singular hypersurface. The method of [PaPr2] relies on the computation of suitable characteristic cycles [LeMe], [BMM], whereas our approach is based on Verdier's specialization [Ve1, Ve2], [Sch2].

5.5. Another view on Milnor class

Assume, as before, that codim $X = 1$. Then monodromic functions in $N_X Z \setminus X$ are determined by functions on X. Any section of $N_X Z \setminus X$ gives us a local identification

$$CF(X) \simeq CF_{mon}(C_X Z \setminus X) .$$

In this case the vanishing cycle transformation passes to a map

$$\mu_i : CF(Z) \to CF(X) .$$

It satisfies the formula

$$\Phi_i(\alpha) = \pi^* \mu_i(\alpha) - k_* \mu_i(\alpha) .$$

Then

$$\begin{aligned} k^* c_*(\Phi_i(\alpha)) &= k^* c_*(\pi^* \mu_i(\alpha) - k_*(\mu_i(\alpha))) \\ &= k^* c_* \pi^* \mu_i(\alpha) - k^* c_* k_* \mu_i(\alpha) \\ &= k^*(c^*(T_\pi) \cap \pi^* c_* \mu_i(\alpha)) - k^* k_* c_* \mu_i(\alpha) \\ &= c^*(N_X Z) \cap c_* \mu_i(\alpha) - c^1(N_X Z) \cap c_* \mu_i(\alpha) \\ &= (c^*(N_X Z) - c^1(N_X Z)) \cap c_*(\mu_i(\alpha)) = c_*(\mu_i(\alpha)) . \end{aligned}$$

Hence

$$\mathcal{M}(X \subset Z; \alpha) = c^*(N_X Z)^{-1} \cap k^* c_*(\Phi_i(\alpha)) = c^*(N_X Z)^{-1} \cap c_*(\mu_i(\alpha)) .$$

For higher codimension complete intersection there exists a constructible function $\mu_i(\alpha) \in CF(X)$ such, that

$$\Phi_i(\alpha) = \pi^* \mu_i(\alpha) - k_* \mu_i(\alpha) + \text{correction terms.}$$

The correction terms are supported in lower dimensions. Then

$$k^* c_* \Phi_i(\alpha) = (c^*(N_X Z) - c^{\text{top}}(N_X Z)) \cap c_*(\mu_i(\alpha)) + \text{correction terms}$$

and

$$\begin{aligned} \mathcal{M}(X \subset Z; \alpha) &= c^*(N_X Z)^{-1} \cap k^* c_*(\Phi_i(\alpha)) \\ &= c^*(N_X Z)^{-1} \cap (c^* N_X Z - c^{top} N_X Z) \cap c_*(\mu_i(\alpha)) + \text{correction terms.} \end{aligned}$$

This matter and some related topics are discussed in the two letters of J. Schürmann, reproduced in the following appendix.

6. Appendix: Two letters of J. Schürmann

Letter addressed to Brasselet, Lehmann, Seade, and Suwa, dated December 2001:

Dear colleagues,

I have now looked in more detail at the more complete version of your paper "Milnor-classes of local complete intersections" [BLSS], and I have thought a little bit more about a question of Prof. Suwa.

First I should say that my approach is very difficult for explicit calculations (but it fits better with functorial properties, and in the algebraic context it gives (localized) Milnor-classes in the Chow group). I think a complete comparison of my formula with your corollary 5.13 seems to be very difficult, but I think at least some parts of your formula are related to my approach:

Let me follow the notations of page 18 of [Sch2]: i.e., S_α is a connected component of the singular locus of X, S'_α is an irreducible component of S_α and let me denote the "generic value" of $\Phi_i(1_Y)$ restricted to $N_X Y|S'_\alpha$ by μ'_α. The description

$$\Phi_i(1_Y) = \sum \mu'_\alpha(\pi^* - k_*)(1_{S'_\alpha}) + \beta\,,$$

with β some constructible function which is generically vanishing (and which induces in some sense a correction-term in the calculation of the localized Milnor-class). The sum is over all irreducible components of the singular locus of X, with $\pi : N_X Y \to X$ the projection of the normal bundle and $k : X \to N_X Y$ the zero-section.

So for the constructible function $\mu'_\alpha(\pi^* - k_*)(1_{S'_\alpha})$ one can make the same calculation as on page 8 of [Sch2] (for the codimension one case), and gets by the self-intersection formula the equation

$$A := k^* c_*(\mu'_\alpha(\pi^* - k_*)(1_{S'_\alpha})) = \mu'_\alpha(c^*(N) - c^{top}(N)) \cap c_*(1_{S'_\alpha})\,.$$

So one sees that one gets from each irreducible component the contribution $c^*(N)^{-1} \cap A$ to the (localized) Milnor-class (plus a correction-term coming from β). In particular, for a smooth component S of the singular locus this is just $\mu'_\alpha(c^*(N) - c^{top}(N))\; c^*(N)^{-1}\; c^*(S) \cap [S]$, i.e., the first term of the formula of your corollary 5.13, with the identification of μ'_α as the Milnor-number of a generic transversal slice $X \cap H$!

Here I would expect that one can deduce this calculation of μ'_α from my general approach by some "generic base-change" argument, but the details of this have to be worked out!

To finish this letter I would like to ask you if your calculation of this generic value would also work without the assumption that S is smooth, i.e., at generic points of the smooth part of the singular locus? (so that one gets in this way the top-dimensional "localized Milnor-class" as in the last formula of your corollary 5.13, without the assumption that S is smooth, similarly to the remark on page 18 of [Sch2]).

Best wishes,

Jörg Schürmann

Letter addressed to Aluffi, Brasselet, Kennedy, Pragacz, Suwa, and Weber, dated May 7, 2002:

Dear colleagues,

I have now thought about some questions asked at the conference in Warsaw about my Verdier-Riemann-Roch theorem for Chern (Schwartz-MacPherson) classes.

Enclosed are some remarks and also a question:

(a) Consider a regular embedding $X \to Y$ of spaces and the Verdier specialization for constructible functions $sp : CF(Y) \to CF_{mon}(C_X Y)$. Fix $\alpha \in CF(Y)$, and denote by $sp(\alpha)_{gen} \in CF(X)$ the unique constructible function such that $sp(\alpha) = \pi^*(sp(\alpha)_{gen})$ on a Zariski-dense (conic) open subset of $C_X Y$, with $\pi : C_X Y \to X$ the projection of the normal cone= normal bundle (i.e., write $sp(\alpha)$ as a (locally finite) sum of terms m_i times indicator functions of closed irreducible subcones C_i. Then $sp(\alpha)_{gen}$ is the sum of m_i times the indicator function of $\pi(C_i)$, where we only consider those C_i for which $C_i = \pi^{-1}(\pi(C_i))$).

If the regular embedding is of codimension one, given locally by one equation $g = 0$, then one knows $sp(\alpha)_{gen} = \psi_g(\alpha)$ as in (SP6) of [Sch2] (with ψ_g the nearby cycle functor of Deligne on the level of constructible functions). Here is now the generalization to the higher codimension case: Suppose X is locally given by $f_1 = \ldots = f_n = 0$ for a regular sequence $f_1, \ldots, f_n$. Then one gets:

$$sp(\alpha)_{gen} = \psi_{f_n}(\ldots(\psi_{f_1}(\alpha))\ldots) \tag{1}$$

with the right iterated nearby cycle functor as in [McPa2].

The proof of (1) is by induction similarly to an argument of Verdier on p. 204 in Astérisque 36–37 (1976) !

(b) The formula (1) shows in particular, that this iterated nearby cycle functor (on the level of constructible functions) does not (!!) depend on the order of the local tuple $f_1, \ldots, f_n$ (as in the more general context studied by McCrory-Parusiński). So in the case of a regular embedding one can speak of the Euler characteristic of "the Milnor-fiber" of the function $f = (f_1, \ldots, f_n)$ (weighted by α), which for an isolated singularity (of Y and f) reduces to the usual notion! In particular, this gives a proof of my claim on p.18 of [Sch2] that for isolated complete intersection singularities the localized Milnor-class is given by the "usual Milnor-number" (up to signs) as in the paper [Suw] of Suwa!

(c) This (iterated) nearby cycle functor commutes with restriction to transversal slices (with respect to suitable Whitney-stratification) as for example proved in Part V, lemma 3.5 of my book [Sch1] In particular, I get for the generic value μ_α on p.18 of [Sch2] the local description: (Euler-characteristic of the Milnor-fiber of f restricted to a suitable normal slice) -1 . And this implies my expected formula for the top localized Milnor-class. In particular, my top localized Milnor-class is the same as the one of [BLSS] !

(d) There is one important additional case, where one gets more information on $sp(\alpha)$, and not only on its generic value (which I have not mentioned in

Warsaw). This comes from the fact that sp for constructible functions commutes with exterior products (!!). So one can for example describe completely $sp(1_Y)$ for a cartesian product of hypersurfaces (and then my approach to Milnor-classes implies the formulas of Ohmoto-Yokura [OhYo] for this special case !

Consider two closed embeddings $X \to Y$ and $X' \to Y'$. Then one has $C_X Y \times C_{X'} Y' = C_{X \times X'}(Y \times Y')$ and

$$sp(\alpha) \times sp(\beta) = sp(\alpha \times \beta)! \tag{2}$$

For the proof of (2) I use that ψ_f commutes with fiber-products (which follows easily from corresponding Milnor-fibrations and the Künneth-formula), and the following property. Let $M(X,Y), M(X',Y')$ and $M(X \times X', Y \times Y')$ be the corresponding deformation spaces to the normal cones. Then I can define a natural morphism:

$$M(X \times X', Y \times Y') \to M(X,Y) \times_{A^1} M(X',Y') \tag{3}$$

of the first deformation space to the fiber-product of the other ones (which is an isomorphism of the underlying sets) such that over $\{0\}$ one gets the natural isomorphism $C_{X \times X'}(Y \times Y') = C_X Y \times C_{X'} Y'$, and over $A^* = A^1 \setminus \{0\}$ the natural isomorphism

$$Y \times Y' \times A^* = (Y \times A^*) \times_{A^*} (Y' \times A^*).$$

This natural map (3) is induced by the ruled join construction of the two blow-up spaces

$$Bl_{X \times \{0\}}(Y \times A^1) \times Bl_{X' \times \{0\}}(Y' \times A^1) \leftarrow \text{Join}$$

and the natural closed embedding

$$Bl_{X \times X' \times \{0\}}(Y \times Y' \times A^1) \to \text{Join}\,.$$

Here I would expect that this map (3) is an isomorphism, but I didn't checked this (since it is not important for my argument).

References

[Al1] P. Aluffi: *Chern classes for singular hypersurfaces.* Trans. A.M.S. 351(10), 3949–4026 (1999).

[Al2] P. Aluffi: *Shadows of blow-up algebras.* preprint. math.AG/0201228, to appear in Tôhoku Math. J.

[BFM] P. Baum, W. Fulton, R. MacPherson: *Riemann-Roch for singular varieties.* Publ. Math. I.H.E.S., 45, 101–145 (1975).

[Br] J.-P. Brasselet: *Existence des classes de Chern en théorie bivariante.* Astérisque 101–102, 7–22 (1981).

[BLSS] J.-P. Brasselet, D. Lehmann, J. Seade, T. Suwa: *Milnor classes of local complete intersections.* Trans. A.M.S. 354(4), 1351–1371 (2002).

[BrSw] J.-P. Brasselet, M.-H. Schwartz: *Sur les classes de Chern d'un ensemble analytique complexe.* Astérisque 82–83, 93–147 (1981).

[BMM] J. Briançon, P. Maisonobe, M. Merle: *Localisation de systèmes différentiels, stratifications de Whitney et condition de Thom.* Inv. Math. 117, 531–550 (1994).

[BDK] J.-L. Brylinski, A. Dubson, M. Kashiwara: *Formule d'indice pour les modules holonomes et obstruction d'Euler locale.* C. R. Acad. Sci. Paris 293, 573–576 (1981).

[Ch] S.S. Chern: *Characteristic classes of Hermitian manifolds.* Ann. of Math. 47, 85–121 (1946).

[Du] A. Dubson: *Calcul des Invariants numériques des singularités et applications.* preprint SFB Theor. Math. Bonn (1981).

[Fu] J.H.G. Fu: *Curvature measures and Chern classes of singular varieties.* J. Differ. Geom. 39, 251–280 (1994).

[FuMc] J.H.G. Fu, C. McCrory: *Stiefel-Whitney classes and the conormal cycle of a singular variety.* Trans. A.M.S. 349, 809–835 (1997).

[Fl] W. Fulton: *Intersection Theory.* Springer-Verlag, Berlin (1984).

[FlMa] W. Fulton, R. MacPherson: *Categorical framework for study of singular spaces.* Mem. A.M.S. 243, chap. 6 (1981).

[Gi] V. Ginzburg: *Characteristic cycles and vanishing cycles.* Inv. Math. 84, 327–402 (1986).

[G-S] G. González-Sprinberg: *L'obstruction locale d'Euler et le théorème de MacPherson.* Astérisque 82–83, 7–32 (1981).

[GoMa] M. Goresky, R. MacPherson: *Stratified Morse theory.* Springer-Verlag (1988).

[GrMa] M. Grinberg, R. MacPherson: *Euler characteristics and Lagrangian intersections.* In: "Symplectic geometry and topology", IAS/Park City Math. Ser. 7, A.M.S., 265–293 (1999).

[Gr] A. Grothendieck: *Récoltes et Semailles*, Preprint, Montpellier (1984–86).

[Ka1] M. Kashiwara: *Index theorem for a Maximally Overdetermined System of Linear Differential Equations.* Proc. Japan Acad. 49, 803–804 (1973).

[Ka2] M. Kashiwara: *Systems of Microdifferential Equations.* Progress in math. 34, Birkhäuser-Verlag.

[KaSp] M. Kashiwara, P. Schapira: *Sheaves on manifolds.* Springer-Verlag (1990).

[Ke] G. Kennedy: *MacPherson's Chern classes of singular algebraic varieties.* Comm. in Alg. 18, 2821–2839 (1990).

[Kw] M. Kwieciński: *Formule du produit pour les classes caractéristiques de Chern-Schwartz-MacPherson et homologie d'intersection.* C. R. Acad. Sci. Paris Sér. I Math. 314 no.8, 625–628 (1992).

[La] G. Laumon: *Comparaison de characteristiques d'Euler-Poincaré en cohomogie l-adique.* C. R. Acad. Sci. Paris 292, 209–212 (1981).

[LeMe] D.T. Lê, Z. Mebkhout: *Variétés caractéristiques et variétés polaires* C. R. Acad. Sci. Paris 296, 129–132 (1983).

[Ma] R. MacPherson: *Chern classes for singular algebraic varieties.* Ann. of Math. 100, 423–432 (1974).

[Mat] J.N. Mather *Stratifications and mappings.* In: "Dynamical systems" Proc. Sympos., Univ. Bahia, Salvador, Academic Press, 195–232 (1973).

[McPa1] C. McCrory, A. Parusiński: *Algebraically constructible functions.* Ann. Sci. Ec. Norm. Sup. 30, 527–552 (1997).

[McPa2] C. McCrory, A. Parusiński: *Complex monodromy and the topology of real algebraic sets.* Comp. Math. 106, 211–233 (1997).

[MiSt] J. Milnor, J. Stasheff: *Characteristic classes.* Princeton University Press (1974).

[OhYo] T. Ohmoto, S. Yokura: *Product formulas for the Milnor class.* Bull. Pol. Acad. Sci. Math. 48, 387–401 (2000).

[PaPr1] A. Parusiński, P. Pragacz: *A formula for the Euler characteristic of singular hypersurfaces.* J. Algebr. Geom. 4, 337–351 (1995).

[PaPr2] A. Parusiński, P. Pragacz: *Characteristic classes of hypersurfaces and characteristic cycles.* J. Algebr. Geom. 10, 63–79 (2001).

[Sa] C. Sabbah: *Quelques remarques sur la géométrie des espaces conormaux.* Astérisque 130, 161–192 (1985).

[Sp] P. Schapira: *Operations on constructible functions.* J. Pure Appl. Algebra 72, 83–93 (1991).

[ScVi] W. Schmid, K. Vilonen: *Characteristic cycles of constructible sheaves.* Inv. Math. 124, 451–502 (1996).

[Sch1] J. Schürmann: *Topology of singular spaces and constructible sheaves.* "Monografie Matematyczne" New Series vol. 63, Birkhäuser-Verlag (2003).

[Sch2] J. Schürmann: *A generalized Verdier-type Riemann-Roch theorem for Chern-Schwartz-MacPherson classes*, preprint. math.AG/0202175.

[Su] D. Sullivan: *Combinatorial invariants of analytic spaces.* In: "Proceedings of Liverpool Singularities Symposium I", Springer LNM 192, 165–168 (1971).

[Suw] T. Suwa: *Classes de Chern des intersections complètes locales* C. R. Acad. Sci. Paris 324, 67–70 (1996).

[Sw] M.-H. Schwartz: *Classes caractéristiques définies par une stratification d'une variété analytique complexe.* C. R. Acad. Sci. Paris 260, 3262–3264 (1965).

[Ve1] J.-L. Verdier: *Spécialisation des classes de Chern.* Astérisque 82-83, 149–159 (1981).

[Ve2] J.-L. Verdier: *Spécialisation de faisceaux et monodromie modérée.* Astérisque 101-102, 332–364 (1983).

[Yo] S. Yokura: *On a Verdier-type Riemann-Roch for Chern-Schwartz-MacPherson class.* Topology and its Appl. 94, 315–327 (1999).

Jörg Schürmann
Westf. Wilhelms-Universität
SFB 478 "Geometrische Strukturen in der Mathematik"
Hittorfstr. 27
D 48149 Münster, Germany
e-mail: jschuerm@math.uni-muenster.de

Trends in Mathematics:
Topics in Cohomological Studies of Algebraic Varieties, 203–270

Algebraic K-theory of Schemes

Marek Szyjewski

Abstract. The present notes contain the foundations of algebraic K-theory together with a series of explicit computations of the K-groups of fields and some classical varieties. Our goal is to provide an introduction to a more advanced reading, as well as to convince the reader that such a study may be useful and interesting. The exposition is by no means complete nor self-contained. We hope nevertheless, that the covered part of the theory is sufficient for effective computations in algebraic geometry.

The organization of these notes follows the historical development of algebraic K-theory in the 2nd half of the XXth century. We give a brief outline of the theory of the Grothendieck groups $K_0(X)$, $K_0'(X)$. We also develop the higher K-theory (of Milnor and Quillen) of fields and compute the K-groups of finite fields. Next, the Quillen's definition of K-groups as homotopy groups is given and their properties are discussed. Some instructive examples are also included. Moreover, we compute the higher algebraic K-groups of projective bundles, Brauer-Severi varieties and quadrics. In the end we apply these techniques to compute the Chow ring of a split quadric and to prove "Hilbert 90" for $K_2\left(F(\sqrt{a})\right)$ following Merkurjev's proof.

1. Introduction

Following [30, Introduction], we remind that algebraic K-theory has two components: the classical theory and the higher one. The classical theory centers around the Grothendieck group K_0 of a category and uses algebraic presentations, while the higher algebraic K-theory requires topological or homological machinery to define. We introduce as much of this machinery as is necessary to state definitions, and no more.

For the Grothendieck group K_0 we use the approach with relations for exact sequences, which is the version used in algebraic geometry. Thus we omit other versions: group completion construction in symmetric monoidal categories (e.g., projective modules over a ring or vector bundles over compact spaces) and the one with relations for weak equivalences in Waldhausen categories (e.g., category of chain complexes).

There are four basic constructions for higher algebraic K-theory: the $+$-construction for rings, Quillen's Q-construction for exact categories, the group completion construction (for symmetric monoidal categories), and Waldhausen's wS. construction (for categories with cofibrations and weak equivalences). All these constructions give the same K-theory of a ring, but are useful in various distinct settings. To avoid topological details, we use Q-construction as the fundamental one; the $+$-construction is only mentioned in Section 3.

The organization of these notes follows the historical development of algebraic K-theory in the 2nd half of the XXth century. Section 2 contains a brief outline of the theory of the Grothendieck groups $K_0(X)$, $K_0'(X)$: definitions, multiplicative structure, contravariant properties, Cartan map, the Fundamental Theorem (homotopy property), localization, cycle map, filtrations, and a proof of the Projective Bundle Theorem for K_0.

In Section 3 we develop the higher K-theory of rings, notably (of Milnor and Quillen) of fields and compute the K-groups of finite fields. Note that if X is a variety over a field F, then "compute the higher K-theory of X" means "express $K_n(X)$ in terms of $K_m(A_i)$ for some F-algebras $A_1, \dots, A_r$", e.g.,

- for affine space $X = \mathbb{A}_F^r$
$$K_n(X) = K_n(F);$$
for a complement $Y = \mathbb{A}_F^r \backslash \mathbb{A}_F^{r-1}$ to hyperplane
$$K_n(Y) = K_n(F) \oplus K_{n-1}(F);$$
(Fundamental Theorem 15);
- for projective space $X = \mathbb{P}_F^r$
$$K_n(X) \cong K_n(F)^{r+1}$$
(Projective Bundle Theorem 33);
- for d-dimensional smooth quadric hypersurface X given by equation $q = 0$ in $\mathbb{P}_F^{d+1}$
$$K_n(X) \cong K_n(F)^d \oplus K_n(C_0(q)),$$
where $C_0(q)$ is the even part of the Clifford algebra of the quadratic form q (Swan Theorem 46);
- for a Brauer-Severi variety X corresponding to central simple algebra A of dimension r^2
$$K_n(X) \cong K_n(F) \oplus K_n(A) \oplus K_n(A^{\otimes 2}) \oplus \cdots \oplus K_n(A^{\otimes r-1})$$
(Theorem 34).

Thus using higher algebraic K-theory of a variety X one should know what is $K_n(A)$ for $n > 0$.

The next two sections are based on Quillen's fundamental paper [24]: the definition of K-groups as homotopy groups of the Q-construction is given and their properties are discussed. These are functoriality with respect to exact functors, reduction by resolution, devissage and localization (Theorem 14). Some instructive examples not contained in [24] are also included. In Section 5 we discuss connection between K'-theory and Chow groups. The Brown-Gersten-Quillen spectral sequence

$$E_1^{p,q}(X) = \coprod_{x \in X_p} K_{-p-q}\left(k(x)\right) \Longrightarrow K'_{-n}(X)$$

(Theorem 20) and $\mathcal{K}'$-cohomology

$$E_2^{p,q}(X) = H^p(X, \mathcal{K}'_{-q})$$

(Proposition 23) appear to be efficient tools for explicit computations. The abovementioned connection with Chow groups is $H^p(X, \mathcal{K}'_p) = E_2^{p,-p}(X) = CH^p(X)$ (Theorem 21). Here also we include instructive examples. In Section 6 we compute the higher algebraic K-groups of projective bundles, Brauer-Severi varieties (after [24]) and quadrics (after Swan [28]). In the last section we apply these techniques to compute the Chow ring of a split quadric (here we follow the author's approach [29]) and to prove "Hilbert 90" for $K_2\left(F(\sqrt{a})\right)$ following Merkurjev's proof.

These notes are slightly expanded version of a series of lectures at Algebraic Geometry, Algebra and Applications Conference and Summer School in September 2003, held in Borovetz (Bulgaria). I am grateful to the organizers of this conference for their invitation. I am also grateful to Grzegorz Banaszak, Piotr Krasoń and Piotr Pragacz, who read carefully the manuscript and helped to convert it into a paper.

2. Grothendieck groups

There are numerous connections between properties of a scheme/variety X and properties of the categories of sheaves of $\mathcal{O}_X$-modules on X. We are interested in two categories:

- the category $\mathcal{M}(X)$ of coherent $\mathcal{O}_X$-modules on X,
- the category $\mathcal{P}(X)$ of vector bundles (locally free $\mathcal{O}_X$-modules of finite rank) on X.

If $X = \operatorname{Spec}(R)$ is an affine scheme, the first one is equivalent to the category of finitely generated R-modules, while the second one is equivalent to its subcategory of projective R-modules.

In the 50's of the XXth century Alexander Grothendieck studied additive functions on these categories: for an abelian group G, a function $f : \mathcal{C} \to G$ is called *additive*, iff for every exact sequence

$$0 \to A \to B \to C \to 0$$

in $\mathcal{C}$ the equality

$$f(B) = f(A) + f(C)$$

holds.

In 1957, in the manuscript [8] along with an algebraic proof of (a generalized) Riemann-Roch-Hirzebruch Theorem, Grothendieck introduced a universal target groups for additive functions. Given an additive category $\mathcal{C}$ embedded in an abelian category $\mathcal{A}$, the *Grothendieck group* $K_0(\mathcal{C})$ of $\mathcal{C}$ is a factor of the free abelian group generated by isomorphic classes $[A]$ of objects of $\mathcal{C}$ modulo subgroup generated by all expressions

$$[B] - [A] - [C]$$

for every sequence

$$0 \to A \to B \to C \to 0$$

in $\mathcal{C}$, which is exact in $\mathcal{A}$. Thus $K_0(\mathcal{C})$ has generators: one generator $[A]$ for each $A \in \mathrm{Ob}(\mathcal{C})$ and every element of $K_0(\mathcal{C})$ may be expressed in the form $[A] - [B]$ (in many ways). Moreover, split sequences are exact independently of the embedding in an abelian category, so the addition rule may be written as

$$[A] + [B] = [A \oplus B].$$

In this way, to a scheme X there are associated two K-groups:

$$\begin{aligned} G_0(X) \text{ or } K_0'(X) &= K_0(\mathcal{M}(X)), \\ K_0(X) &= K_0(\mathcal{P}(X)) \end{aligned}$$

(this is modern notation; Grothendieck's was $K_.(X)$ and $K^{\cdot}(X)$ - see eg. [3]. Moreover, notation $K_0(X)$ and $K^0(X)$ is in use, e.g., [6], or even $K(X)$ and $K_1(X)$ - [10].) These groups have generators $[A]$ for $A \in \mathrm{Ob}(\mathcal{M}(X))$ (resp. $A \in \mathrm{Ob}(\mathcal{P}(X))$) which are subject to relations

$$[B] = [A] + [C]$$

associated with sequences $0 \to A \to B \to C \to 0$ which are exact in $\mathcal{M}(X)$.

The function

$$A \longmapsto [A]$$

from $\mathrm{Ob}(\mathcal{M}(X))$ (resp. $A \in \mathrm{Ob}(\mathcal{P}(X))$) to $K_0'(X)$ (resp. $K_0(X)$) is universal additive function, in the sense that every additive function factors through it.

In the affine case, $X = \mathrm{Spec}(R)$, it is customary to simplify the notation

$$\begin{aligned} K_0'(\mathrm{Spec}(R)) &= K_0'(R), \\ K_0(\mathrm{Spec}(R)) &= K_0(R). \end{aligned}$$

It is known that it is impossible to classify vector bundles up to isomorphism on all varieties; in contrast, in many cases one can compute K-groups.

Example 1. *If $R = F$ is a field ($X = \mathrm{Spec}(F)$ is a point), then every finitely generated F-module is free, and these modules are classified by dimension. Thus*

$$K_0'(F) = K_0(F) \cong \mathbb{Z}$$

and the isomorphism is induced by the additive function $A \longmapsto \dim A$.

Example 2. *If $R = D$ is a skew-field (a division ring), then finitely generated left D-modules are free. If A is such a module, then every basis of A has the same number of elements. So two finitely generated D-modules are isomorphic iff they have the same rank (or dimension). It follows that*

$$K_0'(D) = K_0(D) \cong \mathbb{Z}$$

and the isomorphism is induced by the additive function $A \longmapsto \dim A$.

Exercise 1. *Prove that for a (not necessarily commutative) ring R and finitely generated projective left R-modules A, B*

$$[A] = [B] \text{ in } K_0(R) \text{ iff there exists}$$
$$\text{a f.g. projective } C \text{ such that } A \oplus C \cong B \oplus C.$$

Exercise 2. *Prove that, in general, for a category $\mathcal{C}$ embedded in an abelian category $\mathcal{A}$,*

$$[A] = [B] \text{ in } K_0(\mathcal{C}) \text{ iff there exist objects } C', C, C'' \text{ and exact sequences}$$
$$0 \to C' \to C \to A \oplus C'' \to 0 \text{ and } 0 \to C' \to C \to B \oplus C'' \to 0.$$

2.1. Definitions, multiplicative structure, contravariant properties, Cartan map

If $\mathcal{F}$ is a vector bundle, then it is easy to check locally that

$$\mathcal{F} \otimes - : \mathcal{M}(X) \to \mathcal{M}(X)$$

is an exact functor (takes exact sequences to exact sequences). In fact, one has:

Proposition 1. *The formula*

$$[A] \cdot [B] = [A \otimes B]$$

defines a multiplication rules

$$\begin{array}{rcl} K_0(X) \otimes_{\mathbb{Z}} K_0(X) & \longrightarrow & K_0(X), \\ K_0(X) \otimes_{\mathbb{Z}} K_0'(X) & \longrightarrow & K_0'(X). \end{array}$$

With this multiplication rule, $K_0(X)$ is a ring with unit element $1 = [\mathcal{O}_X]$, and $K_0'(X)$ is a $K_0(X)$-module. □

Example 3. *For a field F, the map $\dim : K_0(X) \to \mathbb{Z}$ is a ring isomorphism.*

In the "categorical" context to study dependence of K-groups on functors one must restrict to additive functors which preserve exactness of sequences (exact functors). To do this we will introduce the notion of exact category below. In our set-up, where exactness is defined by embedding into an abelian category, one must use exact functors defined on ambient abelian categories. It is obvious, that such a functor $f : \mathcal{A}_1 \to \mathcal{A}_2$ which takes $\mathcal{C}_1$ to $\mathcal{C}_2$ defines a homomorphism

$$f : K_0(\mathcal{C}_1) \to K_0(\mathcal{C}_2).$$

Example 4. *The inclusion functor $\mathcal{P}(X) \to \mathcal{M}(X)$ preserves exactness. Thus there is a homomorphism*

$$K_0(X) \longrightarrow K_0'(X)$$

which acts naturally on generators ($[A] \mapsto [A]$.) This is a homomorpism of $K_0(X)$-modules.

This homomorphism is called the *Cartan homomorphism.*

Example 5. *If $f : X \to Y$ is a morphism of schemes, then the pull-back functor $f^* : \mathcal{M}(Y) \to \mathcal{M}(X)$ is exact and takes $\mathcal{P}(Y)$ to $\mathcal{P}(X)$. Thus K_0 and K_0' are functors from schemes to abelian groups.*

Example 6. *(Reduction by resolution) We keep the notation following Example 3. Assume that:*

1) *for every object A in $\mathcal{C}_2$ there is a finite resolution of A by objects of the form $f(B)$, $B \in \mathcal{C}_1$,*
2) *if for $A, C \in \mathrm{Ob}(\mathcal{C}_2)$ there is a positive number n and an exact sequence*

$$0 \to C \to f(B_{n-1}) \to \cdots \to f(B_0) \to A \to 0,$$

then there exists $B_n \in \mathrm{Ob}(\mathcal{C}_1)$ such that $C \cong f(B_n)$.

Then the homomorphism $f : K_0(\mathcal{C}_1) \to K_0(\mathcal{C}_2)$ is surjective. In fact, if there is an exact sequence

$$0 \to f(B_n) \to \cdots \to f(B_0) \to A \to 0,$$

then to show that

$$[A] = \sum_i (-1)^i f[B_i]$$

one needs only a standard inductive argument.

Example 7. *Consider $X = \mathrm{Spec}(\mathbb{Z})$. Locally free abelian groups are simply free abelian groups, so the additive function* rank *defines a ring isomorphism $K_0(\mathbb{Z}) \cong \mathbb{Z}$. Every finitely generated abelian group A has a resolution*

$$0 \to B_1 \to B_0 \to A \to 0$$

with free abelian groups B_0, B_1. Thus the inclusion functor $\mathcal{P}(\mathbb{Z}) \to \mathcal{M}(\mathbb{Z})$ induces a surjective homomorphism $K_0(\mathbb{Z}) \to K_0'(\mathbb{Z})$ which again splits by means of rank function. Thus $K_0'(\mathbb{Z}) = K_0(\mathbb{Z}) \cong \mathbb{Z}$.

Another example: if $f : X \to Y$ is a proper morphism of locally noetherian schemes and every object in $\mathcal{M}(X)$ has a finite resolution by coherent sheaves with no higher derived images, then there is a well-defined group homomorphism

$$f_* : K_0'(X) \to K_0'(Y)$$

such that $f_*[\mathcal{F}] = [f_*\mathcal{F}]$ provided $\mathcal{F}$ has no higher derived images.

Proposition 2. *If X is a separated noetherian regular scheme, then the Cartan map $K_0(X) \longrightarrow K_0'(X)$ is an isomorphism.* □

2.2. Fundamental Theorem (homotopy property), localization, cycle map, filtrations

In the affine case $X = \mathrm{Spec}(R)$, where R is a commutative ring, every element of $K_0(R)$ may be written in the following special form:

$$[A] - [B] = [A \oplus C] - [B \oplus C] = [R^n] - [B \oplus C] = n - [B \oplus C]$$

where C is such that $A \oplus C \cong R^n$. Moreover, an equality $[A] = [B]$ is equivalent to existence of a C such that $A \oplus C \cong B \oplus C$ (Exercise 1), and C may be chosen to be a free module (we say that A and B are *stably isomorphic*). The class $[R^1]$ of rank 1 free module is the unit of the ring $K_0(R)$, so $K_0(R) = \mathbb{Z}$ means that R^1 is a generating module and every finitely generated projective R-module is stably free. It is not very difficult to prove that every finitely generated projective module over a polynomial ring $F[x_1, x_2, \ldots, x_n]$ is stably free, so $K_0(F[x_1, x_2, \ldots, x_n]) = \mathbb{Z}$. Serre in 1955 posed the question if every finitely generated projective module is in this case actually a free module. This fact was proven independently by Quillen and Suslin in 1973.

A weaker statement – the equality $K_0(F[x_1, x_2, \ldots, x_n]) = \mathbb{Z}$ – has two generalizations.

Theorem 3 (Fundamental Theorem). *If R is a regular ring, then the inverse image functors $M \longmapsto R[t] \otimes_R M$, $M \longmapsto R[t, t^{-1}] \otimes_R M$ induce isomorphisms*

$$K_0(R[t]) = K_0(R) = K_0(R[t, t^{-1}]).$$

If R is a noetherian ring, then analogously

$$K_0'(R[t]) = K_0'(R) = K_0'(R[t, t^{-1}]).$$

Proof. [2, Theorem XII.3.1], [2, Theorem XII.4.1]. □

Theorem 4 (Homotopy property). *If $f : X \to Y$ is a morphism of smooth varieties such that all fibers $f^{-1}(y)$ are affine spaces, then inverse image functor f^* induces a ring isomorphism $K_0(Y) = K_0(X)$.*

Proof. [3, Exposé IX, Prop. 1.6] □

The fact that a (locally free) sheaf on an open subset $U \subset X$ extends to a (locally free) sheaf on whole X can be stated as follows:

Theorem 5 (Localization). *Let U be an open subset of a scheme X and let Z be the reduced closed complement of U in X with the reduced scheme structure. Denote by $j : U \hookrightarrow X$ and $i : Z \hookrightarrow X$ the inclusions. The following sequence is exact:*

$$K_0'(Z) \xrightarrow{i_*} K_0'(X) \xrightarrow{j^*} K_0'(U) \longrightarrow 0$$

Proof. [3, Exposé IX, Prop. 1.1] □

One of the reasons to introduce higher algebraic K-theory was to extend this sequence to a long exact localization sequence. We achieve this in Proposition 18 below.

Every closed subscheme Z of X defines an element $[\mathcal{O}_Z] \in K_0'(X)$. If $\mathcal{Z}^r(X)$ is the group of cycles of codimension r in X, and

$$\mathcal{Z}^\bullet(X) = \bigoplus_{r=0}^{\dim X} \mathcal{Z}^r(X)$$

then there is a group homomorphism

$$\begin{array}{ccc} \mathcal{Z}^\bullet(X) & \to & K_0'(X) \\ [Z] & \longmapsto & [\mathcal{O}_Z], \end{array}$$

called the *cycle map*. In the affine case it is easy to see that the image of the cycle map generates $K_0'(X)$ – every finitely generated R-module has a filtration with factors isomorphic to $R/\mathfrak{p}$ for suitable $\mathfrak{p} \in \operatorname{Spec} R$. This fact remains valid for smooth varieties over a field. Thus the homomorphism $\mathcal{Z}^\bullet(X) \to K_0'(X)$ is surjective ([3, Exposé X, Corollaire 1.1.4].)

There is the topological filtration

$$F^p K_0'(X) = \left\{ \alpha \in K_0'(X) : \begin{array}{c} \alpha \text{ is in the kernel of restriction} \\ K_0'(X) \to K_0'(\mathcal{O}_{X,x}) \text{ to the generic} \\ \text{point } x \text{ of every subvariety of codimension } < p \end{array} \right\},$$

and consecutive factors of this filtration are connected with the cycle groups $\mathcal{Z}^r(X)$.

Multiplication in $K_0(X)$ extends to multiplication in $K_0'(X)$:

$$[\mathcal{F}] \cdot [\mathcal{G}] = \sum (-1)^i \left[\operatorname{Tor}_i^{\mathcal{O}_X}(\mathcal{F}, \mathcal{G})\right].$$

For this multiplication, if cycles Z_1, Z_2 intersect properly, then

$$[\mathcal{O}_{Z_1}] \cdot [\mathcal{O}_{Z_2}] = [\mathcal{O}_{Z_1 \cap Z_2}].$$

2.3. Projective Bundle Theorem

Computation of the Grothendieck group of a projective bundle ([3, Exposé VI]) was crucial for the theory and its applications. We present here a modern proof.

Theorem 6. *If $\mathcal{E}$ is a vector bundle of rank n on a quasicompact scheme Y,*

$$p : X = \mathbb{P}_Y(\mathcal{E}) \to Y$$

is the projective bundle associated to $\mathcal{E}$, $\mathcal{O}_X(-1)$ is the tautological line bundle, then, via $p^* : K_0(Y) \to K_0(X)$, the $K_0(Y)$-module $K_0(X)$ is free with the basis $[\mathcal{O}_X], [\mathcal{O}_X(-1)], \ldots, [\mathcal{O}_X(1-n)]$.

Proof. (cf. [21, Theorem 2.1] or [22, Theorem 2.1]) Define a sheaf $\mathcal{J}$ on X by the exact sequence:

$$0 \to \mathcal{J} \to p^*(\mathcal{E}) \to \mathcal{O}_X(1) \to 0.$$

The sheaf $\mathcal{J}$ is locally free of finite rank, since $p^*(\mathcal{E})$ and $\mathcal{O}_X(1)$ are such. Moreover,

$$\begin{aligned} p_*\mathcal{O}_X(1) &= \mathcal{E}^{\check{}} = \mathrm{Hom}_{\mathcal{O}_\mathbf{Y}}(\mathcal{E},\mathcal{O}_\mathbf{Y}), \\ p_*\mathcal{J}^{\check{}} &= \mathcal{E}, \end{aligned}$$

where $\mathcal{J}^{\check{}} = \mathrm{Hom}_{\mathcal{O}_\mathbf{X}}(\mathcal{J},\mathcal{O}_\mathbf{X})$. Therefore

$$H^0\left(X\times_Y X, \mathcal{O}_X(1)\boxtimes\mathcal{J}^{\check{}}\right) = H^0\left(Y,\mathcal{E}^{\check{}}\otimes\mathcal{E}\right) = \mathrm{End}_Y(\mathcal{E}),$$

where

$$\mathcal{F}\boxtimes\mathcal{G} = p_1^*(\mathcal{F})\otimes_{\mathcal{O}_{X\times_Y X}} p_2^*(\mathcal{G})$$

is a sheaf on $X\times_Y X$, $\mathcal{F},\mathcal{G}$ are sheaves on X, and p_1, p_2 are the projections $X\times_Y X\to X$. Consider the global section s of $\mathcal{O}_X(1)\boxtimes\mathcal{J}^{\check{}}$ which corresponds to the identity endomorphism of $\mathcal{E}$. Since the restriction to the diagonal $\Delta : X\to X\times_Y X$ is locally induced by the evaluation map $\mathcal{E}^{\check{}}\otimes\mathcal{E}\to\mathcal{O}_\mathbf{X}$, the section s vanishes exactly along the diagonal, which yields the Koszul resolution

$$\begin{aligned} 0 &\to \mathcal{O}_X(1-n)\boxtimes\bigwedge^{n-1}\mathcal{J}\to\cdots \\ &\to \mathcal{O}_X(-2)\boxtimes\bigwedge^2\mathcal{J}\to\mathcal{O}_X(-1)\boxtimes\mathcal{J}\to\mathcal{O}_X\boxtimes\mathcal{O}_X\to\mathcal{O}_{\Delta(X)}\to 0. \end{aligned}$$

By the projection formula, for every bundle $\mathcal{F}$ on X

$$\begin{aligned} [\mathcal{F}] &= p_{1*}\left((p_2^*[\mathcal{F}])\cdot\left[\mathcal{O}_{\Delta(X)}\right]\right) = p_{1*}\left((p_2^*[\mathcal{F}])\cdot\sum_{i=0}^{n-1}(-1)^i\left[\mathcal{O}_X(-i)\boxtimes\bigwedge^i\mathcal{J}\right]\right) \\ &= \sum_{i=0}^{n-1}(-1)^i p^*\left(p_*\left(\left[\bigwedge^i\mathcal{J}\right]\cdot[\mathcal{F}]\right)\right)[\mathcal{O}_X(-i)] \\ &= \sum_{i=0}^{n-1}(-1)^i p_*\left(\left[\bigwedge^i\mathcal{J}\right]\cdot[\mathcal{F}]\right)\cdot[\mathcal{O}_X(-i)], \end{aligned}$$

which implies that the map

$$\begin{aligned} \varphi &: \bigoplus_{i=0}^{n-1}K_0(Y)\to K_0(X) \\ \varphi(a_0,\ldots,a_{n-1}) &= \sum_{i=0}^{n-1}a_i\,[\mathcal{O}_X(-i)] \end{aligned}$$

is a surjective homomorphism of $K_0(Y)$-modules.

On the other hand, it is known that

$$\begin{aligned} R^m p_*\mathcal{O}_X(i) &= 0 \text{ for every } m>0, i\geq 0 \\ p_*\mathcal{O}_X(i) &= \mathrm{Sym}^i\,\mathcal{E}^{\check{}} \text{ for every } i\geq 0 \\ R^m p_*\mathcal{O}_X(i) &= 0 \text{ for every } m\geq 0, 1-n\leq i\leq -1. \end{aligned}$$

Thus for the map

$$\begin{aligned}\psi &: K_0(X) \to \bigoplus_{i=0}^{n-1} K_0(Y)\\ \psi([\mathcal{F}]) &= (p_*[\mathcal{F}], p_*[\mathcal{F}(-1)], \ldots, p_*[\mathcal{F}(1-n)]),\end{aligned}$$

the composition $\psi \circ \varphi : K_0(Y)^n \to K_0(Y)^n$ is given by an upper triangular matrix with units on the diagonal, which implies that φ is injective. □

Remark 1. *We shall prove a more general version of the Projective Bundle Theorem in Theorem* 33 *below.*

3. K-theory of fields

3.1. $K_1(F)$ and $K_2(F)$ (Matsumoto Theorem)

With some effort, proper notions of $K_1(R)$, $K_2(R)$ were found even without assumption that R is commutative. Milnor's book [17] is an excellent, easy and self-contained exposition of this part of development of the theory.

The group $\mathrm{GL}_n(R)$ may be identified with a subgroup of $\mathrm{GL}_{n+1}(R)$ of block diagonal matrices with 1 in the right lower corner:

$$A \longmapsto \left(\begin{array}{c|c} A & 0 \\ \hline 0 & 1 \end{array}\right).$$

Definition 1.

$$\begin{aligned}\mathrm{GL}(R) &= \varinjlim \mathrm{GL}_n(R)\\ \mathrm{E}(R) &= \varinjlim \mathrm{E}_n(R)\end{aligned}$$

where $\mathrm{E}_n(R)$ *is a subgroup of* $\mathrm{GL}_n(R)$ *generated by all elementary matrices.*

Recall that:

- a group G is a *perfect group* iff G has trivial center and G is equal to its commutator subgroup,
- a group is perfect iff it has the universal central extension,
- the group $E(R)$ is perfect,

(see [17, § 5])

Definition 2. *The Steinberg group* $\mathrm{St}(R)$ *is a universal central extension of the group* $\mathrm{E}(R)$.

Definition 3.

$$\begin{aligned}K_1(R) &= H_1(\mathrm{GL}(R), \mathbb{Z}) = \mathrm{GL}(R)/[\mathrm{GL}(R), \mathrm{GL}(R)] = \mathrm{GL}(R)/\mathrm{E}(R)\\ K_2(R) &= H_2(\mathrm{E}(R), \mathbb{Z}) = \ker(\mathrm{St}(R) \twoheadrightarrow \mathrm{E}(R))\\ K_3(R) &= H_3(\mathrm{St}(R), \mathbb{Z})\end{aligned}$$

If R is a commutative ring, there is a homomorphism

$$R^* \xrightarrow{\sim} \mathrm{GL}_1(R) \hookrightarrow \mathrm{GL}(R) \twoheadrightarrow K_1(R).$$

If R is a commutative ring, then

$$\det : \mathrm{GL}(R) \longrightarrow R^*$$

factors through $K_1(R)$ and splits the homomorphism $R^* \to K_1(R)$.

The group $K_2(R)$ may be characterized as a kernel of universal central extension of perfect group $\mathrm{E}(R)$, which is a subgroup of $\mathrm{GL}(R)$ generated by (images of) elementary matrices and coincides with the commutator subgroup $[\mathrm{GL}(R), \mathrm{GL}(R)]$.

In the case of a field, it is easy to see that $K_1(F)$ is a multiplicative group of F:

$$\det : K_1(F) \xrightarrow{\sim} F^*.$$

In general, the determinant map defines a splitting epimorphism

$$\det : K_1(R) \twoheadrightarrow R^*$$

and the kernel is usually denoted as $SK_1(R)$.

We will use an additive notation for the group $K_1(F)$. An element corresponding to $a \in F^*$ is denoted by $\{a\}$ (it is not a set!), so

$$\begin{aligned} \{-\} &: F^* \xrightarrow{\sim} K_1(F) \\ \{ab\} &= \{a\} + \{b\} \\ \{1\} &= 0. \end{aligned}$$

Every element of $K_1(R)$ arises from an automorphism of a free R-module. In particular, if E/F is an finite field extension, then there are canonical transfer maps

- $N_{E/F} : K_0(E) \to K_0(F)$ which as a group homomorphism coincides with multiplication by the degree $[E : F]$,
- $N_{E/F} : K_1(E) \to K_1(F)$ which coincides with usual norm, since the usual norm of $x \in E^{\cdot}$ is determinant of multiplication $x \cdot - : E \to E$.

Description of $K_2(F)$ is more complicated. A *Steinberg symbol* with values in an abelian group A is a map $f : F^{*\cdot} \times F^* \to A$ which:

- is bimultiplicative, i.e.,

$$f(ab, c) = f(a, c) + f(b, c), \qquad f(a, bc) = f(a, b) + f(a, c)$$

- vanishes on pairs (a, b) such that $a + b = 1$,

$$f(a, 1 - a) = 0.$$

Matsumoto and (independently) Moore proved that $K_2(F)$ is the group of values of universal Steinberg symbol:

Theorem 7 (Matsumoto).

$$K_2(F) \cong F^* \otimes F^*/S$$

where S is a subgroup of $F^* \otimes F^*$ generated by all $a \otimes b$ such that $a + b = 1$.

For a proof see [17]. We use the notation

$$\{a, b\} = a \otimes b \bmod S.$$

Thus the identities

$$\{ab, c\} = \{a, c\} + \{b, c\}, \qquad \{a, bc\} = \{a, b\} + \{a, c\}, \tag{3.1}$$

$$\{1, a\} = \{a, 1\} = 0, \tag{3.2}$$

$$\{a, 1 - a\} = 0 \text{ for } a \neq 1 \tag{3.3}$$

hold in $K_2(F)$. The map

$$\{-, -\} : F^* \times F^* \longrightarrow K_2(F)$$

is called *the universal Steinberg symbol.*

There is a transfer map $N_{E/F} : K_2(E) \to K_2(F)$, but it is not so easy to write formula for it (see Lemma 61 and Proposition 62 below for the case $(E : F) = 2$.)

It is worth to point out that there are natural multiplications

$$\begin{aligned} K_0(R) \otimes K_i(R) &\to K_i(R), \\ K_1(R) \otimes K_1(R) &\to K_2(R); \end{aligned}$$

in the case of a field the latter is

$$\{a\}\{b\} = \{a, b\}$$

and is "graded commutative" (see Exercise 4 on page 216). For a finite field extension, the projection formula

$$N_{E/F}(\{a\} \cdot \{x\}) = \{a\} \cdot \{N_{E/F}(x)\}$$

for $a \in F^*$, $x \in E^*$, holds.

3.2. Milnor K-theory of fields

The Matsumoto Theorem 7 was a motivation for J. Milnor in 1970 ([16]) to define, for a field F, a graded ring

$$K_*^M(F) = K_0(F) \oplus K_1(F) \oplus K_2(F) \oplus K_3^M(F) \cdots \oplus K_n^M(F) \oplus \cdots$$

which has interesting connections with the theory of quadratic forms and Galois cohomology, as an approximation of K-theory.

Definition 4. *Let $T^\bullet(F)$ be the tensor algebra of a $\mathbb{Z}$-module F^**

$$T^\bullet(F) = \bigoplus_{n=0}^{\infty} (F^*)^{\otimes n}$$

and let I be the two-sided ideal of $T^\bullet(F)$ generated by all expressions $a \otimes b$ such that $a + b = 1$. Then

$$K_\bullet^M(F) \stackrel{\text{df}}{=} T^\bullet(F)/I.$$

The ideal I is homogeneous, so $K_\bullet^M(F)$ is a graded ring. Milnor's K-group $K_n^M(F)$ is the nth homogeneous component of $K_\bullet^M(F)$, and $K_n^M(F) = K_n(F)$ for $n = 0, 1, 2$. We usually write

$$\{a_1, a_2, \ldots, a_n\} = a_1 \otimes a_2 \otimes \cdots \otimes a_n \bmod I.$$

Milnor was able to prove the following theorem:

Theorem 8. *If $F(t)$ is a field of rational functions in t, then there are split exact sequences*

$$0 \to K_n^M(F) \to K_n^M(F(t)) \to \bigoplus_\pi K_{n-1}^M(F[t]/\pi) \to 0$$

for $n = 1, 2, \ldots$, where π runs over closed points of the affine line $\operatorname{Spec} F[t]$.

Proof. [16, Theorem 2.3] □

3.3. Quillen K-theory of finite fields

Known values of K-functor for a ring R led D. Quillen to use homotopy theory methods to compute appropriate homology groups.

In 1969 Quillen introduced a "+-construction": given a pointed arcwise connected CW-complex (X, x), and perfect normal subgroup $H \subset \pi_1(X, x)$, there is a map of CW-complexes $f : (X, x) \to (X^+, x^+)$ such that

- f induces an isomorphism $\pi_1(X, x)/H \xrightarrow{\sim} \pi_1(X^+, x^+)$;
- f induces an isomorphism $H_\bullet(X, f^*L) \xrightarrow{\sim} H_\bullet(X^+, L)$ for every local system of coefficients L on X^+.

There exists the *classifying space* $B\,\mathrm{GL}(R)$ of the group $\mathrm{GL}(R)$, such that

$$H_\bullet(\mathrm{GL}(R)) = H_\bullet(B\,\mathrm{GL}(R)) \text{ and } H^\bullet(\mathrm{GL}(R)) = H^\bullet(B\,\mathrm{GL}(R)).$$

Applying the +-construction to this classifying space yields the following definition ([23]):

Definition 5. *For $i > 0$,*

$$K_i(R) = \pi_i(B\,\mathrm{GL}(R)^+).$$

The space $B\,\mathrm{GL}(R)^+$ is connected, so $\pi_0(B\,\mathrm{GL}(R)^+) = 0$. Consider the disjoint union $K_0(R) \times B\,\mathrm{GL}(R)^+$ of copies of connected space $B\,\mathrm{GL}(R)^+$, one for each element of $K_0(R)$. This space has the same higher homotopy groups as $B\,\mathrm{GL}(R)^+$ (higher homotopy group do depend on the connected component of the base point only). Moreover, $\pi_0(K_0(R) \times B\,\mathrm{GL}(R)^+) = K_0(R)$ by the construction, so

$$K_i(R) = \pi_i(K_0(R) \times B\,\mathrm{GL}(R)^+)$$

for all $i \geq 0$. The product $K_0(R) \times B\,\mathrm{GL}(R)^+$ is the first example of so called *K-theory space* – a space which homotopy groups are desired K-groups. Note that such a space is defined only up to homotopy equivalence, since $B\,\mathrm{GL}$ is. The proper term "a homotopy type of spaces" is much longer than "space", so we use not-so-correct notion "K-theory space".

With this definition Quillen was able to compute K-groups of a finite field $\mathbb{F}_q$ with q elements:

Theorem 9. *For $i > 0$*

$$K_{2i}(\mathbb{F}_q) = 0,$$
$$K_{2i-1}(\mathbb{F}_q) = \mathbb{Z}/(q^i - 1).$$

If $\mathbb{F}_q \subset \mathbb{F}_r$, then $K_n(\mathbb{F}_q) \subset K_n(\mathbb{F}_r)$ for all n.

In particular, this shows that Milnor K-theory, although useful, gives no proper values of $K_n(F)$ for $n > 1$.

Exercise 3.

1. *Simplify the expression* $\dfrac{1-x}{1-x^{-1}}$.
2. *Use this to show that* $\{x, -x\} = 0$ *in* $K_2(F)$.
3. *Deduce that* $\{x, x\} = \{x, -1\}$.
4. *Deduce that* $\{x, y\} + \{y, x\} = 0$.

Exercise 4. *Assume that F is a finite field.*

1. *Show that for every $a, b \in F^*$, the equation $ax^2 + by^2 = 1$ has a solution in F.*
2. *Use this and the fact that the group F^* is cyclic to prove that $K_2(F) = 0$.*
3. *Deduce that $K_n^M(F) = 0$ for $n > 1$.*

4. Quillen Q-construction

Homotopy theory turns out to be a proper setting for developing K-theory. For a suitable category (symmetric monoidal or exact or Waldhausen), (the homotopy type of) a topological space is attached; K-theory groups are homotopy groups of this K-theory space. We discuss the first general Quillen definition of higher algebraic K-theory of an exact category in some detail.

4.1. Exact categories, Q-construction, classifying space of a category

4.1.1. Exact categories.

Definition 6. Exact category $\mathfrak{M} = (\mathfrak{M}, \mathfrak{A}, \mathfrak{E})$ *is an additive category $\mathfrak{M}$, having a set of isomorphism classes of objects, embedded as a full subcategory in abelian category $\mathfrak{A}$, closed under extensions in $\mathfrak{A}$, with a family $\mathfrak{E}$ of exact (in $\mathfrak{A}$) sequences*

$$0 \to M' \overset{\alpha}{\to} M \overset{\beta}{\to} M'' \to 0 \tag{4.1}$$

(called admissible exact sequences*) satisfying the following conditions:*

1. all split exact sequences of objects of $\mathfrak{M}$ are in $\mathfrak{E}$; if 4.1 is in $\mathfrak{E}$, then α is a kernel of β in $\mathfrak{M}$ and β is the cokernel of α in $\mathfrak{M}$;
   We say that a monomorphism (an epimorphism) is an *admissible monomorphism* (an *admissible epimorphism*) iff it occurs in an admissible exact sequence.

2. a composition of admissible epimorphisms (monomorphisms) is an admissible epimorphism (monomorphism)

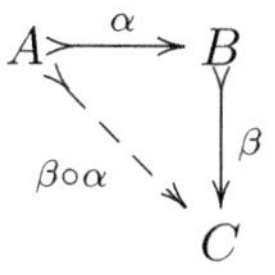

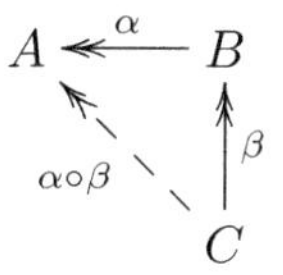

a (co)base change of an admissible epimorphism (monomorphism) is an admissible epimorphism (monomorphism)

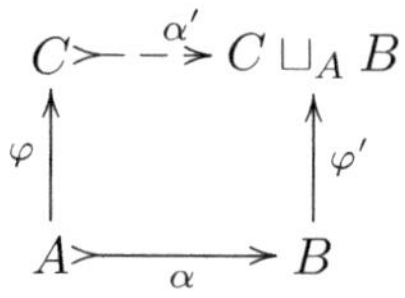

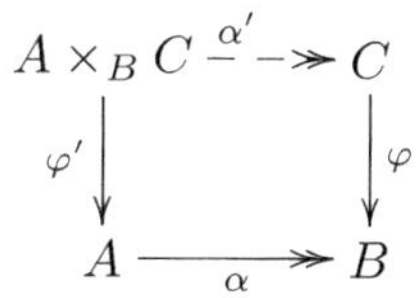

3. if $M \to M''$ possesses a kernel in $\mathfrak{M}$ and the composition $N \to M \to M''$ is an admissible epimorphism, then $M \to M''$ is an admissible epimorphism; the dual statement for monomorphisms holds true.

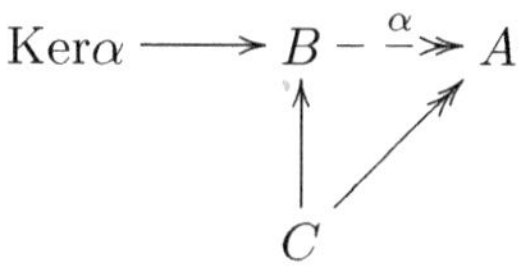

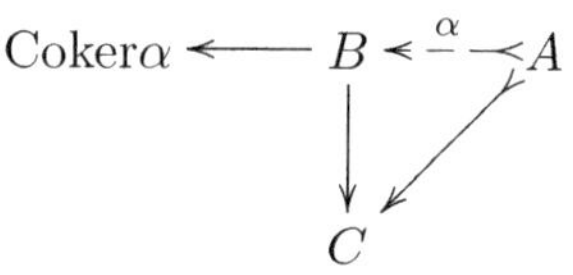

For example, given an admissible exact sequence (4.1) and a map $N \xrightarrow{\varphi} M''$ the sequence

$$0 \to M' \xrightarrow{\overline{\alpha}} M \times_{M''} N \xrightarrow{p_2} N \to 0$$

is an admissible exact sequence.

It is now known that condition (3) is a consequence of (1) - (2) (Keller, [11, Appendix A]).

Moreover, for an exact category $(\mathfrak{M}, \mathfrak{A}, \mathfrak{E})$ there exists a functor $\mathfrak{M} \hookrightarrow \mathfrak{A}'$ which embeds $\mathfrak{M}$ as a full, closed under extensions subcategory into an abelian category $\mathfrak{A}'$ such that $\mathfrak{E}$ is the class of all short exact sequences in $\mathfrak{M}$ which are exact in $\mathfrak{A}'$. So there is another way to define an exact category.

We will use arrows $\rightarrowtail$ and $\twoheadrightarrow$ to denote admissible monomorphisms and admissible epimorphisms respectively.

An admissible subobject is a source of an admissible monomorphism.

There is a notion of an *exact functor* between exact categories: it is an additive functor, which takes admissible exact sequences to admissible exact sequences.

Some examples:

Example 8. $\mathfrak{M} = \mathfrak{A}$, $\mathfrak{E}$ *is the family of all exact sequences.*

Example 9. $\mathfrak{M} = \mathfrak{A}$, $\mathfrak{E}$ *is the family of all split exact sequences.*

Example 10. *$\mathfrak{A}$ is the category of finitely generated left R-modules, $\mathfrak{M}$ is the full subcategory of projective modules, $\mathfrak{E}$ is the family of all sequences that are exact in $\mathfrak{A}$.*

Example 11. *X is a scheme, $\mathfrak{A}$ is the category of coherent $\mathcal{O}_X$-modules, $\mathfrak{M}$ is the full subcategory of locally free modules, $\mathfrak{E}$ is the family of all sequences that are exact in $\mathfrak{A}$.*

4.1.2. The Q-construction. For an exact category $\mathfrak{M} = (\mathfrak{M}, \mathfrak{A}, \mathfrak{E})$ Quillen defined another category, denoted by $Q\mathfrak{M}$ (called the *Quillenization* of $\mathfrak{M}$.) Given an exact category $(\mathfrak{M}, \mathfrak{A}, \mathfrak{E})$ the category $Q\mathfrak{M}$ has the same objects as $\mathfrak{M}$, and a morphism from M to M' in $Q\mathfrak{M}$ is a class of diagrams

$$M \overset{\beta}{\twoheadleftarrow} N \overset{\alpha}{\rightarrowtail} M'$$

up to isomorphism, which induces the identities on M and M'.

The composition of morphisms $M \overset{\beta}{\twoheadleftarrow} N \overset{\alpha}{\rightarrowtail} M'$ and $M' \overset{\delta}{\twoheadleftarrow} N' \overset{\gamma}{\rightarrowtail} M''$ in the category $Q\mathfrak{M}$ is defined by the fiber product as follows. Given two composable morphisms $M \overset{\beta}{\twoheadleftarrow} N \overset{\alpha}{\rightarrowtail} M'$, $M' \overset{\beta'}{\twoheadleftarrow} N' \overset{\gamma}{\rightarrowtail} M''$, the diagram

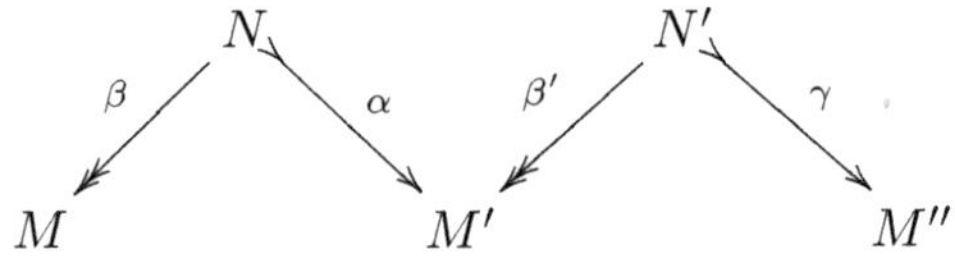

is completed to the diagram

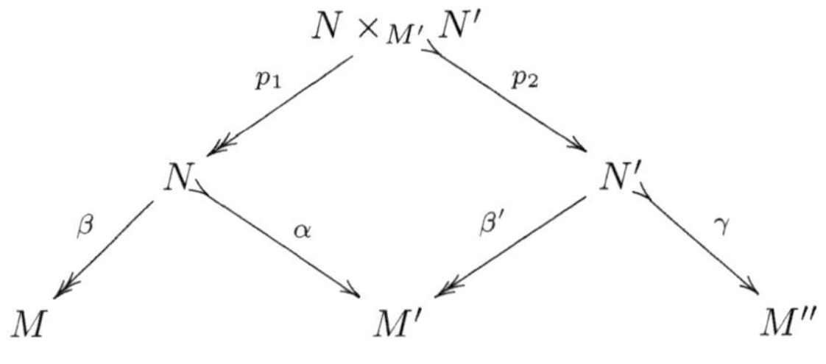

which yields the composition

$$M \overset{\beta\circ p_1}{\twoheadleftarrow} N \times_{M'} N' \overset{\gamma\circ p_2}{\rightarrowtail} M''.$$

It is clear that composition is well-defined and associative. Since the isomorphism classes of diagrams $M \overset{\beta}{\twoheadleftarrow} N \overset{\alpha}{\rightarrowtail} M'$ form a set for each M, M', the $Q\mathfrak{M}$ is a well defined category.

Each admissible monomorphism $N \overset{\alpha}{\rightarrowtail} M$ gives rise to a morphism $\alpha_! : N \to M$ in $Q\mathfrak{M}$ represented by the diagram

$$\alpha_! : N \overset{1_N}{\twoheadleftarrow} N \overset{\alpha}{\rightarrowtail} M$$

and morphisms of this type are called *injective.*

Dually, each admissible epimorphism $M \overset{\beta}{\twoheadleftarrow} N$ defines a morphism $\beta^! : M \to N$ in $Q\mathfrak{M}$ represented by the diagram

$$\beta^! : M \overset{\beta}{\twoheadleftarrow} N \overset{1_N}{\rightarrowtail} N$$

and these morphisms are called *surjective.*

Note that $M \overset{\beta}{\twoheadleftarrow} N \overset{\alpha}{\rightarrowtail} M' = \alpha_! \circ \beta^!$. There is a dual decomposition: given a map $M \overset{\beta}{\twoheadleftarrow} N \overset{\alpha}{\rightarrowtail} M'$ the fiber product

$$\begin{array}{ccc} N & \overset{\alpha}{\rightarrowtail} & M' \\ {\scriptstyle\beta}\downarrow & & \downarrow{\scriptstyle\delta} \\ M & \underset{\gamma}{\rightarrowtail} & M \times_N M' \end{array}$$

defines a decomposition $M \overset{\beta}{\twoheadleftarrow} N \overset{\alpha}{\rightarrowtail} M' = \delta^! \circ \gamma_!$. Conversely, for a diagram $M \overset{\gamma}{\rightarrowtail} N' \overset{\delta}{\twoheadleftarrow} M'$ the fiber sum $N = M \amalg_{N'} M'$

$$\begin{array}{ccc} M \amalg_{N'} M' & \overset{\alpha}{\rightarrowtail} & M' \\ {\scriptstyle\beta}\downarrow & & \downarrow{\scriptstyle\delta} \\ M & \underset{\gamma}{\rightarrowtail} & N' \end{array}$$

defines a morphism $M \overset{\beta}{\twoheadleftarrow} N \overset{\alpha}{\rightarrowtail} M' = \alpha_! \circ \beta^!$ with the decomposition. In fact, this is an alternative way to define morphisms in $Q\mathfrak{M}$. It may be convenient to regard a morphism in the category $Q\mathfrak{M}$ as a bicartesian (cartesian and cocartesian simultaneously) square

$$\begin{array}{ccc} N & \overset{\alpha}{\rightarrowtail} & M' \\ {\scriptstyle\beta}\downarrow & \square & \downarrow{\scriptstyle\delta} \\ M & \overset{\gamma}{\rightarrowtail} & N' \end{array}$$

and agree that usually we omit one corner of the square for short. The $\square$ sign indicates that marked square is bicartesian.

Proposition 10.

a) *If α and α' are composable monomorphisms in $\mathfrak{M}$, then $(\alpha \circ \alpha')_! = \alpha_! \circ \alpha'_!$ in $Q\mathfrak{M}$;*

b) *if β and β' are composable epimorphisms in $\mathfrak{M}$, then $(\beta' \circ \beta)^! = \beta^! \circ \beta'^!$ in $Q\mathfrak{M}$;*

c) $(1_M)_! = (1_M)^! = 1_M$;

d) *for a bicartesian square*

$$\begin{array}{ccc} N & \overset{\alpha}{\rightarrowtail} & M' \\ {\scriptstyle\beta}\downarrow & \square & \downarrow{\scriptstyle\delta} \\ M & \underset{\gamma}{\rightarrowtail} & N' \end{array}$$

with admissible arrows in $\mathfrak{M}$ *the equality* $\alpha_! \circ \beta^! = \delta^! \circ \gamma_!$ *holds in* $Q\mathfrak{M}$. □

4.1.3. The universal property of the Q-construction. Suppose given a category $\mathfrak{C}$, for each object M in $\mathfrak{M}$ an object hM of $\mathfrak{C}$, and for each $N \overset{\alpha}{\rightarrowtail} M'$ (resp. $M \overset{\beta}{\twoheadleftarrow} N$) in $\mathfrak{M}$ a map $\alpha_! : hN \to hM'$ (resp. $\beta^! : hM \to hN$) such that the properties a), b), c), d) of Proposition 10 hold. Then it is clear that this data induce a unique functor $F : Q\mathfrak{M} \to \mathfrak{C}$, $F(M) = hM$ compatible with the operations $\alpha \mapsto \alpha_!$, $\beta \mapsto \beta^!$.

This universal property of the Q-construction shows that an exact functor $F : \mathfrak{M} \to \mathfrak{M}'$ between exact categories induces a functor $Q\mathfrak{M} \to Q\mathfrak{M}'$, $M \mapsto FM$, $\alpha_! \mapsto (F\alpha)_!$, $\beta^! \mapsto (F\beta)^!$. Also for the dual category $\mathfrak{M}^o$ of an exact category $\mathfrak{M}$ there is an isomorphism of categories

$$Q\mathfrak{M}^o = Q\mathfrak{M}$$

such that the injective arrows in the former correspond to the surjective arrows in the latter and conversely.

4.1.4. Isomorphisms. For an isomorphism $N \overset{\alpha}{\rightarrowtail} M$, two maps $\alpha_! : N \to M$ and $\alpha^{-1!} : N \to M$ are equal since there is commutative diagram

$$\begin{array}{ccccc} N & \overset{1_N}{\twoheadleftarrow} & N & \overset{\alpha}{\rightarrowtail} & M \\ \| & & \downarrow{\scriptstyle\alpha} & & \| \\ N & \overset{\alpha^{-1}}{\twoheadleftarrow} & M & \overset{1_M}{\rightarrowtail} & M \end{array}$$

Conversely, a map in $Q\mathfrak{M}$ which is both injective and surjective is an isomorphism and it is of the form $\alpha_! = \alpha^{-1!}$ for a unique isomorphism α in $\mathfrak{M}$.

4.1.5. Zero maps. Let us denote $0^M : 0 \rightarrowtail M$ and $0_M : M \twoheadrightarrow 0$ unique maps in the additive category $\mathfrak{M}$. The set $\mathrm{Mor}_{Q\mathfrak{M}}(0, M)$ is in $1-1$ correspondence with the set of admissible subobjects of M (i.e., admissible monomorphisms $N \overset{\alpha}{\rightarrowtail} M$ up to automorphism of N over M), since each such a morphism by the definition is given by diagram

$$0^M_\alpha : 0 \overset{0_N}{\twoheadleftarrow} N \overset{\alpha}{\rightarrowtail} M.$$

Thus the set $\mathrm{Mor}_{Q\mathfrak{M}}(0, M)$ of morphisms from 0 to M is partially ordered with the smallest element $0^M_!$ and the greatest element $0^!_M$. There are decompositions

$0_!^M = \alpha_! \circ 0_!^N$ and $0_\alpha^M = \alpha_! \circ 0_N^!$ since each such a morphism is, by definition, given by the diagram:

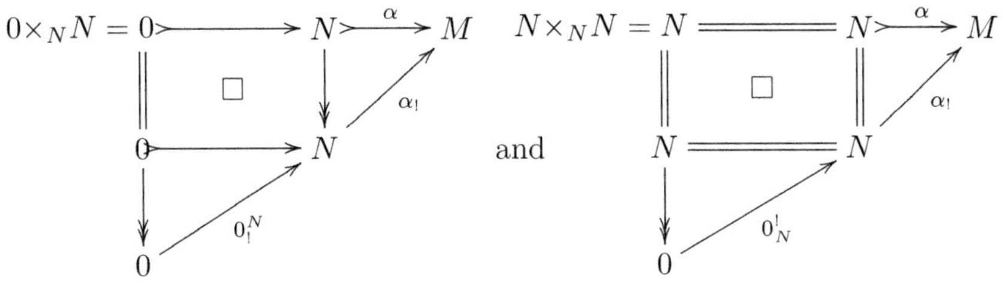

Dually, given admissible epimorphism $M \overset{\beta}{\twoheadrightarrow} M''$ with kernel $N \overset{\alpha}{\rightarrowtail} M$, there are decompositions $0_M^! = \beta^! \circ 0_{M''}^!$ and $0_\alpha^M = \beta^! \circ 0_!^{M''}$ given by commutative diagrams:

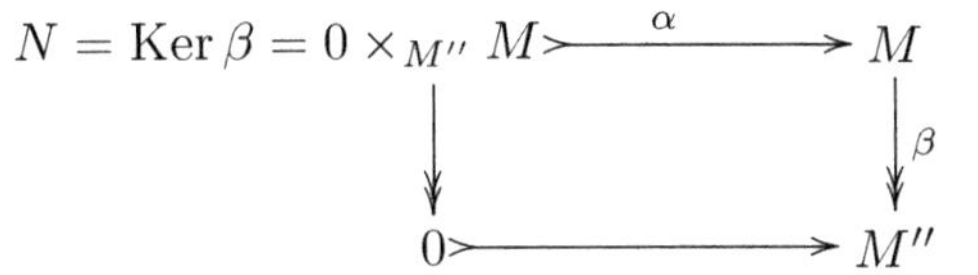

Proposition 11. *An admissible exact sequence* 4.1 *produces following commutative diagram in* $Q\mathfrak{M}$*:*

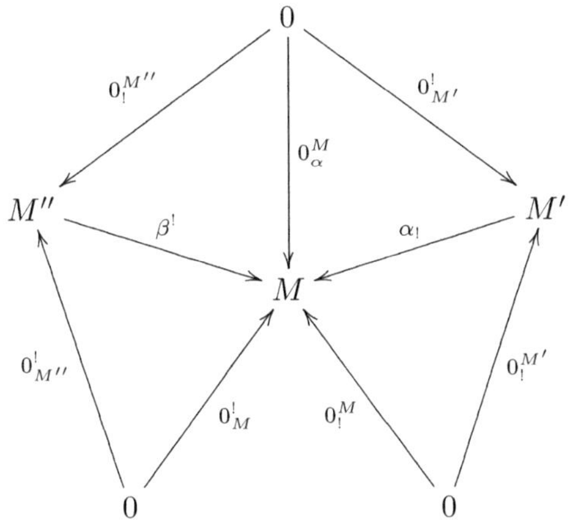

Proof. There are decompositions $0_!^M = \alpha_! \circ 0_!^{M'}$ and $0_\alpha^M = \alpha_! \circ 0_{M'}^!$. □

4.2. Higher K-groups and their elementary properties

The classifying space $B\mathfrak{C}$ of a small category $\mathfrak{C}$ is the geometric realization of the nerve $N\mathfrak{C}$ of this category. The nerve $N\mathfrak{C}$ of a small category $\mathfrak{C}$ is a simplicial set whose

1. p-simplices are diagrams in $\mathfrak{C}$ of the form

$$X_0 \to X_1 \to \cdots \to X_p$$

2. the ith face of the simplex is obtained by deleting the object X_i and composing arrows which come to and go from X_i;
3. the ith degeneracy of the simplex is obtained by inserting $X_i \stackrel{1_{X_i}}{\to} X_i$ in place of X_i.

Quillen [24, I.2., Theorem 1 p. 18] proved the following

Theorem 12. *The fundamental group $\pi_1(B(Q\mathfrak{M}), 0)$ is canonically isomorphic to the Grothendieck group $K_0(\mathfrak{M})$.* □

It is obvious from this proof that the class $[M] \in K_0(\mathfrak{M})$ corresponds under this isomorphism to the loop $|0^!_M| - |0^M_!|$ formed from paths $|0^!_M|$ corresponding to 1-simplex $0^!_M : 0 \to M$ and $|0^M_!|$ corresponding to 1-simplex $0^M_! : 0 \to M$.

This theorem was a motivation for the following definition:

Definition 7. *The higher algebraic K-theory groups of an exact category $\mathfrak{M}$ are the homotopy groups of the classifying space of the category $Q\mathfrak{M}$:*

$$K_i(\mathfrak{M}) = \pi_{i+1}(B(Q\mathfrak{M}), 0).$$

It is known, that this definition agrees with the earlier definition of K-theory of a ring, that used the $+$-construction. In particular K-groups of a finite field are known as well as K_0, K_1 and K_2 of arbitrary field. The last theorem on page 228 of [7] states that for a ring R the loop space $\Omega Q\mathcal{P}(\mathrm{Spec}(R))$ is homotopy equivalent to $K_0(R) \times B\,\mathrm{GL}(R)^+$ (cf. [30, IV§ 5, Theorem 5.1].)

Example 12. *A geometrical realization of the simplicial set, associated to the commutative diagram of an admissible exact sequence* 4.1 *consists of the following four 2-simplices*

$$0 \xrightarrow{0^{M'}_!} M' \xrightarrow{\alpha_!} M, \qquad 0 \xrightarrow{0^{M''}_!} M'' \xrightarrow{\beta^!} M,$$

$$0 \xrightarrow{0^!_{M'}} M' \xrightarrow{\alpha_!} M, \qquad 0 \xrightarrow{0^!_{M''}} M'' \xrightarrow{\beta^!} M,$$

visible in diagram 11. *This simplicial complex is obtained from the diagram by gluing together the three* 0 *vertices* (*Fig.* 1). *It may be realized as a parachute form* (*Fig.* 2) *or a sphere with a hole of a clover-leaf shape* (*Fig.* 3) – *boundary of each part of the leaf is a loop corresponding to one of objects M', M, M'' with such an orientation that sum of the loop of the subobject and the loop of the factor is the loop of M.*

Example 13. *A long exact sequence, e.g.,*

$$0 \to A \xrightarrow{\alpha} B \xrightarrow{\beta} C \xrightarrow{\gamma} D \to 0$$

may be split into two short exact sequences by $E = \mathrm{Im}\,\beta = \ker\gamma$ and resulting complexes are glued together along the boundary of the hole corresponding to E into a sphere with four-lobed hole.

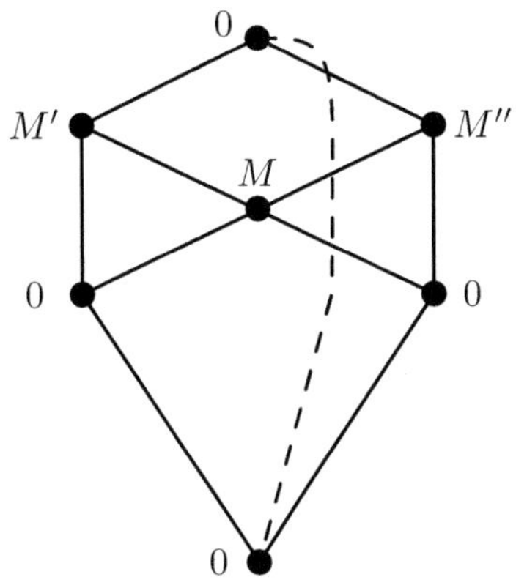

FIGURE 1. Connecting 0 vertices

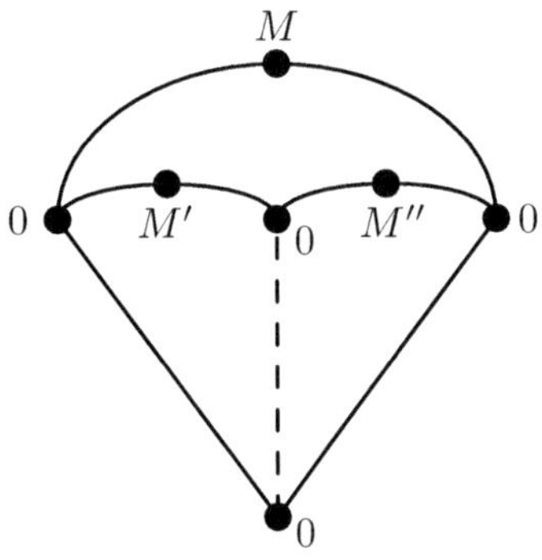

FIGURE 2. Parachute

Example 14. *Given two short exact sequences with the same subobject, a total object and a factor (so-called double short exact sequence) there are two clover-holed spheres with common boundary of the holes. This defines an element of* $\pi_2(B(Q\mathfrak{M}), 0)$. *A. Nenashev proved ([19], [20]) that for arbitrary exact category* $\mathfrak{M}$, *every element of* $K_1(\mathfrak{M})$ *is given by a double short exact sequence in this way. Moreover, he gave a family of defining relations for* $K_1(\mathfrak{M})$ *in terms of double short exact sequences.*

Any exact functor $f : \mathfrak{M} \to \mathfrak{M}'$ induces a functor $Qf : Q\mathfrak{M} \to Q\mathfrak{M}'$, and hence a homomorphism of K-groups which we denote by

$$f_* : K_\bullet(\mathfrak{M}) \to K_\bullet(\mathfrak{M}').$$

A natural transformation $Qf \to Qg$ of functors induces a homotopy between maps $B(Qf)$ and $B(Qg)$, so naturally equivalent exact functors $f, g : \mathfrak{M} \to \mathfrak{M}'$ induce the same homomorphism $f_* = g_*$.

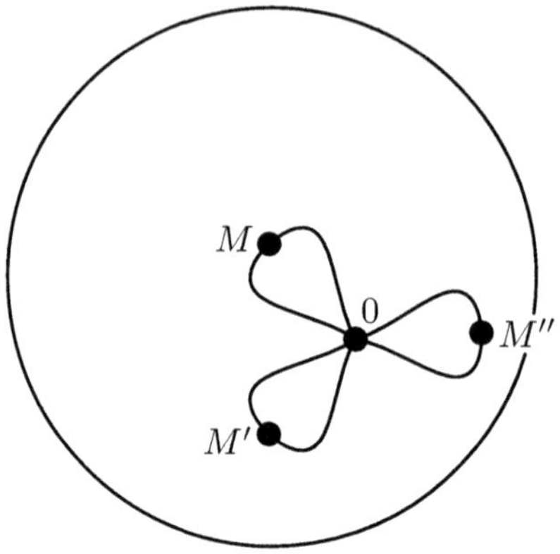

FIGURE 3. Clover-holed sphere

Example 15. *Given two exact categories $\mathfrak{M}, \mathfrak{M}'$ one may convert the product $\mathfrak{M} \times \mathfrak{M}'$ into exact category in the obvious way. Clearly $Q(\mathfrak{M} \times \mathfrak{M}') = Q(\mathfrak{M}) \times Q(\mathfrak{M}')$ and $(\mathrm{pr}_1, \mathrm{pr}_2)$ is an isomorphism*

$$K_i(\mathfrak{M} \times \mathfrak{M}') = K_i(\mathfrak{M}) \oplus K_i(\mathfrak{M}').$$

Example 16. *Let*

$$f' \rightarrowtail f \twoheadrightarrow f''$$

be an exact sequence of exact functors from an exact category $\mathfrak{M}$ to an exact category $\mathfrak{N}$. Then

$$f_* = f'_* + f''_*$$

as homomorphisms $K_i(\mathfrak{M}) \to K_i(\mathfrak{N})$.

This may be generalized as follows. We say that a filtration

$$0 = f_0 \subset f_1 \subset \cdots \subset f_n = f : \mathfrak{M} \to \mathfrak{N}$$

is an *admissible filtration* iff $f_{p-1}(X) \to f_p(X)$ is an admissible monomorphism for all X and p. This implies that there exist quotient functors f_p/f_q for all $q \leq p$ and if f_p/f_{p-1} are exact, then all quotients f_p/f_q are exact.

Example 17. *If $f : \mathfrak{M} \to \mathfrak{N}$ is an exact functor between exact categories equipped with an admissible filtration $0 = f_0 \subset f_1 \subset \cdots \subset f_n = f$ such that the consecutive quotients functors f_p/f_{p-1} are exact, then*

$$f_* = \sum_{p=1}^{n} (f_p/f_{p-1})_*.$$

Example 18. *If there is an exact sequence*

$$0 \to f_0 \to f_1 \to \cdots \to f_n \to 0$$

of exact functors $\mathfrak{M} \to \mathfrak{N}$, *then*

$$\sum_{p=0}^{n}(-1)^p(f_p)_* = 0.$$

Example 19. *Let* X *be a ringed space,* $\mathcal{P}(X)$ *– the category of vector bundles on* X *with the usual notion of an exact sequence,* $K_i(X) = K_i(\mathcal{P}(X))$*. Given* E *in* $\mathcal{P}(X)$*, there is an exact functor:*

$$E \otimes_{\mathcal{O}_X} - : \mathcal{P}(X) \to \mathcal{P}(X)$$

which induces a homomorphism $(E \otimes_{\mathcal{O}_X} -)_* : K_i(X) \to K_i(X)$*. If*

$$0 \to E' \to E \to E'' \to 0$$

is an exact sequence of vector bundles, then

$$(E \otimes_{\mathcal{O}_X} -)_* = (E' \otimes_{\mathcal{O}_X} -)_* + (E'' \otimes_{\mathcal{O}_X} -)_*.$$

Thus there is a product

$$\begin{aligned} K_0(X) \otimes_{\mathbb{Z}} K_i(X) &\to K_i(X) \\ [E] \otimes x &\longmapsto (E \otimes_{\mathcal{O}_X} -)_*(x). \end{aligned}$$

In fact, there are products $K_i(X) \otimes_{\mathbb{Z}} K_j(X) \to K_{i+j}(X)$ which generalize the above and make $K_\bullet(X)$ into a graded-commutative ring but the construction requires more machinery.

Example 20. *If* X *is a scheme, then there are exact functors*

$$\begin{aligned} E \otimes_{\mathcal{O}_X} - &: \mathcal{P}(X) \to \mathcal{P}(X) \\ E \otimes_{\mathcal{O}_X} - &: \mathcal{M}(X) \to \mathcal{M}(X) \end{aligned}$$

which yield a $K_0(X)$*-module structure on both* $K_.(X)$ *and* $K'_.(X)$*.*

4.3. Devissage and localization in abelian categories

Technical theorems on homotopy theory of categories – Quillen's Theorem A and B ([24, § 7]) – yield results on K-theory of abelian categories.

Let $\mathcal{A}$ be an abelian category having a set of isomorphism classes of objects, with all short exact sequences admissible. Let moreover $\mathcal{B}$ be a non-empty full subcategory closed under taking subobjects, quotient objects and finite products in $\mathcal{A}$, with all short exact sequences admissible. Thus $\mathcal{B}$ is an abelian category and the inclusion functor $\mathcal{B} \to \mathcal{A}$ is exact.

Then $Q\mathcal{B}$ is a full subcategory of $Q\mathcal{A}$ consisting of those objects which are also objects of $\mathcal{B}$.

Theorem 13 (Devissage). *Suppose that every object* A *of abelian category* $\mathcal{A}$ *has a finite filtration*

$$0 = A_0 \subset A_1 \subset \cdots \subset A_n = A$$

such that A_j/A_{j-1} *is in* $\mathcal{B}$ *for each* j*. Then the inclusion* $Q\mathcal{B} \to Q\mathcal{A}$ *is a homotopy equivalence, so* $K_i(\mathcal{B}) \cong K_i(\mathcal{A})$*.*

Proof. [24, § 5, Theorem 4]. □

Example 21. *Let $\mathcal{A}$ be an abelian category such that:*

- *the isomorphism classes of objects of $\mathcal{A}$ form a set,*
- *every object of $\mathcal{A}$ has finite length (has a finite filtration with simple objects as consecutive factors.)*

Then

$$K_i(\mathcal{A}) \cong \coprod_{j \in J} K(D_j)$$

where $\{X_j : j \in J\}$ is a set of representatives for the isomorphism classes of simple objects in $\mathcal{A}$, and D_j is the skew field $\mathrm{End}(X_j)^{op}$.

Example 22. *If I is a nilpotent two-sided ideal in a noetherian ring R, then $K_i'(R/I) \cong K_i'(R)$.*

The second result – long exact homotopy sequence – is the main tool for producing long exact sequence of localization.

Theorem 14 (Localization). *Let $\mathcal{B}$ be a Serre subcategory of an abelian category $\mathcal{A}$, let $\mathcal{A}/\mathcal{B}$ be the associated quotient abelian category, and let $\varepsilon : \mathcal{B} \to \mathcal{A}$, $\sigma : \mathcal{A} \to \mathcal{A}/\mathcal{B}$ denote the canonical functors. Then there is a long exact sequence*

$$\cdots \xrightarrow{\sigma_*} K_1(\mathcal{A}/\mathcal{B}) \xrightarrow{\partial} K_0(\mathcal{B}) \xrightarrow{\varepsilon_*} K_0(\mathcal{A}) \xrightarrow{\sigma_*} K_0(\mathcal{A}/\mathcal{B}) \to 0$$

Proof. [24, § 5, Theorem 5]. □

This is a homotopy long exact sequence of a (homotopy) fibration $B(Q\mathcal{B}) \to B(Q\mathcal{A}) \to B(Q(\mathcal{A}/\mathcal{B}))$; it is natural in $\mathcal{A}, \mathcal{B}$.

Example 23. *If R is a Dedekind domain (regular normal domain of dimension 1) with quotient field $R_{(0)}$, then there is a long exact sequence*

$$\cdots \to K_{i+1}\left(R_{(0)}\right) \to \coprod_{\mathfrak{m} \in \mathrm{Specm} R} K_i(R/\mathfrak{m}) \to K_i(R) \to K_i\left(R_{(0)}\right) \to \cdots \qquad \square$$

5. $K'_\bullet$ of noetherian schemes

The category $\mathcal{M}(X)$ of coherent $\mathcal{O}_X$-modules on a noetherian scheme X is an abelian category, so the Devissage and Localization Theorems apply. We list shortly basic results, following [24, § 7].

5.1. Fundamental Theorem

Theorem 15. *If R is a noetherian ring, then there are canonical isomorphisms*

$$K'_i(R[t]) \cong K'_i(R),$$
$$K'_i(R[t,t^{-1}]) \cong K'_i(R) \oplus K'_{i-1}(R)$$

If R is regular, there are canonical isomorphisms

$$K_i(R[t]) \cong K_i(R),$$
$$K_i(R[t,t^{-1}]) \cong K_i(R) \oplus K_{i-1}(R).$$

Proof. [24, § 6, Theorem 8]. □

In general, if a noetherian separated scheme X is regular, then the Cartan homomorphism $K_i(X) \to K'_i(X)$ is an isomorphism.

5.2. Functoriality, localization, homotopy property

If $f : X \to Y$ is a morphism of schemes, then the inverse image functor $f^* : \mathcal{P}(Y) \to \mathcal{P}(X)$ is exact and induces homomorphism of K-groups which is denoted $f^* : K_i(Y) \to K_i(X)$.

If $f : X \to Y$ is a flat morphism of schemes, then the inverse image functor $f^* : \mathcal{M}(Y) \to \mathcal{M}(X)$ is exact and induces homomorphism of K-group which is denoted $f^* : K'_i(Y) \to K'_i(X)$.

In both cases the formula

$$(fg)^* = g^* f^*$$

holds.

Thus K_i becomes a contravariant functor from noetherian separated schemes to abelian groups and K'_i becomes a contravariant functor on the subcategory of noetherian separated schemes and flat morphisms. Some homotopy theory is needed to prove that K_i takes filtered projective limits of schemes with affine transition maps to filtered injective limits of abelian groups, and K'_i takes such a limits with flat affine transition maps to appropriate limits of abelian groups.

Let $f : X \to Y$ be a proper morphism, so that the higher derived image functors $\mathrm{R}^i f_*$ carry coherent sheaves on X to coherent sheaves. Let $\mathcal{F}(X,f)$ denote the full subcategory of $\mathcal{M}(X)$ consisting of sheaves F such that $\mathrm{R}^i f_*(F) = 0$ for $i > 0$. The functor f_* induces a homomorphism

$$f_* : K'_i(\mathcal{F}(X,f)) \to K'_i(Y).$$

In the case when f is finite, in particular, when f is a closed immersion, $\mathcal{F}(X,f) = \mathcal{M}(X)$.

In the case when X has an ample line bundle, every coherent sheaf on X can be embedded in an object of $\mathcal{F}(X,f)$, which implies that the inclusion $\mathcal{F}(X,f) \to \mathcal{M}(X)$ induces an isomorphism on K'-groups.

In both cases a homomorphism $f_* : K'_i(X) \to K'_i(Y)$ is defined.

Proposition 16 (Projection Formula). *If for $f : X \to Y$ both $f_* : K_i'(X) \to K_i'(Y)$ and $f^* : K_i'(Y) \to K_i'(X)$ are defined, then for $x \in K_0(X)$, $y \in K_i'(Y)$ the equality:*

$$f_*\left(x \cdot f^*(y)\right) = f_*(x) \cdot y$$

in $K_i'(Y)$ holds.

Proof. [24, § 7, Proposition 2.10]. □

Let $\iota : Z \to X$ be a closed subscheme of X, and let $\mathcal{I}$ be the coherent sheaf of ideals in $\mathcal{O}_X$ defining Z. The functor $\iota_* : \mathcal{M}(Z) \to \mathcal{M}(X)$ allows us to identify coherent sheaves on Z with coherent sheaves on X annihilated by $\mathcal{I}$.

Proposition 17. *If $\mathcal{I}$ is nilpotent, then $\iota_* : K_i'(Z) \to K_i'(X)$ is an isomorphism. In particular $K_i'(X_{red}) \cong K_i'(X)$.* □

Now we extend the exact sequence of Localization Theorem 5:

Proposition 18 (Localization). *Let $j : U \to X$ be the open complement of Z in X. The there is a long exact sequence*

$$\cdots \to K_{i+1}'(U) \overset{\partial}{\to} K_i'(Z) \overset{\iota_*}{\to} K_i'(X) \overset{j^*}{\to} K_i'(U) \to \cdots .$$

Proof. The functor $j^* : \mathcal{M}(X) \to \mathcal{M}(U)$ induces an equivalence of $\mathcal{M}(U)$ with the quotient category $\mathcal{M}(X)/\mathcal{B}$, where $\mathcal{B}$ is the Serre subcategory consisting of coherent sheaves with support in Z. Devissage implies that $\iota_* : \mathcal{M}(Z) \to \mathcal{B}$ induces isomorphism on K-groups, so the desired exact sequence results from 14. □

We get, as an immediate application, the following generalization of the Fundamental Theorem 15:

Proposition 19 (Homotopy Property). *Let $f : P \to X$ be a flat map whose fibres are affine spaces. Then $f^* : K_i'(X) \to K_i'(P)$ is an isomorphism.*

Proof. It follows by noetherian induction, starting from a closed point and passing to generic point, since the affine case is proved – see [24, § 7, Proposition 4.7]. □

5.3. Filtration, K-cohomology, Brown-Gersten-Quillen spectral sequence, Chow ring

Let $\mathcal{M}_p(X)$ denote the Serre subcategory of $\mathcal{M}(X)$ consisting of those sheaves whose support is of codimension $\geq p$. It is clear that

$$K_i\left(\mathcal{M}_p(X)\right) = \varinjlim K_i'(Z)$$

where Z runs over closed subsets of codimension $\geq p$. Moreover subcategories $\mathcal{M}_p(X)$ are preserved by flat inverse image functor and by filtered projective limits with affine flat transition maps.

Theorem 20. *Let X_p be the set of points of codimension p in X. There is the following Brown-Gersten-Quillen spectral sequence (shortly: BGQ spectral sequence)*

$$E_1^{p,q}(X) = \coprod_{x \in X_p} K_{-p-q}\left(k(x)\right) \Longrightarrow K_{-n}'(X)$$

which is convergent when X has finite dimension. BGQ spectral sequence is contravariant for flat morphisms. If $X = \varprojlim X_i$, where $i \mapsto X_i$ is a filtered projective system with flat affine transition morphisms, then the BGQ spectral sequence for X is the inductive limit of the BGQ spectral sequences for the X_i.

This sequence is concentrated in the range $p \geq 0$, $p + q \leq 0$; if $\dim X = d$ then the E_1-term looks as follows:

$$\begin{array}{ccccccc}
\coprod_{x \in X_0} K_0(k(x)) & & & & & & \\
\coprod_{x \in X_0} K_1(k(x)) & \longrightarrow & \coprod_{x \in X_1} K_0(k(x)) & & & & \\
\vdots & & \vdots & & & & \\
\coprod_{x \in X_0} K_d(k(x)) & \longrightarrow & \coprod_{x \in X_1} K_{d-1}(k(x)) & \longrightarrow & \cdots & \longrightarrow & \coprod_{x \in X_d} K_0(k(x)) \\
\coprod_{x \in X_0} K_{d+1}(k(x)) & \longrightarrow & \coprod_{x \in X_1} K_d(k(x)) & \longrightarrow & \cdots & \longrightarrow & \coprod_{x \in X_d} K_1(k(x)) \\
\vdots & & \vdots & & & & \vdots
\end{array}$$

Proof. Consider the filtration

$$\mathcal{M}(X) = \mathcal{M}_0(X) \supset \mathcal{M}_1(X) \supset \cdots$$

of $\mathcal{M}(X)$ by Serre subcategories. There is an equivalence

$$\mathcal{M}_p(X)/\mathcal{M}_{p+1}(X) \cong \coprod_{x \in X_p} \bigcup_n \mathcal{M}\left(\mathcal{O}_{X,x} / (\mathrm{rad}\,(\mathcal{O}_{X,x}))^n\right)$$

which yields an isomorphism

$$K_i\left(\mathcal{M}_p(X)/\mathcal{M}_{p+1}(X)\right) \cong \coprod_{x \in X_p} K_i\left(k(x)\right)$$

where $k(x)$ is the residue field at x.

Localization exact sequences

$$\to K_i\left(\mathcal{M}_{p+1}(X)\right) \to K_i\left(\mathcal{M}_p(X)\right) \to \coprod_{x \in X_p} K_i\left(k(x)\right) \to K_{i-1}\left(\mathcal{M}_{p+1}(X)\right) \to$$

form an exact couple:

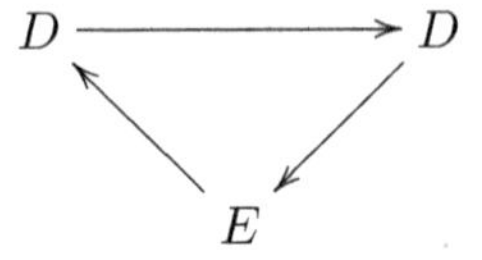

where

$$D = \coprod_{i,p} K_i\left(\mathcal{M}_p(X)\right), \qquad E = \coprod_{i,p} \coprod_{x \in X_p} K_i\left(k(x)\right)$$

and give rise to the BGQ spectral sequence in a standard way (see eg. [13, XI.5]). □

The filtration, arising in $K'_n(X)$ from the BGQ spectral sequence is the one given by codimension of support: $F^{p+1}K'_n(X)$ consists of elements of $K'_n(X)$ which vanish at generic points of subschemes of codimension $\leq p$.

One of geometric applications of BGQ spectral sequence is the following:

Theorem 21. *Let X be regular scheme of finite dimension over a field. Then there is a canonical isomorphism*

$$E_2^{p,p}(X) = CH^p(X).$$

Proof. [24, § 7, Proposition 5.14]. □

In fact, the group $\coprod_{x \in X_p} K_0\left(k(x)\right) = \coprod_{x \in X_p} \mathbb{Z}$ is a group of codimension p cycles on X, and one may check that image of $\coprod_{x \in X_{p-1}} K_1\left(k(x)\right) \to \coprod_{x \in X_p} K_0\left(k(x)\right)$ (or $\coprod_{x \in X_{p-1}} k(x)^* \to \coprod_{x \in X_p} \mathbb{Z}$) is exactly the group of cycles linearly equivalent to 0.

For $y \in X_{p-1}$ and $x \in X_p$ the yx component of differential $\coprod_{y \in X_{p-1}} k(y)^* \to \coprod_{x \in X_p} \mathbb{Z}$ assigns to $f \in k(y)^*$ the multiplicity of its 0 in x.

Another property of the BGQ spectral sequence is connected with the so called Gersten Conjecture.

Proposition 22. *The following conditions are equivalent:*

(i) *For every $p \geq 0$ the inclusion $\mathcal{M}_{p+1}(X) \to \mathcal{M}_p(X)$ induces 0 on the K-groups;*

(ii) *For all q, $E_2^{p,q} = 0$ if $p \neq 0$ and the edge homomorphism $K'_{-q} \to E_2^{0,q}$ is an isomorphism;*

(iii) *For every n the sequence*

$$0 \to K'_n(X) \to \coprod_{x \in X_0} K_n\left(k(x)\right) \to \coprod_{x \in X_1} K_{n-1}\left(k(x)\right) \to \cdots \tag{5.1}$$

is exact. □

Proposition 23 (Gersten). *Let $\mathcal{K}'_n$ denote the Zariski sheaf on X associated to the presheaf $U \mapsto K'_n(U)$. Assume that* Spec $(\mathcal{O}_{X,x})$ *satisfies the equivalent conditions of Proposition* 22 *for all $x \in X$. Then there is a canonical isomorphism*

$$E_2^{p,q}(X) = H^p(X, \mathcal{K}'_{-q}).$$

Proof. We use the sequences (5.1) for different open subsets of X and we sheafify them to get a sequence of sheaves

$$0 \to \mathcal{K}'_n \to \coprod_{x \in X_0} (i_x)_* (K_n(k(x))) \to \coprod_{x \in X_1} (i_x)_* (K_{n-1}(k(x))) \to \cdots$$

where $i_x : \text{Spec}\,(k(x)) \to X$ denotes the canonical map. The stalk of this sequence over x is the sequence (5.1) for Spec $(\mathcal{O}_{X,x})$, which is exact by hypothesis, whence it is a flasque resolution of $\mathcal{K}'_n$. □

We refer to $H^p(X, \mathcal{K}'_n)$ as *K-cohomology groups of X.*

The Gersten Conjecture is the following:

Conjecture 24. *The conditions of Proposition* 22 *are satisfied for the spectrum of a regular local ring.*

Some partial cases of this conjecture are proved. In particular Quillen proved it even for a semilocal ring obtained from a finite type algebra over a field by localizing with respect to a finite set of regular points ([24, Theorem 5.11]). So $E_2^{p,q}(X) = H^p(X, \mathcal{K}'_{-q})$ holds for a smooth variety over a field. This is a tool for effective computation of $E_2^{p,q}(X)$, e.g.,

- $H^p(-, \mathcal{K}'_{-q})$ has a homotopy property: if $f : X \to Y$ is a flat map whose fibers are affine spaces, then f^* induces an isomorphism $H^p(X, \mathcal{K}'_{-q}) \cong H^p(Y, \mathcal{K}'_{-q})$.
- if Z is a closed subset of X of pure codimension d, then

$$0 \to \cdots \to 0 \to \coprod_{x \in Z_0} K_{n-d}(k(x)) \to \coprod_{x \in Z_1} K_{n-d-1}(k(x)) \to \cdots$$

  is a subcomplex of

$$\coprod_{x \in X_0} K_n(k(x)) \to \coprod_{x \in X_1} K_{n-1}(k(x)) \to$$

  and the factor complex is

$$\coprod_{x \in (X \setminus Z)_0} K_n(k(x)) \to \coprod_{x \in (X \setminus Z)_1} K_{n-1}(k(x)) \to$$

  which yields excision property: $H^p(X, \mathcal{K}'_n) = H^p(X \setminus Z, \mathcal{K}'_n)$ for $p = 0, 1, \ldots, d-2$ and there is an exact excision sequence:

$$\begin{aligned} 0 \quad &\to \quad H^{d-1}(X, \mathcal{K}'_n) \to H^{d-1}(X \setminus Z, \mathcal{K}'_n) \to H^0(Z, \mathcal{K}'_{n-d}) \to \\ &\to \quad H^d(X, \mathcal{K}'_n) \to H^d(X \setminus Z, \mathcal{K}'_n) \to H^1(Z, \mathcal{K}'_{n-d}) \to \cdots \end{aligned}$$

Other important properties of BGQ spectral sequence are that if X is a variety over a field F, then $E_\bullet^{-,-}(X)$ is a spectral sequence of $K_\bullet(F)$-modules. In particular the constants – elements of $K_\bullet(F)$ – are in kernel of every differential. Thus if a term is constant, $E_r^{p,q}(X) = K_{-p-q}(F)$, then all differentials starting from $E_r^{p,q}(X)$ are trivial.

Example 24. *Affine space.*

If $X = \mathbb{A}_F^n$ is the affine space, then $E_2^{p,q}(X)$ looks as follows:

$$\begin{array}{ccccc}
K_0(F(X)) & 0 & 0 & \cdots & 0 \\
 & & & & \\
K_1(F) & 0 & 0 & \cdots & 0 \\
 & & & & \\
\vdots & \vdots & \vdots & & \vdots \\
 & & & & \\
K_n(F) & 0 & 0 & \cdots & 0 \\
 & & & & \\
\vdots & \vdots & \vdots & & \vdots
\end{array}$$

and the BGQ spectral sequence degenerates from the E_2-term on - all differentials in all E_n, $n \geq 2$, are trivial. □

Example 25. *Projective space.*

If $X = \mathbb{P}_F^n$ is the projective space, then for a hyperplane $Z = \mathbb{P}_F^{n-1}$ there is an exact excision sequence:

$$\begin{array}{rl}
0 \;\rightarrow & H^0(\mathbb{P}_F^n, \mathcal{K}'_{-q}) \rightarrow H^0(\mathbb{A}_F^n, \mathcal{K}'_{-q}) \rightarrow H^0(\mathbb{P}_F^{n-1}, \mathcal{K}'_{-q-1}) \rightarrow \\
\rightarrow & H^1(\mathbb{P}_F^n, \mathcal{K}'_{-q}) \rightarrow H^1(\mathbb{A}_F^n, \mathcal{K}'_{-q}) \rightarrow H^1(\mathbb{P}_F^{n-1}, \mathcal{K}'_{-q-1}) \rightarrow \cdots
\end{array}$$

which in fact is:

$$\begin{array}{rl}
0 \;\rightarrow & H^0(\mathbb{P}_F^n, \mathcal{K}'_{-q}) \rightarrow K_{-q}(F) \rightarrow H^0(\mathbb{P}_F^{n-1}, \mathcal{K}'_{-q-1}) \rightarrow \\
\rightarrow & H^1(\mathbb{P}_F^n, \mathcal{K}'_{-q}) \rightarrow 0 \rightarrow H^1(\mathbb{P}_F^{n-1}, \mathcal{K}'_{-q-1}) \rightarrow \cdots
\end{array}$$

Moreover, the homomorphism $H^0(\mathbb{A}_F^n, \mathcal{K}'_{-q}) \rightarrow H^0(\mathbb{P}_F^{n-1}, \mathcal{K}'_{-q-1})$ is induced by a differential, so it is trivial and

$$E_2^{0,q}(\mathbb{P}_F^n) = K_{-q}(F), \qquad E_2^{p,q}(\mathbb{P}_F^n) = E_2^{p-1,q-1}(\mathbb{P}_F^{n-1}) \text{ for } p > 0.$$

By induction on n we get

$$E_2^{p,q}(\mathbb{P}_F^n) = K_{-p-q}(F).$$

The BGQ spectral sequence for a projective space degenerates from E_2 onwards. In the E_2-term we have:

$$E_2(\mathbb{P}^n_F):\quad \begin{array}{c|ccccc} K_0(F) & 0 & \cdots & 0 & 0 \\ \hline K_1(F) & K_0(F) & & 0 & 0 \\ \vdots & \vdots & \ddots & \vdots & \vdots \\ K_n(F) & K_{n-1}(F) & \cdots & K_1(F) & K_0(F) \\ K_{n+1}(F) & K_n(F) & \cdots & K_2(F) & K_1(F) \\ \vdots & \vdots & & \vdots & \vdots \end{array}$$

It follows that there is a canonical isomorphism of $K_\bullet(F)$-modules

$$K_\bullet(\mathbb{P}^n_F) = K'_\bullet(\mathbb{P}^n_F) = K_\bullet(F)^{n+1}$$

compatible with the topological filtration: here in lower right block there is the image of $E_2(\mathbb{P}^{n-1}_F)$. Thus $F^1K_\bullet(\mathbb{P}^n_F)$ is the image of $K_\bullet(\mathbb{P}^{n-1}_F)$, where $\mathbb{P}^{n-1}_F$ is embedded in $\mathbb{P}^n_F$ as a hyperplane. The $K_0(F)$, at $p = 1, q = -1$ is the Picard group, unit element of this group is the class $\left[\mathcal{O}_{\mathbb{P}^{n-1}_F}\right]$ of the structural sheaf of a hyperplane. □

Example 26. *A quadric.*

Let Q^d be a d-dimensional projective quadric over F defined by equation $q_d = 0$, where

$$\begin{aligned} q_{2k} &= x_0y_0 + x_1y_1 + \cdots + x_ky_k, \\ q_{2k+1} &= z^2 + x_0y_0 + x_1y_1 + \cdots + x_ky_k. \end{aligned}$$

Consider the hyperplane section $Z^{d-1} : y_k = 0$ and its open complement $U^d = Q^d \backslash Z^{d-1}$. U^d is an affine space - the spectrum of $F\left[\frac{x_0}{y_k}, \frac{y_0}{y_k}, \cdots, \frac{x_k}{y_k}\right]/(\frac{q_{2k}}{y_k})$ or $F\left[\frac{z}{y_k}, \frac{x_0}{y_k}, \frac{y_0}{y_k}, \cdots, \frac{x_k}{y_k}\right]/(\frac{q_{2k+1}}{y_k})$, which is a polynomial ring in d variables. The excision property yields

$$H^0(Q^d, \mathcal{K}'_n) = K_n(F), \qquad H^p(Q^d, \mathcal{K}'_n) = H^{p-1}(Z^{d-1}, \mathcal{K}'_{n-1}) \text{ for } p > 0.$$

Z^{d-1} is a projective cone over Q^{d-2} with the vertex pt $= (0 : 0 : \cdots : 0 : 1 : 0)$. Z^{d-1} with the vertex removed is a bundle of affine lines over Q^{d-2}, so by the homotopy property

$$H^{p-1}(Z^{d-1}\backslash \text{pt}, \mathcal{K}'_{n-1}) = H^{p-1}(Q^{d-2}, \mathcal{K}'_{n-1}).$$

Starting the induction with $Q^0 = \text{pt} \amalg \text{pt}$ and $Q^1 \cong \mathbb{P}^1_F$ we obtain

- for odd d

$$E_2^{p,q}(Q^d) = H^p(Q^d, \mathcal{K}'_{-q}) = K_{-p-q}(F),$$

- for $d = 2k$

$$\begin{aligned} E_2^{p,q}(Q^d) &= H^p(Q^d, \mathcal{K}'_{-q}) = K_{-p-q}(F) \text{ for } p \neq k, \\ E_2^{k,q}(Q^d) &= H^k(Q^d, \mathcal{K}'_{-q}) = K_{-k-q}(F) \oplus K_{-k-q}(F) \text{ for } p = k \end{aligned}$$

and the BGQ spectral sequence degenerates from the E_2-term on.

In particular, we see that for $i \leq d$,

$$CH^i(Q^d) = \mathbb{Z} \text{ for } 2i \neq d, \qquad CH^k(Q^{2k}) = \mathbb{Z} \oplus \mathbb{Z}. \qquad \square$$

6. $K_\bullet$ of certain varieties

There are several constructions of K-theory spaces other than Q-construction, like Gillet-Grayson G-construction or Giffen K-construction (see [30, Chapter IV], i.e., `http://math.rutgers.edu./~weibel/Kbook.IV.dvi`.) If one wants to obtain results like $K_3(\mathbb{Z}) \cong \mathbb{Z}/48\mathbb{Z}$ ([12]) or $K_4(\mathbb{Z}) = 0$, then this topological machinery is required. If one wants K-theory as a tool for algebraic geometry of varieties, then it is reasonable to use above basic properties as axioms and simply do computations. So there will be no homotopy theory in the following.

6.1. Regular sheaves and projective bundles

We recall here the computation from § 8 of [24] of K-theory ring of a projective bundle, for two reasons: firstly, this gives methods for explicit computations, secondly this provides a model for computations of K-theory of other varieties.

Let T be a scheme (not necessarily noetherian or separated), let $\mathcal{E}$ be a vector bundle of rank r over T, and let $X = \mathbb{P}(\mathcal{E}) = \operatorname{Proj}(\operatorname{Sym}\mathcal{E})$ be associated projective bundle, where $\operatorname{Sym}\mathcal{E}$ is the symmetric algebra of $\mathcal{E}$ over $\mathcal{O}_T$. Let $\mathcal{O}_X(1)$ be the canonical line bundle on X and $f : X \to T$ the structural map.

Lemma 25.

(i) *For every quasi-coherent sheaf $\mathcal{F}$ of $\mathcal{O}_X$-modules, $R^q f_*(\mathcal{F})$ is a quasi-coherent sheaf of $\mathcal{O}_T$-modules which is zero for $q \geq r$.*

(ii) *For any quasi-coherent sheaf $\mathcal{F}$ of $\mathcal{O}_X$-modules and vector bundle $\mathcal{G}$ on T, one has*

$$R^q f_*(\mathcal{F}) \otimes_{\mathcal{O}_T} \mathcal{G} = R^q f_*(\mathcal{F} \otimes_{\mathcal{O}_X} f^*\mathcal{G}).$$

(iii) *For any quasi-coherent sheaf $\mathcal{N}$ of $\mathcal{O}_X$-modules, one has*

$$R^q f_*(\mathcal{O}_X(n) \otimes_{\mathcal{O}_T} \mathcal{N}) = \begin{cases} 0 & \text{for } q \neq 0, r-1 \\ \operatorname{Sym}^n \mathcal{E} \otimes_{\mathcal{O}_T} \mathcal{N} & \text{for } q = 0 \\ \left(\operatorname{Sym}^{-r-n} \mathcal{E}\right)\check{} \otimes_{\mathcal{O}_T} \mathcal{N} & \text{for } q = r-1 \end{cases} \qquad \square$$

(iv) *If $\mathcal{F}$ is a coherent sheaf of $\mathcal{O}_X$-modules, and T is affine, then $\mathcal{F}$ is a quotient of $\left(\mathcal{O}_X(-1)^{\otimes n}\right)^k$ for some n, k.*

The following notion is usually attributed to Mumford ([18, Lecture 14]):

Definition 8. *A quasi-coherent sheaf $\mathcal{F}$ of $\mathcal{O}_X$-modules on X is m-regular iff*

$$R^i f_*(\mathcal{F}(m-i)) = 0$$

for all $i > 0$. "0-regular" is simply "regular". $\mathcal{F}$ is m-regular iff $\mathcal{F}(m)$ is regular.

Mumford himself attributes this notion and its properties to Castelnuovo.

The long exact sequence of higher derived images yields immediately the following Lemma.

Lemma 26. *Let $0 \to \mathcal{F}' \to \mathcal{F} \to \mathcal{F}'' \to 0$ be an exact sequence of quasi-coherent sheaves of $\mathcal{O}_X$-modules.*

(i) *If $\mathcal{F}'(n)$ and $\mathcal{F}''(n)$ are m-regular, so is $\mathcal{F}(n)$.*
(ii) *If $\mathcal{F}(n)$ and $\mathcal{F}'(n+1)$ are m-regular, so is $\mathcal{F}''(n)$.*
(iii) *If $\mathcal{F}(n+1)$ and $\mathcal{F}''(n)$ are m-regular, and if $f_*(\mathcal{F}(n)) \to f_*(\mathcal{F}''(n))$ is onto, then $\mathcal{F}'(n+1)$ is m-regular.* □

Lemma 27. *If $\mathcal{F}$ is regular, then $\mathcal{F}(n)$ is regular for all $n \geq 0$.*

Proof. From the canonical epimorphism $f^*\mathcal{E} \to \mathcal{O}_X(1)$ one gets an epimorphism

$$\mathcal{O}_X(-1) \otimes_{\mathcal{O}_X} f^*(\mathcal{E}) \to \mathcal{O}_X \tag{6.1}$$

and Koszul resolutions

$$0 \to \mathcal{O}_X(-r) \otimes_{\mathcal{O}_X} f^*\left(\bigwedge^r \mathcal{E}\right) \to \cdots \to \mathcal{O}_X(-1) \otimes_{\mathcal{O}_X} f^*(\mathcal{E}) \to \mathcal{O}_X \to 0, \tag{6.2}$$

$$0 \to \mathcal{F}(-r) \otimes_{\mathcal{O}_X} f^*\left(\bigwedge^r \mathcal{E}\right) \to \cdots \to \mathcal{F}(-1) \otimes_{\mathcal{O}_X} f^*(\mathcal{E}) \to \mathcal{F} \to 0. \tag{6.3}$$

Assuming $\mathcal{F}$ to be regular, one sees that $\left(\mathcal{F}(-p) \otimes_{\mathcal{O}_X} f^*\left(\bigwedge^p \mathcal{E}\right)\right)(p)$ is regular. If (6.3) is split into short exact sequences

$$0 \to \mathcal{Z}_p \to \mathcal{F}(-p) \otimes_{\mathcal{O}_X} f^*\left(\bigwedge^p \mathcal{E}\right) \to \mathcal{Z}_{p-1} \to 0,$$

then by decreasing induction on p we obtain that the sheaf $\mathcal{Z}_p(p+1)$ is also regular. Thus the regularity of $\mathcal{F}$ implies that $\mathcal{Z}_0(1) = \mathcal{F}(1)$ is regular. □

Lemma 28. *If $\mathcal{F}$ is regular, then the canonical map $f^*f_*(\mathcal{F}) \to \mathcal{F}$ is surjective.*

Proof. As above there is an exact sequence

$$0 \to \mathcal{Z}_1 \to \mathcal{F}(-1) \otimes_{\mathcal{O}_X} f^*(\mathcal{E}) \to \mathcal{F} \to 0$$

with regular $\mathcal{Z}_1(2)$. Thus $R^1f_*(\mathcal{Z}_1(n)) = 0$ for $n > 1$, so the canonical map $f_*(\mathcal{F}(n-1)) \otimes_{\mathcal{O}_T} \mathcal{E} \to f_*(\mathcal{F}(n))$ is surjective for $n \geq 1$. Hence the canonical map of $\operatorname{Sym}\mathcal{E}$-modules

$$f_*(\mathcal{F}) \otimes_{\mathcal{O}_T} \operatorname{Sym}\mathcal{E} \to \coprod_{n \geq 0} f_*(\mathcal{F}(n))$$

is surjective. The Lemma follows by taking the associated sheaves. □

Now we shall describe a recursive procedure which gives the canonical resolution of a regular sheaf. This procedure is the key point here and in other computations below. But first suppose that a coherent sheaf $\mathcal{F}$ of $\mathcal{O}_X$-modules admits a resolution

$$0 \to f^*(\mathcal{T}_{r-1})(1-r) \to \cdots \to f^*(\mathcal{T}_0) \to \mathcal{F} \to 0$$

where $\mathcal{T}_i$ are coherent sheaves of $\mathcal{O}_T$-modules. Then $\mathcal{F}$ has to be regular. Moreover, the above exact sequence can be viewed as a resolution of the zero sheaf by acyclic objects for the δ-functor $\mathcal{F} \longmapsto R^q f_*(\mathcal{F}(n))$, where n is any fixed nonnegative integer. Applying f_* we get an exact sequence

$$0 \to \operatorname{Sym}^{n-r+1} \mathcal{E} \otimes_{\mathcal{O}_T} \mathcal{T}_{r-1} \to \cdots \to \operatorname{Sym}^n \mathcal{E} \otimes_{\mathcal{O}_T} f_*(\mathcal{F}) \to 0$$

for $n \geq 0$. In particular, we have the exact sequences:

$$0 \to \mathcal{T}_n \to \mathcal{E} \otimes_{\mathcal{O}_T} \mathcal{T}_{n-1} \to \cdots \to f_*(\mathcal{F}(n)) \to 0$$

for $n = 0, 1, \ldots, r-1$ which can be used to show that the sheaves $\mathcal{T}_n$ are determined by $\mathcal{F}$ up to canonical isomorphisms.

Recursive construction of a canonical resolution:

Conversely, given a coherent sheaf $\mathcal{F}$ of $\mathcal{O}_X$-modules, we inductively define a sequence of coherent sheaves of $\mathcal{O}_X$-modules $\mathcal{Z}_n = \mathcal{Z}_n(\mathcal{F})$ and a sequence of coherent sheaves of $\mathcal{O}_T$-modules $\mathcal{T}_n = \mathcal{T}_n(\mathcal{F})$ as follows. Starting with $\mathcal{Z}_{-1} = \mathcal{F}$, let

$$\begin{aligned} \mathcal{T}_n &= f_*(\mathcal{Z}_{n-1}(n)), \\ \mathcal{Z}_n &= \ker(f^*(\mathcal{T}_n)(-n) \to \mathcal{Z}_{n-1}). \end{aligned} \tag{6.4}$$

$\mathcal{Z}_n$ and $\mathcal{T}_n$ are additive functors of $\mathcal{F}$.

Assume now that $\mathcal{F}$ is regular. We see by induction that $\mathcal{Z}_n(n+1)$ is regular. This is clear for $n = -1$. The exact sequences

$$0 \to \mathcal{Z}_n(n) \to f^*(\mathcal{T}_n) \to \mathcal{Z}_{n-1}(n) \to 0 \tag{6.5}$$

allow one to complete the induction. In addition we have

$$f_*(\mathcal{Z}_n(n)) = 0 \text{ for } n \geq 0 \tag{6.6}$$

because the functor f_* maps $f^*(\mathcal{T}_n) \to \mathcal{Z}_{n-1}(n)$ onto an isomorphism. The functor f_* is exact on the category of regular coherent sheaves of $\mathcal{O}_X$-modules, so $\mathcal{F} \longmapsto \mathcal{T}_n(\mathcal{F})$ is an exact functor.

Example 27. *Projective space*

Let F be a field, $T = \operatorname{Spec} F$, $X = \mathbb{P}^m_F$. The sheaf $\mathcal{F} = \mathcal{O}_X(1)$ is regular.

$$\mathcal{T}_0 = f_*(\mathcal{F}) = H^0(X, \mathcal{O}_X(1))$$

is a vector space of dimension $m+1$. Exact sequence (6.5) for $n = 0$

$$0 \to \mathcal{Z}_0 \to \mathcal{O}_X^{m+1} \to \mathcal{O}_X(1) \to 0$$

defines $\mathcal{Z}_0$; for the next step we twist this sequence by 1:

$$0 \to \mathcal{Z}_0(1) \to \mathcal{O}_X(1)^{m+1} \to \mathcal{O}_X(2) \to 0$$

to obtain that $\dim H^0(X, \mathcal{Z}_0(1)) = (m+1)\binom{m+1}{1} - \binom{m+2}{2} = \binom{m+1}{2}$. Thus

$$\mathcal{T}_1 = f_*(\mathcal{Z}_0(1)) = H^0(X, \mathcal{Z}_0(1)) = F^{\binom{m+1}{2}}.$$

Exact sequence (6.5) for $n = 1$ is

$$0 \to \mathcal{Z}_1(1) \to \mathcal{O}_X^{\binom{m+1}{2}} \to \mathcal{Z}_0(1) \to 0.$$

For convenience we glue it with the exact sequence for $n = 0$ twisted by 1:

$$0 \to \mathcal{Z}_1(1) \to \mathcal{O}_X^{\binom{m+1}{2}} \to \mathcal{O}_X(1)^{m+1} \to \mathcal{O}_X(2) \to 0.$$

Thus next twist by 1

$$0 \to \mathcal{Z}_1(2) \to \mathcal{O}_X(1)^{\binom{m+1}{2}} \to \mathcal{O}_X(2)^{m+1} \to \mathcal{O}_X(3) \to 0,$$

yields

$$\dim H^0(X, \mathcal{Z}_0(2)) = \binom{m+1}{2}(m+1) - (m+1)\binom{m+2}{2} + \binom{m+3}{3} = \binom{m+1}{3},$$

$$\mathcal{T}_2 = f_*(\mathcal{Z}_1(2)) = H^0(X, \mathcal{Z}_1(2)) = F^{\binom{m+1}{3}},$$

so the next step is

$$0 \to \mathcal{Z}_2(2) \to \mathcal{O}_X^{\binom{m+1}{3}} \to \mathcal{O}_X(1)^{\binom{m+1}{2}} \to \mathcal{O}_X(2)^{m+1} \to \mathcal{O}_X(3) \to 0.$$

The final result is $\mathcal{Z}_m = 0$ and

$$0 \to \mathcal{O}_X \to \mathcal{O}_X(1)^{\binom{m+1}{m}} \to \cdots$$
$$\cdots \to \mathcal{O}_X(m-1)^{\binom{m+1}{2}} \to \mathcal{O}_X(m)^{m+1} \to \mathcal{O}_X(m+1) \to 0.$$

Untwisting yields the resolution of $\mathcal{O}_X(1)$:

$$0 \to \mathcal{O}_X(-m) \to \mathcal{O}_X(1-m)^{\binom{m+1}{m}} \to \cdots$$
$$\cdots \to \mathcal{O}_X(-1)^{\binom{m+1}{2}} \to \mathcal{O}_X^{\binom{m+1}{1}} \to \mathcal{O}_X(1) \to 0.$$

According to Proposition 6 above, $K_0(X)$ is a free abelian group with a basis $1 = [\mathcal{O}_X]$, $\xi = [\mathcal{O}_X(-1)], \xi^2, \ldots, \xi^{m-1}, \xi^m$. Here we have an identity

$$\xi^{-1} = \binom{m+1}{1} + \binom{m+1}{2}(-\xi) + \cdots + \binom{m+1}{m}(-\xi)^{m-1} + (-\xi)^m$$

which yields the ring structure:

$$K_0(X) - Z[\xi] / \left((1-\xi)^{m+1}\right). \qquad \square$$

Going back to the general case we see that the key point is that the recursive process described above terminates, i.e., that $\mathcal{Z}_{r-1} = 0$. From 6.5 we get an exact sequence

$$R^{q-1}f_*(\mathcal{Z}_{n+q-1}(n)) \to R^q f_*(\mathcal{Z}_{n+q}(n)) \to R^q f_*(f^*(\mathcal{T}_{n+q})(-q))$$

which allows one to prove by induction on q, starting from 6.6, that

$$R^q f_* \left(\mathcal{Z}_{n+q}(n)\right) = 0 \quad \text{for} \quad q, n \geq 0.$$

This shows that $\mathcal{Z}_{r-1}(r-1)$ is regular, since $R^q f_*$ is zero for $q \geq r$. Thus $f_* \left(\mathcal{Z}_{r-1}(r-1)\right) = 0$ and $\mathcal{Z}_{r-1}(r-1)$ by Lemma 28.

We have proved the following.

Proposition 29. *Any regular coherent sheaf of $\mathcal{O}_X$-modules $\mathcal{F}$ has the canonical resolution of the form*

$$0 \to f^* \left(\mathcal{T}_{r-1}(\mathcal{F})\right)(1-r) \to \cdots \to f^* \left(\mathcal{T}_1(\mathcal{F})\right)(1) \to f^* \left(\mathcal{T}_1(\mathcal{F})\right) \to \mathcal{F} \to 0 \quad (6.7)$$

where the $\mathcal{T}_i(\mathcal{F})$ are coherent sheaves of $\mathcal{O}_T$-modules determined up to a unique isomorphism by $\mathcal{F}$. Moreover $\mathcal{F} \longmapsto \mathcal{T}_i(\mathcal{F})$ is an exact functor from the category of regular coherent sheaves of $\mathcal{O}_X$-modules to the category of coherent sheaves of $\mathcal{O}_T$-modules. □

We state now three lemmas.

Lemma 30. *Assume T is quasi-compact. Then for any vector bundle $\mathcal{F}$ on X there exists an integer n_0 such that for all coherent sheaves $\mathcal{N}$ of $\mathcal{O}_T$-modules, one has*

(a) $R^q f_* \left(\mathcal{F}(n) \otimes_{\mathcal{O}_X} f^*(\mathcal{N})\right) = 0$ *for* $q > 0$,
(b) $f_* \left(\mathcal{F}(n)\right) \otimes_{\mathcal{O}_T} \mathcal{N} \xrightarrow{\sim} f_* \left(\mathcal{F}(n) \otimes_{\mathcal{O}_X} f^*(\mathcal{N})\right)$,
(c) $f_* \left(\mathcal{F}(n)\right)$ *is a vector bundle on* T.

Proof. [24, § 8, Lemma 1.12]. □

Lemma 31. *If $\mathcal{F}$ is a vector bundle on X such that $R^q f_* \left(\mathcal{F}(n)\right) = 0$ for $q > 0$ and $n \geq 0$, then $f_* \left(\mathcal{F}(n)\right)$ is a vector bundle on T for all $n \geq 0$.*

Proof. [24, § 8, Lemma 1.13]. □

Lemma 32. *If $\mathcal{F}$ is a regular vector bundle on X, then $T_i(\mathcal{F})$ is a vector bundle on T for each i.*

Proof. [24, § 8, Lemma 1.14]. □

Now we can prove main result of this section (Theorem 2.1 of [24, § 8]).

Theorem 33. *Let $\mathcal{E}$ be a vector bundle of rank r over a scheme T and $X = \mathbb{P}(\mathcal{E})$ the associated projective bundle. If T is quasi-compact, then one has isomorphisms*

$$\left(K_q(T)\right)^r \xrightarrow{\sim} K_q(X), \qquad (a_0, a_1, \ldots, a_{r-1}) \longmapsto \sum_{i=0}^{r-1} t^i \cdot f^* a_i$$

where $t \in K_0(X)$ is the class of the canonical line bundle $\mathcal{O}_X(-1)$, product $\cdot$ is the multiplication $K_0(X) \otimes K_q(X) \to K_q(X)$ defined in Example 19, *and $f : X \to T$ is the structural map.*

Proof. Let $\mathfrak{P}_n = \mathfrak{P}_n(X)$ denote the full subcategory of $\mathcal{P}(X)$ consisting of vector bundles $\mathcal{F}$ such that $R^q f_*(\mathcal{F}(k)) = 0$ for $q \neq 0$ and $k \geq n$. Let $\mathfrak{R}_n = \mathfrak{R}_n(X)$ denote the full subcategory of $\mathcal{P}(X)$ consisting of n-regular vector bundles. Each of this subcategories is closed under extensions, so its K-groups are defined.

We prove that inclusions induce isomorphisms $K_q(\mathfrak{R}_n) \cong K_q(\mathfrak{P}_n) \cong K_q(\mathcal{P}(X)) = K_q(X)$. To see this change the exact sequence (6.2) into

$$0 \to \mathcal{O}_X \to (f^*(\mathcal{E}))\,(1) \to \cdots \to \left(f^*\left(\bigwedge^r \mathcal{E}\right)\right)(r) \to 0$$

and tensor it by $\mathcal{F}$:

$$0 \to \mathcal{F} \to f^*(\mathcal{E}) \otimes_{\mathcal{O}_X} \mathcal{F}(1) \to \cdots \to f^*\left(\bigwedge^r \mathcal{E}\right) \otimes_{\mathcal{O}_X} \mathcal{F}(r) \to 0.$$

For each $p > 0$, $\mathcal{F} \longmapsto \mathcal{F}(p) \otimes_{\mathcal{O}_X} f^*\left(\bigwedge^p \mathcal{E}\right)$ is an exact functor from $\mathfrak{P}_n$ to $\mathfrak{P}_{n-1}$, hence it induces a homomorphism $u_p : K_p(\mathfrak{P}_n) \to K_p(\mathfrak{P}_{n-1})$. It is clear, that $\sum_{p>0}(-1)^{p-1}u_p$ is an inverse to the map induced by the inclusion $\mathfrak{P}_{n-1} \to \mathfrak{P}_n$. Thus we have $K_p(\mathfrak{P}_{n-1}) \xrightarrow{\sim} K_p(\mathfrak{P}_n)$ for all n. By Lemma 30 (a), $\mathcal{P}(X)$ is the union of the $\mathfrak{P}_n$'s, so $K_p(\mathfrak{P}_n) \cong K_p(\mathcal{P}(X))$ for all n. The proof that $K_p(\mathfrak{R}_n) \cong K_p(\mathcal{P}(X))$ is similar.

Put $U_n(\mathcal{N}) = (f^*(\mathcal{N}))\,(-n)$ for $\mathcal{N} \in \mathcal{P}(T)$. For $0 \leq n < r$, U_n is an exact functor from $\mathcal{P}(T)$ to $\mathfrak{P}_0$, hence it induces a homomorphism $u_n : K_p(\mathcal{P}(T)) \to K_p(\mathfrak{P}_0)$. To prove the theorem it suffices to show that the homomorphism

$$u : K_p(\mathcal{P}(T))^r \to K_p(\mathfrak{P}_0), \qquad u\,(a_0, a_1, \ldots, a_{r-1}) = \sum_{n=0}^{r-1} u_n(a_n),$$

is an isomorphism.

From Lemma 31 we know that $V_n(\mathcal{F}) = f_*(\mathcal{F}(n))$ is an exact functor from $\mathfrak{P}_0$ to $\mathcal{P}(T)$ for $n \geq 0$, hence we have a homomorphism

$$v : K_p(\mathfrak{P}_0) \to K_p(\mathcal{P}(T))^r, \qquad v(x) = (v_0(x), v_1(x), \ldots, v_{r-1}(x)),$$

where v_n is a homomorphism induced by V_n. Since

$$V_n \circ U_m(\mathcal{N}) = f_*\left((f^*(\mathcal{N}))\,(n-m)\right) = \mathrm{Sym}^{n-m}(\mathcal{E}) \otimes_{\mathcal{O}_T} \mathcal{N},$$

it follows that the composition vu is described by a triangular matrix with 1's on the diagonal. Therefore vu is an isomorphism, so u is injective.

On the other hand, $\mathcal{T}_n$ is an exact functor from $\mathfrak{R}_0$ to $\mathcal{P}(T)$, hence we have a homomorphism

$$t : K_p(\mathfrak{R}_0) \to K_p(\mathcal{P}(T))^r, \qquad t(x) = \left(t_0(x), -t_1(x), \ldots, (-1)^{r-1}t_{r-1}(x)\right),$$

where t_n is induced by $\mathcal{T}_n$. The composition ut is the map $K_p(\mathfrak{R}_0) \to K_p(\mathfrak{P}_0)$ induced by the inclusion $\mathfrak{R}_0 \to \mathfrak{P}_0$. Since ut is an isomorphism u is surjective. This concludes the proof. $\square$

6.2. Brauer-Severi varieties

If a variety X is in some sense similar to a projective space, then K-theory of X differs slightly from the K-theory of projective space. To show similarities and differences we discuss Brauer-Severi varieties and quadric hypersurfaces.

Brauer-Severi variety is a twisted form of a projective space. Let k be a field and let $\overline{k}$ be its algebraic closure.

Definition 9. *A k-variety P is a* Brauer-Severi variety *over k if $P_{\overline{k}} \cong \mathbb{P}^{r-1}_{\overline{k}}$.*

(see [1].) For example a conic is a Brauer-Severi variety (Example 35 below.) So a Brauer-Severi variety is similar to a projective space.

Brauer-Severi varieties over a fixed field k are classified by central simple k-algebras. Since this connection algebras-varieties is relevant, we give here some details.

Recall that a k-algebra A is *central* iff k is its center; it is *simple* iff it has no proper two-sided ideals. By Wedderburn Theorem ([25, Chapter 8, Theorem 1.5]) every central simple k-algebra is a matrix algebra over a division ring, which is central over k. In particular every central simple $\overline{k}$-algebra is a matrix algebra.

Given a central simple algebra A of rank r^2 over a field k, a Brauer-Severi variety over k may be defined directly as follows. For a field extension $k \subset K$ let P_K be the set of all left ideals L of $A \otimes_k K$ of dimension r. Let P be the k-variety which set of K-points is P_K for every K. It is clear that P possesses the structure of an algebraic variety over k. Indeed, picking a fixed basis for A over k, one embeds P as a closed subvariety of a Grassmannian $\mathrm{Gr}(r, r^2)$ of r-spaces in $A \otimes_k K$, defined by the relations stating that each L is a left ideal of A.

If A is a division ring, then P is a variety without rational points.

If $A = M_r(k)$ is a matrix algebra and $e_i = e_{ii}$ are the usual idempotents (e_{ij} is a matrix which differs from zero matrix by exactly one 1 at place i, j), then each left ideal L may be decomposed as $L = e_1 L \oplus \cdots \oplus e_r L$, when viewed as a k-module. Matrices in $e_i L$ differ form zero matrix by ith row. Since $e_{ji} e_i L = e_j L$, $\dim_k e_i L = 1$ for each $1 \leq i \leq r$. Thus all i-rows of matrices in $e_i L$ form a line in k^r, and for $i = 1, 2, \ldots, r$ this is the same line - the common image $k^r \cdot x$ of all nonzero $x \in L$. Choose $x \neq 0$ in $e_1 L$; then x may be written as $x = \sum a_j e_{1j}$ for some $(a_1, \ldots, a_r) \in k^r$, where at least one $a_i \neq 0$. For another choice of x, we have $x' = \lambda x$. It follows that each left ideal L of A corresponds to a point $(a_1 : \ldots : a_r)$ of $\mathbb{P}^{r-1}_k$, which is the common image of rank 1 matrices in L. On the other hand, pick $(a_1 : \ldots : a_r)$ in $\mathbb{P}^{r-1}_k$ and let $l = \sum a_j e_{1j}$. Then we associate to it the left ideal L it generates in $M_r(k)$, which is $L = kl + ke_{21}l + \cdots + ke_{r1}l$. Thus the Brauer Severi-variety associated to $M_r(k)$ is just a projective $(r-1)$-space over k. In particular, a Brauer-Severi variety over an algebraically closed field is a projective space, and a Brauer-Severi variety over a field k becomes isomorphic to a projective space over $\overline{k}$.

Brauer-Severi varieties may also be defined by descent. By descent theory, if $G = \mathrm{Gal}(\overline{k}/k)$, then the pointed set of isomorphism classes of k-varieties P with the property that $P_{\overline{k}} \cong \mathbb{P}^{r-1}_{\overline{k}}$ is isomorphic to $H^1(G, PGL_r)$, because $\mathrm{Aut}P = PGL_r$. On the other hand, each k-automorphism of $M_r(k)$ is inner (Noether-Skolem Theorem, [25, Chapter 8, Theorem 4.2]), $\mathrm{Aut}M_r(k) = PGL_r(k)$. Now, let A be a central simple k-algebra and $\overline{k}$ the algebraic closure of k. Then $A \otimes_k \overline{k} = M_r(\overline{k})$. So by descent theory the pointed set of isomorphism classes of central simple k-algebras of rank r^2 is isomorphic to $H^1(G, PGL_r)$, too. We associate to a k-variety P as above the central simple algebra A of the same class in $H^1(G, PGL_r)$. If one looks at the explicit action of PGL_r on $\mathbb{P}^{r-1}$ and the variety of left ideals of rank r of M_r, then it is easy to see that both constructions yield the same result.

Example 28. *Conic and quaternion algebra.*

Let $r = 2$. Then there exists a quadratic extension $F = k\left(\sqrt{a}\right)$ of k such that $P_F \cong \mathbb{P}^1_F$ and $A \otimes_k F \cong M_2(F)$. One may check explicitly that if P and A are defined by the same cocycle, then there exists $b \in k^*$ such that A has a basis $1, i, j, k$ such that

$$i^2 = a, \quad j^2 = b, \quad ij = -ji = k.$$

Such A is the *quaternion algebra* $\left(\frac{a,b}{k}\right)$; these are classified up to isomorphism by isomorphism classes of quadratic form (their reduced norm – see the final section below for a definition)

$$Nrd(x+yi+zj+tk) = (x+yi+zj+tk)(x-yi-zj-tk) = x^2-ay^2-bz^2+abz^2. \quad (6.8)$$

This is a very special quadratic form (two-fold Pfister form), its isomorphism class is uniquely defined by isomorphism class of its subform

$$Nrd(yi + zj + tk) = -(yi + zj + tk)^2 = -ay^2 - bz^2 + abz^2$$

(in the quadratic form theory it is called a *pure subform of* (6.8); a quaternion of the form $yi + zj + tk$ is called a *pure quaternion.*) We show that Brauer-Severi variety P is the projective conic defined by the equation $ay^2 + bz^2 - abz^2 = 0$. A left ideal L is of the form $A\alpha$ for $\alpha \in A$; if $N(\alpha) \neq 0$, then α is invertible, so $N(\alpha) = 0$ for $\dim L = 2$; if $\alpha = x + yi + zj + tk$ and $x \neq 0$ and, say, $x^2 - ay^2 \neq 0$, then $x - yi$ is invertible and

$$\begin{aligned} L = Ai\,(x - yi)\,(x + yi + zj + tk) &= Ai\left(x^2 - ay^2 + (xz - byt)\,j + (xt - yz)\,k\right) \\ &= A\left(\left(x^2 - ay^2\right)i + a\,(xt - yz)\,j + (xz - byt)\,k + (xt - yz)\,k\right) \\ &= A\,(y'i + z'j + t'k)\,. \end{aligned}$$

One easily checks that

$$A\,(y'i + z'j + t'k) = A\,(y''i + z''j + t''k) \text{ iff } (y' : z' : t') = (y'' : z'' : t'') \qquad \square$$

There are relative versions of the above notions. For a scheme T we introduce the following definitions:

- A sheaf $\mathcal{A}$ of $\mathcal{O}_T$-algebras of rank r^2 over T is an *Azumaya algebra* iff $\mathcal{A}$ is locally isomorphic to the sheaf $M_r\left(\mathcal{O}_T\right)$ with respect to étale topology (cf. [9, § 5].)

- A T-scheme $f : X \to T$ is a *Brauer-Severi scheme* of relative dimension $r-1$ iff it is locally isomorphic to the projective space $\mathbb{P}_T^{r-1}$ with respect to étale topology on T.
  (cf. op. cit., § 8.)

By descent, the connection between central simple algebras and Brauer-Severi varieties remains valid for Azumaya algebras and Brauer-Severi schemes – for details see [9, § 8]. Moreover, Quillen ([24, § 8.4]) has generalized the Projective Bundle Theorem 33 to this situation.

So let $f : X \to T$ be a Brauer-Severi scheme of relative dimension $r-1$ corresponding to the Azumaya algebra $\mathcal{A}$ over T.

If there exists a line bundle $\mathcal{L}$ on X which restricts to $\mathcal{O}(-1)$ on each geometric fibre, then one has $X = \mathbb{P}(\mathcal{E})$, where $\mathcal{E}$ is the vector bundle $f_*\mathcal{L}^{\vee}$ on T. In general such a line bundle $\mathcal{L}$ exists only locally for the étale topology on X. However, we shall now show that there is a canonical vector bundle of rank r on X which restricts to $\mathcal{O}(-1)^r$ on each geometric fibre.

Let the group scheme $GL(r,T)$ act on $\mathcal{O}_T^r$ in the standard way, and put $Y = \mathbb{P}_T^{r-1} = \mathbb{P}(\mathcal{O}_T^r)$. The induced action on Y factors through the projective group $PGL(r,T) = GL(r,T)/\mathbb{G}_{m,T}$. Since the multiplicative group $\mathbb{G}_{m,T}$ acts trivially on the vector bundle $\mathcal{O}_Y(-1) \otimes_{\mathcal{O}_Y} g^*(\mathcal{O}_T^r)$, where $g : Y \to T$ is the structural map, the group $PGL(r,T)$ operates on this vector bundle compatibly with its action on Y. As X is locally isomorphic to Y with respect to the étale topology on T and $PGL(r,T)$ is the group of automorphisms of Y over T, one knows that X is the bundle over T with fibre Y associated to a torsor A under $PGL(r,T)$ locally trivial for the étale topology. Thus by a faithfully flat descent, the bundle $\mathcal{O}_Y(-1) \otimes_{\mathcal{O}_Y} g^*(\mathcal{O}_T^r)$ on Y gives rise to a vector bundle $\mathcal{J}$ on X of rank r.

It is clear that the construction of $\mathcal{J}$ is compatible with the base change, and that $\mathcal{J} = \mathcal{O}_X(-1) \otimes_{\mathcal{O}_X} f^*(\mathcal{E})$ if $X = \mathbb{P}\mathcal{E}$. In the general case, there is a cartesian square

$$\begin{array}{ccc} X' & \xrightarrow{h'} & X \\ {\scriptstyle f'}\downarrow & & \downarrow{\scriptstyle f} \\ T' & \xrightarrow[h]{} & T \end{array}$$

where h is faithfully flat (e.g., an étale surjective map over which $\mathcal{A}$ becomes trivial) such that $X' = \mathbb{P}(\mathcal{E})$ for some vector bundle $\mathcal{E}$ of rank r on T', and further

$$h'^*(\mathcal{J}) = \mathcal{O}_{X'}(-1) \otimes_{\mathcal{O}_{X'}} f'^*(\mathcal{E}).$$

Let $\mathcal{A}$ be the sheaf of (non-commutative) $\mathcal{O}_T$-algebras given by

$$\mathcal{A} = f_*\left(\mathrm{End}_{\mathcal{O}_X}(\mathcal{J})\right)^{op}$$

where 'op' denotes the opposed ring structure. As h is flat, we have $h^*f_* = f'_*h'^*$. Hence

$$\begin{aligned} h^*(\mathcal{A})^{op} &= h^*f_*\left(\mathrm{End}_{\mathcal{O}_X}(\mathcal{J})\right) = f'_*h'^*\left(\mathrm{End}_{\mathcal{O}_X}(\mathcal{J})\right) \\ &= f'_*\left(\mathrm{End}_{\mathcal{O}_X}\left(\mathcal{O}_{X'}(-1)\otimes_{\mathcal{O}_{X'}} f'^*(\mathcal{E})\right)\right) \\ &= f'_*f'^*\left(\mathrm{End}_{\mathcal{O}_{T'}}(\mathcal{E})\right) = \mathrm{End}_{\mathcal{O}_{T'}}(\mathcal{E}). \end{aligned}$$

Thus $\mathcal{A}$ is an Azumaya algebra of rank r^2 over T. Moreover one has

$$f^*\mathcal{A} = \mathrm{End}_{\mathcal{O}_X}(\mathcal{J})^{op}$$

as one verifies by pulling back both sides to X'.

Let $\mathcal{J}_n$ (resp. $\mathcal{A}_n$) be the n-fold tensor product of $\mathcal{J}$ on X (resp. $\mathcal{A}$ on T), so that $\mathcal{A}_n$ is an Azumaya algebra of rank $(r^n)^2$ such that

$$\mathcal{A}_n = f_*\left(\mathrm{End}_{\mathcal{O}_X}(\mathcal{J}_n)\right)^{op}, \qquad f^*\mathcal{A}_n = \mathrm{End}_{\mathcal{O}_X}(\mathcal{J}_n)^{op}.$$

Let $\mathcal{P}(T,\mathcal{A}_n)$ be the category of vector bundles on T which are left modules for $\mathcal{A}_n$. Since $\mathcal{J}_n$ is a right $f^*(\mathcal{A}_n)$-module, which locally on X is a direct summand of $f^*(\mathcal{A}_n)$, we have an exact functor

$$\mathcal{J}_n\otimes_{\mathcal{A}_n} - : \mathcal{P}(T,\mathcal{A}_n)\to\mathcal{P}(X), \qquad \mathcal{M}\longmapsto \mathcal{J}_n\otimes_{f^*(\mathcal{A}_n)} f^*(\mathcal{M})$$

and hence an induced map of K-groups.

Theorem 34. *If T is quasicompact, one has isomorphisms (for all i):*

$$\coprod_{n=0}^{r-1} K_i(\mathcal{A}_n) \xrightarrow{\sim} K_i(X), \qquad (x_0,x_1,\ldots,x_{r-1})\longmapsto \sum_{n=0}^{r-1}(\mathcal{J}_n\otimes_{\mathcal{A}_n} -)_*(x_n).$$

This is actually a generalization of 33 because if two Azumaya algebras $\mathcal{A}$, $\mathcal{B}$ represent the same element of the Brauer group of T, then the categories $\mathcal{P}(T,\mathcal{A})$ and $\mathcal{P}(T,\mathcal{B})$ are equivalent (Morita equivalence), and hence have isomorphic K-groups. Thus $K_i(\mathcal{P}(T,\mathcal{A}_n)) = K_i(T)$ for all n if X is the projectivization of some vector bundle.

A proof of 34 is a modification of the proof of 33. One calls a sheaf $\mathcal{F}$ of $\mathcal{O}_X$-modules to be a regular sheaf if its inverse image on $X' = \mathbb{P}(\mathcal{E})$ is regular. For a regular $\mathcal{F}$ one constructs a sequence

$$0\to\mathcal{J}_{r-1}\otimes_{\mathcal{A}_{r-1}}\mathcal{T}_{r-1}(\mathcal{F})\to\cdots\to\mathcal{O}_X\otimes_{\mathcal{O}_X} f^*(\mathcal{T}_0(\mathcal{F}))\to\mathcal{F}\to 0 \tag{6.9}$$

recursively by

$$\begin{aligned} \mathcal{T}_n(\mathcal{F}) &= f_*\left(\mathrm{Hom}_{\mathcal{O}_X}(\mathcal{J}_n,\mathcal{Z}_{n-1}(\mathcal{F}))\right), \\ \mathcal{Z}_n(\mathcal{F}) &= \ker\left(\mathcal{J}_n\otimes_{\mathcal{A}_n}\mathcal{T}_n(\mathcal{F})\to\mathcal{Z}_{n-1}(\mathcal{F})\right) \end{aligned}$$

starting with $\mathcal{Z}_{-1}(\mathcal{F}) = \mathcal{F}$. It is easy to see that this sequencc, when lifted to X', coincides with the canonical resolution (6.7) of inverse image of $\mathcal{F}$ on X'. Since X' is faithfully flat over X, (6.9) is a resolution of $\mathcal{F}$.

We note also that there is a canonical epimorphism $\mathcal{J} \twoheadrightarrow \mathcal{O}_X$ obtained by descending (6.1), and hence a Koszul complex

$$0 \to \bigwedge^r \mathcal{J} \to \cdots \to \mathcal{J} \to \mathcal{O}_X \to 0$$

is an exact sequence of vector bundles on X, corresponding to (6.2). Therefore it is clear that all of the tools used in the proof of Projective Bundle Theorem are available in the situation under consideration.

This result was used to compute K-cohomology of Brauer-Severi variety in [15]. The field of fractions of Brauer-Severi variety is a generic splitting field for associated central simple algebra, and K-cohomology computation was used to prove that for arbitrary field F and natural number n prime to characteristic of F, there is a natural isomorphism $K_2(F)/nK_2(F) \cong H^2(F, \mu_n^{\otimes 2})$. We show below in 7.2 a particular case of the key argument leading this result, to exhibit the flavor of that kind of applications of the theory.

6.3. Quadric hypersurfaces

We give here a simplified account of Richard Swan paper [28]. The setup of [28] is very general: quadric is defined over a ring R - for a finitely generated projective R-module M with nonsingular quadratic form $q : M \to R$ the quadric $X(q)$ is $\operatorname{Proj}(\operatorname{Sym}(M^*))/(q)$ (here $M^* = \operatorname{Hom}_R(M, R)$.) The main Theorem 9.1 describes K-theory of the category $\mathcal{P}(X(q), \Lambda)$ of vector bundles on $X(q)$ with left action of a generalized Azumaya algebra Λ. With such a general result it is possible to compute the K-theory of products of affine quadrics.

We content ourselves with the K-theory of $\mathcal{P}(X(q))$ of smooth quadric hypersurface $X(q)$ over a field F with characteristic different from 2.

Let us outline one of the key ingredients – the classical notion of Clifford algebra of a quadratic form.

6.3.1. Clifford algebras. The standard reference is [4]. [5, Chapter II § 7] contains all details we need. A more recent source is [25, Chapter 8].

Let V be a vector space over a field F and let $q : V \to F$ be a quadratic form. The Clifford algebra $C(q)$ is a universal object for quadratic algebras of (V, q), i.e., F-algebras containing V as a subspace with the property

$$v^2 = q(v) \text{ for } v \in V.$$

Thus if $T(V) = \coprod_{n=0}^{\infty} V^{\otimes n}$ is the tensor algebra of V with natural grading, I is the two-sided ideal of $T(V)$ generated by all expressions $v \otimes v - q(v)$ for $v \in V$,

Definition 10. *The* Clifford algebra $C(q)$ *of* (V, q) *is*

$$C(q) = T(V)/I.$$

If $T_0(V) = \coprod_{n=0}^{\infty} V^{\otimes 2n}$ *is the even part of* $T(V)$*, then the even Clifford algebra* $C_0(q)$ *of* (V,q) *is*

$$C_0(q) = T_0(V)/I.$$

$C_0(q)$ is a subalgebra of $C(q)$ and $C(q)$ is a direct sum of two $C_0(q)$-modules

$$C(q) = C_0(q) \oplus C_1(q)$$

each of rank 1; this decomposition defines a $\mathbb{Z}_2$-grading in $C(q)$.

It is clear that product of vectors with more that $\dim V$ factors may be written as a shorter product, so if $v_1, v_2, \ldots, v_n$ is a base of V, then set of all ordered products of distinct v_i form a base of $C(q)$; in particular $\dim C(q) = 2^{\dim V}$.

Example 29. *If* $q = 0$ *is a zero form, then* $C(q)$ *is the exterior algebra of* V*.*

It is clear that for an orthogonal direct sum of quadratic forms $(V,q) = (V_1,q_1) \oplus (V_2,q_2)$

$$\begin{aligned} C(q) &= C(q_1) \otimes_F C(q_2), \\ C_0(q) &= C_0(q_1) \otimes_F C_0(q_2) \oplus C_1(q_1) \otimes_F C_1(q_2). \end{aligned}$$

A vector $v \in V$ is called *isotropic* iff $q(v) = 0$ and is called *anisotropic* iff $q(v) \neq 0$. An anisotropic vector v is an invertible element of $C(q)$ and if v is anisotropic, then $C_1(q) = C_0(q)v = vC_0(q)$.

The identity $v^2 = q(v)$ for $v \in V$ yields in a standard way the formula

$$uw + wu = q(u+w) - q(u) - q(w) = 2B(u,w)$$

where B is the symmetric bilinear form associated to q.

Given an anisotropic vector v, any vector w may be decomposed into sum of two components: one parallel to v and one perpendicular to v. If $w = tv + u$ is such a decomposition, then

$$-vwv^{-1} = -\left(tvvv^{-1} + vuv^{-1}\right) = -tv + u$$

is the reflection with respect to the hyperplane $v^{\perp}$ orthogonal to v.

Example 30. *If* $\dim V = 2$ *and* q *has a matrix diag*(a,b) *with respect to the basis* $\{u,v\}$ *of* V*, then* $C(q)$ *has the basis* $\{1,u,v,uv\}$ *with multiplication table* $u^2 = a$*,* $v^2 = b$*,* $vu = -uv$ *so the Clifford algebra is a quaternion algebra:* $C(q) \cong \left(\frac{a,b}{F}\right)$*.*

Example 31. *If* $\dim V = 3$ *and* q *has a matrix diag*(a,b,c) *with respect to the basis* $\{u,v,w\}$*, then* $C_0(q)$ *has basis* $\{1,uv,uw,-avw\}$ *with multiplication table* $(uv)^2 = -ab$*,* $(uw)^2 = -ac$*,* $(uw)(uv) = -(uv)(uw)$*, so* $C_0(q) \simeq \left(\frac{-ab,-ac}{F}\right)$*. The "determinant" element* $\delta = uvw$ *is in the centre of* $C(q)$*,* $\delta^2 = -abc$*; so* $F[\delta] \cong F\left(\sqrt{-abc}\right)$*. Hence*

$$C(q) = C_0(q) \oplus C_0(q)\delta \cong \left(\frac{-ab,-ac}{F\left(\sqrt{-abc}\right)}\right).$$

In particular if $-abc$ *is a square in* F*, then* $F\left(\sqrt{-abc}\right) \cong F \times F$ *and* $C(q) \cong \left(\frac{-ab,-ac}{F}\right) \times \left(\frac{-ab,-ac}{F}\right)$.

Exercise 5. *Let* $\{v_1, v_2, \ldots, v_n\}$ *be an orthogonal basis of* (V, q) *(i.e., a basis of* V*, which is orthogonal with respect to* q*. Let further* $\delta = v_1 \cdot v_2 \cdot \cdots \cdot v_n$*.*

1. δ *commutes in* $C(q)$ *with every vector iff* n *is odd and* δ *anticommutes with every vector if* n *is even.*
2. $\delta^2 = (-1)^{\frac{n(n-1)}{2}} \det(q)$ *(this is the* discriminant *of the quadratic form* q*.)*
3. *If* $\{w_1, w_2, \ldots, w_n\}$ *is another orthogonal base, then* $w_1 \cdot w_2 \cdot \cdots \cdot w_n$ *differs from* δ *by a scalar factor.*

In general for a non-singular q and V of even dimension, $C(q)$ is a central simple F-algebra and is a tensor product of quaternion algebras; for a non-singular q and V of odd dimension, $C_0(q)$ is a central simple F-algebra and is a tensor product of quaternion algebras.

We record here three useful facts:

- for a scalar factor t, quadratic forms q and tq have isomorphic even Clifford algebras (quadratic forms q and tq are usually called *similar* forms);
- if q is an orthogonal direct sum $(V, q) = (V_1, q_1) \oplus (F, x \mapsto -x^2)$, then $C_0(q) \cong C(q_1)$;
- A pair of isotropic $e, f \in V$ such that $B(e, f) = \frac{1}{2}$ gives two orthogonal idempotents $\frac{1}{2}(1 \pm ef)$ in $C_0(q)$ (in this case $Fe \oplus Ff$ is called a *hyperbolic plane.*)

6.3.2. Cohomology of quadric hypersurfaces. Let $B = F[x_0, x_1, \ldots, x_{d+1}]$ and let as usual $\mathbb{P}_F^{d+1} = \operatorname{Proj} B$. Let $q \in B_k$ be a nonzero (homogeneous of degree k) polynomial.

Lemma 35. *Let* $X \subset \mathbb{P}_F^{d+1}$ *be a hypersurface defined by* $q \in B_k$*; i.e.,* $X = \operatorname{Proj} A$ *where* $A = B/(q)$*. Then* $H^p(X, \mathcal{O}_X(n)) = 0$ *for* $p \neq 0, d$ *and, for* $d \geq 1$*,*

(1) $H^0(X, \mathcal{O}_X(n)) = A_n$
(2) $H^d(X, \mathcal{O}_X(n)) = A^*_{k-2-d-n}$.

If $d = 0$*, we have an exact sequence*

$$0 \to A_n \to H^0(X, \mathcal{O}_X(n)) \to A^*_{k-2-n} \to 0.$$

Here $A_n^* = \operatorname{Hom}_F(A_n, F)$ is a dual space.

Proof. The exact sequence

$$0 \to B(n-k) \xrightarrow{q\cdot} B(n) \to A(n) \to 0$$

induces an exact sequence of sheaves

$$0 \to \mathcal{O}_{\mathbb{P}_F^n}(n-k) \xrightarrow{q} \mathcal{O}_{\mathbb{P}_F^n}(n) \to \mathcal{O}_X(n) \to 0.$$

The Lemma follows immediately from the corresponding cohomology sequence and Lemma 25. □

It follows that $\mathcal{O}_X(n)$ is regular for $n \geq k-1$.

Lemma 36. *If X and d are as in Lemma 35 and $\mathcal{F}$ is a vector bundle on X then $H^p(X, \mathcal{F}) = 0$ for $p > d$.*

Proof. There exists an exact sequence

$$0 \to \mathcal{F}' \to \mathcal{O}_X(-n)^r \to \mathcal{F} \to 0.$$

For $p > d+1$ we have $H^p = 0$ as a functor on quasi-coherent sheaves on $\mathbb{P}_F^{d+1}$, so the exact cohomology sequence shows that $H^{d+1}(X, \mathcal{F}) = 0$. □

Lemma 37. *If $\mathcal{F}$ is regular, then the canonical map $f^* f_* (\mathcal{F}) \to \mathcal{F}$ is surjective.*

Proof. The difference between the Lemma and Lemma 28 is that here we have different f^*. Nevertheless the proof remains valid. □

From here on we assume that q is a nonsingular quadratic form (i.e., $k = 2$.)

The first difference between the case of projective bundle and the case of quadric hypersurface is that $\mathcal{O}_X$ is no longer regular, although $\mathcal{O}_X(1)$ still is. One may recursively construct the canonical resolution of a regular sheaf as in the case of projective bundle. Nevertheless, one should remember that $\mathcal{O}_X$ here is different from $\mathcal{O}_X$ in Example 27. The second difference is that the recursive process (6.4) does not terminate.

6.3.3. Generating function for the canonical resolution. We introduce here a computational tool which is not needed to prove the Swan Theorem, but will be helpful for its applications. Let

$$\cdots \to \mathcal{O}_X(-p)^{k_p} \to \cdots \to \mathcal{O}_X(-1)^{k_1} \to \mathcal{O}_X{}^{k_0} \to \mathcal{F} \to 0$$

be the canonical resolution of a regular sheaf $\mathcal{F}$.

Since the functor of global sections is exact on regular sheaves, there is the following recurrence for $k_{p+1} = \dim \Gamma(X, \mathcal{Z}_p(p+1))$ in the process (6.4) of building the canonical resolution:

$$\dim \Gamma(X,\ \mathcal{F}(p+1)) - k_0 \cdot \dim \Gamma(X, \mathcal{O}_X(p+1)) + \cdots \\ + (-1)^{p-1} k_p \cdot \dim \Gamma(X, \mathcal{O}_X(1)) + (-1)^p k_{p+1} = 0. \quad (6.10)$$

Recall that the *Poincaré series* $\Pi_{\mathcal{F}}(t)$ of a sheaf $\mathcal{F}$ is the formal power series

$$\Pi_{\mathcal{F}}(t) := \sum_{i=0}^{\infty} \dim \Gamma(X, \mathcal{F}(i)) \cdot t^i \in \mathbb{Z}[[t]].$$

The Poincaré series $\Pi_X(t)$ of a variety X is the Poincaré series of its structural sheaf:

$$\Pi_X(t) \overset{def}{=} \Pi_{\mathcal{O}_X}(t).$$

In particular if $X = \operatorname{Proj} A$ for a graded algebra A, then $\Pi_X(t)$ is the usual Poincaré series of A.

Example 32. *If S is the projective space, $S = \mathbb{P}^n_F$, then* $\dim \Gamma(S, \mathcal{O}_S(i)) = \binom{n+i}{i}$, *so*

$$P_n(t) := \Pi_S(t) = \sum_{i=0}^{\infty} \binom{n+i}{i} \cdot t^i = (1-t)^{-n-1}.$$

Example 33. *Let φ be a homogeneous polynomial of degree k in homogeneous coordinates in $\mathbb{P}^{d+1}_F = \operatorname{Proj} B$, $B = F[x_0, x_1, \ldots, x_{d+1}]$, $A = B/(\varphi)$, $X = \operatorname{Proj} A$ – a hypersurface $\varphi = 0$ in $\mathbb{P}^{d+1}_F$. Since the exact sequence*

$$0 \to B_n \xrightarrow{\phi\cdot} B_{n+k} \to A_{n+k} \to 0$$

splits for every n, the following equality holds:

$$\Pi_X(t) = P_{d+1}(t) - t^k P_{d+1}(t).$$

Thus $\Pi_S(t) = \dfrac{1-t^k}{(1-t)^{d+2}} = \dfrac{1+t+\ldots+t^{k-1}}{(1-t)^{d+1}}$.

Lemma 38. *For a projective quadric X of dimension d*

$$Q_d(t) := \Pi_X(t) = \frac{1+t}{(1-t)^{d+1}}. \quad \square$$

Proposition 39. *If $0 \to \mathcal{F}' \to \mathcal{F} \to \mathcal{F}'' \to 0$ is an exact sequence of $\mathcal{O}_X$-modules and either $\mathcal{F}'$, $\mathcal{F}''$ are regular or $\mathcal{F}$, $\mathcal{F}'(1)$ are regular, then*

$$\Pi_{\mathcal{F}}(t) = \Pi_{\mathcal{F}'}(t) + \Pi_{\mathcal{F}''}(t).$$

Proof. By Lemma 26, either $\mathcal{F}'$, $\mathcal{F}$, $\mathcal{F}''$ are regular or $\mathcal{F}'(1)$, $\mathcal{F}$, $\mathcal{F}''$ are regular. Hence each exact sequence of sheaves

$$0 \to \mathcal{F}'(i) \to \mathcal{F}(i) \to \mathcal{F}''(i) \to 0$$

induces an exact sequence of the corresponding spaces of global sections. $\square$

The recursive method of finding a canonical resolution of a regular sheaf $\mathcal{F}$ described above, namely the identity (6.10), yields the following identities for the generating function $G_{\mathcal{F}}(t) := \sum_{i=0}^{\infty} k_i t^i$:

$$\Pi_{\mathcal{F}}(t) = G_{\mathcal{F}}(-t) \cdot \Pi_X(t) \qquad \text{and} \qquad G_{\mathcal{F}}(t) = \frac{\Pi_{\mathcal{F}}(-t)}{Q_d(-t)}.$$

Example 34. *The Poincaré series of the sheaf $\mathcal{O}_X(1)$ is*

$$\Pi_{\mathcal{O}_X(1)}(t) = \frac{\Pi_{\mathcal{O}_X}(t) - 1}{t},$$

so

$$G_{\mathcal{O}_X(1)}(t) = \frac{\Pi_{\mathcal{O}_X(1)}(-t)}{Q_d(-t)} = \frac{Q_d(-t)-1}{-tQ_d(-t)} = \frac{\frac{1-t}{(1+t)^{d+1}} - 1}{-t\frac{1-t}{(1+t)^{d+1}}} = \frac{(1+t)^{d+1} - (1-t)}{t(1-t)}.$$

One may easily check that

$$\frac{(1+t)^{d+1}-(1-t)}{t(1-t)}-\frac{2^{d+1}t^{d+1}}{t\,(1-t)}$$

is a polynomial of degree $d-1$, so $k_i = 2^{d+1}$ for $i \geq d$.

6.3.4. The canonical resolution of a regular sheaf.

Lemma 40. *Assume that $H^q(X,\mathcal{F}) = 0$ for $q > d$ and all vector bundles $\mathcal{F}$. Let $\mathcal{Z}_p = \mathcal{Z}_p(\mathcal{F})$ be as in 6.4 above. If $\mathcal{G}$ is d-regular, then* $\operatorname{Ext}^q(\mathcal{Z}_{d-1},\mathcal{G}) = 0$ *for all $q > 0$.*

Proof. The sequence $0 \to \mathcal{Z}_p \to f^*(\mathcal{T}_p)\otimes\mathcal{O}_X(-p) \to \mathcal{Z}_{p-1} \to 0$ gives

$$\cdots \to \operatorname{Ext}^q\left(f^*(\mathcal{T}_p)\otimes\mathcal{O}_X(-p),\mathcal{G}\right) \to \operatorname{Ext}^q\left(\mathcal{Z}_p,\mathcal{G}\right) \to \operatorname{Ext}^{q+1}\left(\mathcal{Z}_{p-1},\mathcal{G}\right)$$
$$\to \operatorname{Ext}^{q+1}\left(f^*(\mathcal{T}_p)\otimes\mathcal{O}_X(-p),\mathcal{G}\right) \to \cdots.$$

Now $\operatorname{Ext}^q\left(f^*(\mathcal{T}_p)\otimes\mathcal{O}_X(-p),\mathcal{G}\right) = \mathcal{T}_p^*\otimes H^q\left(X,\mathcal{G}(p)\right) = 0$ for $q>0$, $p+q\geq d$; so for $q>0$ we have

$$\operatorname{Ext}^q\left(\mathcal{Z}_{d-1},\mathcal{G}\right) \cong \operatorname{Ext}^{q+1}\left(\mathcal{Z}_{d-2},\mathcal{G}\right) \cong \cdots \cong \operatorname{Ext}^{q+d}\left(\mathcal{Z}_{-1},\mathcal{G}\right) = 0 \qquad \square$$

Corollary 41. *Assume $H^q(X,\mathcal{F}) = 0$ for $q > d$ and all vector bundles $\mathcal{F}$. If $0 \to \mathcal{F}' \to \mathcal{F} \to \mathcal{F}'' \to 0$ is an exact sequence of regular sheaves, then*

$$0 \to \mathcal{Z}_{d-1}(\mathcal{F}') \to \mathcal{Z}_{d-1}(\mathcal{F}) \to \mathcal{Z}_{d-1}(\mathcal{F}'') \to 0$$

is split exact.

Proof. Here $\mathcal{Z}_{d-1}(\mathcal{F}')$ is d-regular, so the appropriate Ext^1 is 0. $\square$

Corollary 42. *$K_0(X)$ is torsion-free.*

Proof. Let $K_0(X,\oplus)$ be the K-group of exact category of vector bundles on X with only split exact sequences being admissible. The natural map $K_0(X,\oplus) \to K_0(X)$ has a right inverse given by

$$[\mathcal{F}] \mapsto \sum_{p=0}^{n}(-1)^p\left[\mathcal{O}_X(-p)\otimes f^*\left(\mathcal{T}_p(\mathcal{F})\right)\right] + (-1)^{n+1}\left[\mathcal{Z}_n(\mathcal{F})\right]$$

for n large enough and regular $\mathcal{F}$. Since $K_0(X,\oplus)$ is torsion-free by the Krull-Schmidt Theorem, the same is true for $K_0(X)$. $\square$

Swan's method of computing K-theory of quadric hypersurfaces uses truncation of the canonical resolution of a (-1)-regular sheaf.

If $\mathcal{F}$ is (-1)-regular, then $\mathcal{F}(-1)$ is regular; so we have an epimorphism

$$\mathcal{O}_X\otimes\Gamma\left(X,\mathcal{F}(-1)\right) \twoheadrightarrow \mathcal{F}(-1).$$

Since $\Gamma\left(X,\mathcal{F}(-1)\right)$ has finite dimension m, this is an epimorphism $\mathcal{O}_X^m \twoheadrightarrow \mathcal{F}(-1)$. Therefore there is an exact sequence of vector bundles $0 \to \mathcal{G} \to \mathcal{O}_X(1)^m \to \mathcal{F} \to 0$. If d is as in Lemma 40, then $\mathcal{Z}_{d-1}(\mathcal{G})\oplus\mathcal{Z}_{d-1}(\mathcal{F}) \cong \mathcal{Z}_{d-1}\left(\mathcal{O}_X(1)\right)^m$. Note that

$\mathcal{G}$ is regular, so its canonical resolution is defined. Let $\mathcal{U} = \mathcal{Z}_{d-1}(\mathcal{O}_X(1))$ and let $E = \mathrm{End}_X(\mathcal{U})$ acting on $\mathcal{U}$ from the right. For any vector bundle $\mathcal{W}$, $\mathrm{Hom}_X(\mathcal{U}, \mathcal{W})$ is a left E-module.

Lemma 43. *If $\mathcal{W}$ is a direct summand of $\mathcal{U}^m$ for some $m < \infty$ then*

$$\mathcal{U} \otimes_E \mathrm{Hom}_X(\mathcal{U}, \mathcal{W}) \xrightarrow{\sim} \mathcal{W}.$$

Proof. This is clear for $\mathcal{W} = \mathcal{U}$ and the property is obviously preserved by direct sums and inherited by direct summands. □

Note that if $\mathcal{W}$ is a direct summand of $\mathcal{U}^m$, then $\mathrm{Hom}_X(\mathcal{U}, \mathcal{W})$ is a finitely generated projective E-module since it is a direct summand of $\mathrm{Hom}_X(\mathcal{U}, \mathcal{U}^m) = E^m$.

Definition 11. *For a projective quadric X of dimension d:*

- *the Swan bundle of X is $\mathcal{U} := \mathcal{Z}_{d-1}(\mathcal{O}_X(1))$;*
- *the functor $\mathcal{T}$ from $\mathfrak{R}_{-1}(X)$ to F-modules is $\mathcal{T}(\mathcal{F}) = \mathrm{Hom}_X(\mathcal{U}, \mathcal{Z}_{d-1}(\mathcal{F}))$;*
- *the truncated canonical resolution of $\mathcal{F} \in \mathfrak{R}_{-1}(X)$ is*

$$0 \to \mathcal{U} \otimes_E \mathcal{T}(\mathcal{F}) \to \mathcal{O}_X(1-d) \otimes \mathcal{T}_{d-1}(\mathcal{F}) \to \cdots$$
$$\cdots \to \mathcal{O}_X \otimes \mathcal{T}_0(\mathcal{F}) \to \mathcal{F} \to 0.$$

Each $\mathcal{T}_p$ is an exact functor from $\mathfrak{R}_{-1}(X)$ to $\mathcal{P}(F)$ and $\mathcal{T}$ is also an exact functor.

We omit technical details of Swan's computations with several kinds of resolutions and state the result.

Let $\{v_0, v_1, \ldots, v_{d+1}\}$ be an orthogonal basis of the vector space V; let $\{z_0, z_1, \ldots, z_{d+1}\}$ be the dual basis of V^*. Denote by C_1 the odd part of the Clifford algebra $C(q)$. The subscripts in C_i will be taken $\mod 2$. Put

$$\varphi = \sum_{i=0}^{d+1} z_i \otimes v_i, \qquad \varphi \in \Gamma(X, \mathcal{O}_X(1) \otimes V).$$

The complex

$$\cdots \xrightarrow{\varphi\cdot} \mathcal{O}_X(-n) \otimes C_{n+d+1} \xrightarrow{\varphi\cdot} \mathcal{O}_X(1-n) \otimes C_{n+d} \xrightarrow{\varphi\cdot} \mathcal{O}_X(2-n) \otimes C_{n+d-1} \xrightarrow{\varphi\cdot} \cdots \tag{6.11}$$

is exact and locally splits ([28, Proposition 8.2.(a)].) Moreover, its part for $n > d-1$ coincides with the $n > d-1$ part of the canonical resolution of $\mathcal{O}_X(1)$. Thus if we denote

$$\mathcal{U}_n := \mathrm{Coker}\left(\mathcal{O}_X(-n-2) \otimes C_{n+d+3} \xrightarrow{\varphi\cdot} \mathcal{O}_X(-n-1) \otimes C_{n+d+2}\right)$$

then $\mathcal{U} = \mathcal{U}_{d-1}$ ([28, Corollary 8.6].)

Since the complex (6.11) is – up to twist – periodical with period two, we have

$$\mathcal{U}_{n+2} = \mathcal{U}_n(-2).$$

Consider the exact sequences

$$\mathcal{O}_X(-n-2)\otimes C_{n+d+3} \xrightarrow{\varphi\cdot} \mathcal{O}_X(-n-1)\otimes C_{n+d+2} \to \mathcal{U}_n \to 0$$

for two consecutive values of n; twist the first of them by 1. For any anisotropic vector $w \in V$ the isomorphism given by right multiplication by $1\otimes w$ fits into commutative diagram:

$$\begin{array}{ccccccc}
\mathcal{O}_X(-n-2)\otimes C_{n+d+4} & \xrightarrow{\varphi\cdot} & \mathcal{O}_X(-n-1)\otimes C_{n+d+3} & \longrightarrow & \mathcal{U}_{n+1}(1) & \longrightarrow & 0\\
\cong\big\downarrow \cdot 1\otimes w & & \cong\big\downarrow \cdot 1\otimes w & & & & \\
\mathcal{O}_X(-n-2)\otimes C_{n+d+3} & \xrightarrow{\varphi\cdot} & \mathcal{O}_X(-n-1)\otimes C_{n+d+2} & \longrightarrow & \mathcal{U}_n & \longrightarrow & 0.
\end{array}$$

Thus we have proven the following Lemma:

Lemma 44. *One has*

$$\mathcal{U}_{n+1}\cong\mathcal{U}_n(-1) \quad \textit{and} \quad \mathcal{U}_n\cong\mathcal{U}_0(-n)$$

for an arbitrary integer n. □

There is an exact sequence

$$0\to\mathcal{U}_0 \xrightarrow{\varphi} \mathcal{O}_X\otimes C_0 \to \mathcal{U}_{-1}\to 0 \tag{6.12}$$

where the isomorphism $\cdot(1\otimes w)$ was used to replace $\mathcal{O}_X\otimes C_1$ by $\mathcal{O}_X\otimes C_0$ for even d. In particular

$$\operatorname{rank}(\mathcal{U}) = \frac{1}{2}\dim C_0 = 2^d. \tag{6.13}$$

Lemma 45. $\operatorname{End}_X(\mathcal{U}_n)\cong C_0$ *acts on* $\mathcal{U}_n$ *from the right.*

Proof. [28, Lemma 8.7]. □

Now we can compute the K-theory of the quadric hypersurface X. For $n = 0,1,\dots,d-1$ let $U_n : \mathcal{P}(F)\to\mathcal{P}(X)$ be $U_n(M) := \mathcal{O}_X(-n)\otimes M$; moreover define $U : \mathcal{P}(C_0(q)) \to \mathcal{P}(X)$ by $U(M) := \mathcal{U}\otimes_{C_0(q)} M$. These functors induce maps $u_n : K_i(F)\to K_i(X)$ and $u : K_i(C_0(q))\to K_i(X)$. Therefore we get

$$\overline{u} = (u_0,u_1,\dots,u_{d-1},u) : K_i(F)^d\oplus K_i(C_0(q))\to K_i(X).$$

Theorem 46. *Let $X = X(q)\subset\mathbb{P}_F^{d+1}$ be a quadratic hypersurface of dimension d defined by a nonsingular quadratic space (V,q). Then the map $\overline{u} : K_i(F)^d\oplus K_i(C_0(q))\to K_i(X)$ is an isomorphism.*

Proof. We already know that $K_i(\mathfrak{N}_n)\cong K_i(\mathfrak{P}_n)\cong K_i(\mathcal{P}(X))$. Moreover, we have exact functors $\mathcal{T}_n : \mathfrak{N}_{-1}(X)\to\mathcal{P}(F)$ and $\mathcal{T} : \mathfrak{N}_{-1}(X)\to\mathcal{P}(C_0(q))$, and induced maps $t_n : K_i(X)\to K_i(F)$, $t : K_i(X)\to K_i(C_0(q))$; therefore

$$\overline{t} := (t_0,t_1,\dots,t_{d-1},t)$$

sends $K_i(X)$ to $K_i(F)^d\oplus K_i(C_0(q))$. The truncated canonical resolution shows that $\overline{u}\overline{t}$ is the isomorphism $K_i(\mathfrak{N}_n)\cong K_i(\mathfrak{P}_n)$. This shows that $\overline{u}$ is onto.

Now define functors $W_n : \mathfrak{P}_0(X) \to \mathcal{P}(F)$ by $W_n(\mathcal{F}) = \Gamma(X, \mathcal{F}(n))$ and $W : \mathfrak{P}_0(X) \to \mathcal{P}(C_0(q))$ by $W(\mathcal{F}) = \mathrm{Hom}_X(\mathcal{U}, \mathcal{F}) = \Gamma(X, \mathcal{U}^{\check{}} \otimes \mathcal{F})$. Since

$$\begin{aligned} W_i U_j(M) &= \Gamma\left(X, (\mathcal{O}_X(-j) \otimes M)(i)\right) = \Gamma\left(X, \mathcal{O}_X(i-j)\right) \otimes M, \\ WU(M) &= \mathrm{Hom}_X\left(\mathcal{U}, \mathcal{U} \otimes_{C_0(q)} M\right) = \mathrm{Hom}_X\left(\mathcal{U}, \mathcal{U}\right) \otimes_{C_0(q)} M = M, \\ \Gamma\left(X, \mathcal{U}(n)\right) &= 0 \text{ for } n = 0, 1, \ldots, d-1, \end{aligned}$$

the induced map $\overline{w} : K_i(X) \to K_i(F)^d \oplus K_i(C_0(q))$ has the property that $\overline{wu}$ is given by a triangular matrix with the identity maps on diagonal. Thus $\overline{u}$ is injective. □

Swan gave also an explicit formula for the map induced by an embedding of a nonsingular hyperplane section into X [28, Theorem 10.5], and hence computed the K-theory of a smooth affine quadric. Moreover, Swan computed the K-theory of a cone like $\mathrm{Proj}\, R[x_1, x_2, \ldots, x_n]/\left(q(x_2, \ldots, x_n)\right)$ – [28, Theorem 11.7].

7. Applications

7.1. Chow ring of a split smooth quadric

In theory of quadratic forms a subspace U of a space V with a quadratic form q is said to be an *isotropic subspace* iff U contains a nonzero isotropic vector; a subspace U is said to be a *totally isotropic subspace* iff $q|_U = 0$. This convention is convenient if one is interested in classification of quadratic forms, since nonsingular isotropic space contains a hyperbolic plane. Geometers prefer term "isotropic subspace" for a totally isotropic space. The dimension of maximal totally isotropic subspace is called a *Witt index* of q, or simply an *index* of q. A quadric $q = 0$ is *split* iff the index of q is maximal possible for quadratic forms of given dimension.

We shall apply the results of section 6.3 in the simplest possible case of a split quadric: X is a projective quadric hypersurface over a field F, $\mathrm{char}\, F \neq 2$, defined by the quadratic form of maximal index.

7.1.1. Notation. Consider a vector space V with a basis $\{v_0, v_1, \ldots, v_{d+1}\}$ over a field F, $\mathrm{char}\, F \neq 2$. Denote by $\{z_0, z_1, \ldots, z_{d+1}\}$ the dual basis of $V^* = \mathrm{Hom}_F(V, F)$. Let q be the quadratic form

$$q = \sum_{i=0}^{d+1} (-1)^i z_i^2.$$

Moreover, let $e_i = \frac{1}{2}(v_{2i} - v_{2i+1})$, $f_i = \frac{1}{2}(v_{2i} + v_{2i+1})$ for all possible values of i. Thus $e_0, e_1, \ldots, e_m$ span a maximal totally isotropic subspace and $f_0, f_1, \ldots, f_m$ span a maximal totally isotropic subspace. The index of q is $m+1$, which is maximal possible value for quadratic forms of dimension $d+2$.

- if $d = 2m$, then $e_0, f_0, e_1, f_1, \ldots, e_m, f_m$ form a basis of V with the dual basis $\{x_0, y_0, x_1, y_1, \ldots, x_m, y_m\}$ and

$$q = \sum_{i=0}^{m} x_i y_i.$$

- if $d = 2m+1$, then $f_0, e_1, f_1, \ldots, e_m, f_m, v_{d+1}$ form a basis of V with the dual basis $\{x_0, y_0, x_1, y_1, \ldots, x_m, y_m, z_{d+1}\}$ and

$$q = \sum_{i=0}^{m} x_i y_i + z_{d+1}^2.$$

Note that $v_{2i+1} = f_i - e_i$, and $q(f_i) = q(e_i)$, so that we have:

Lemma 47. *The reflection with respect to the hyperplane ${v_{2i+1}}^{\perp}$ interchanges f_i with e_i and interchanges x_i with y_i.* □

We shall compute table of multiplication in $K_0(X)$ for a d-dimensional projective quadric X defined by the equation $q = 0$ in $\mathbb{P}_F^{d+1}$, i.e., for

$$X = \operatorname{Proj} S(V^*)/(q) \cong \operatorname{Proj} F[z_0, z_1, \ \ldots, z_{d+1}]/(q).$$

7.1.2. Clifford algebra. In the case of $d = 2m+1$ the even part $C_0 = C_0(q)$ of the Clifford algebra $C(q)$ is isomorphic to the matrix algebra $\mathrm{M}_{2^{m+1}}(F)$ (a standard fact),

$$C_0 \cong \mathrm{M}_{2^{m+1}}(F).$$

In particular, the Morita equivalence

$$\begin{array}{rcl} M & \longmapsto & \operatorname{Hom}_F(F^N, F) \otimes_{C_0} M \in \operatorname{Ob}(\mathcal{P}(F)) \\ W & \longmapsto & F^N \otimes_F W \in \operatorname{Ob}(\mathcal{P}(C_0)) \end{array}$$

of categories of C_0-modules and F-modules induces isomorphisms $K_p(C_0) \cong K_p(F)$ for all p.

In the case of even $d = 2m$, the algebra C_0 has the centre $F \oplus F \cdot \delta$, where $\delta = v_0 \cdot v_1 \cdot \cdots \cdot v_{d+1}$ and $\delta^2 = 1$. Thus $\frac{1}{2}(1+\delta)$, $\frac{1}{2}(1-\delta)$ are orthogonal central idempotents of C_0, so

$$C_0 = \frac{1}{2}(1+\delta)C_0 \oplus \frac{1}{2}(1-\delta)C_0$$

where each direct summand is isomorphic to the matrix algebra $\mathrm{M}_{2^m}(F)$ (yet another standard fact).

For every anisotropic vector $w \in V$, the reflection $\alpha \mapsto -w\alpha w^{-1}$ with respect to the hyperplane $w^{\perp}$ induces an automorphism ρ_w of C_0, which interchanges δ with its opposite:

$$\rho_w(\delta) = -\delta. \tag{7.1}$$

Regarding subscripts $i \mod 2$ denote

$$P_i = (1 + (-1)^i \delta) C_0 \text{ for even } d.$$

Lemma 48. *For any anisotropic vector $w \in V$, $\rho_w(P_i) = P_{i+1}$.* □

We are now ready to compute $\mathcal{U}_n\check{}$.

Lemma 49. *$\mathcal{U}_n\check{} \cong \mathcal{U}_n(2n+1)$; in particular, $\mathcal{U}\check{} \cong \mathcal{U}(2d-1)$.*

Proof. We have chosen a basis $\{v_0, v_1, \ldots, v_{d+1}\}$ of V in 7.1.1 above. The set of naturally ordered products of several v_i's with even number of factors forms a basis of C_0. Define a quadratic form Q on C_0 as follows: let the distinct products of elements of the basis $\{v_0, v_1, \ldots, v_{d+1}\}$ be orthogonal to each other and let

$$Q(v_{i_1} \cdot v_{i_2} \cdot \ldots \cdot v_{i_k}) = q(v_{i_1}) \cdot q(v_{i_2}) \cdot \ldots \cdot q(v_{i_k}).$$

The form Q is nonsingular and defines – by scalar extension – a nonsingular symmetric bilinear form Δ on $\mathcal{O}_X \otimes C_0$. Since $(q(v_i))^2 = 1$, a direct computation shows that $\mathrm{Im}(\mathcal{O}_X(-1) \otimes C_1 \xrightarrow{\varphi} \mathcal{O}_X \otimes C_0) = \varphi \cdot \mathcal{U}_0 \cong \mathcal{U}_0$ is a totally isotropic subspace of $\mathcal{O}_X \otimes C_0$. Therefore

$$\mathcal{U}_0 \cong \varphi \cdot \mathcal{U}_0 = (\varphi \cdot \mathcal{U}_0)^{\perp} \cong ((\mathcal{O}_X \otimes C_0)/(\varphi \cdot \mathcal{U}_0))\check{} \cong \mathcal{U}_{-1}\check{}.$$

Thus

$$\mathcal{U}_0\check{} \cong \mathcal{U}_{-1} \cong \mathcal{U}_0(1)$$

and, in general,

$$\mathcal{U}_n\check{} \cong (\mathcal{U}_0(-n))\check{} \cong \mathcal{U}_0\check{}(n) \cong \mathcal{U}_0(n+1) \cong \mathcal{U}_n(2n+1).$$ □

Corollary 50. i) *$[\mathcal{U}\check{}] = [\mathcal{U}(2d-1)]$ and $[\mathcal{U}(d-1)] + [\mathcal{U}(d-1)]\hat{} = 2^{d+1}$ in $K_0(X)$;*
ii) $\mathrm{rank}\,\mathcal{U} = \frac{1}{2}\dim C_0 = 2^d$. □

In case of $d = 2m$ the algebra $\mathrm{End}_X(\mathcal{U}) = C_0$ splits into the direct product of subalgebras defined in 7.1.2 above: $C_0 = P_0 \times P_1$.

Definition 12. *In case of even d:*

$$\mathcal{U}'_n = \mathcal{U}_n \otimes_{C_0} P_0,\ \mathcal{U}''_n = \mathcal{U}_n \otimes_{C_0} P_1,$$
$$\mathcal{U}' = \mathcal{U} \otimes_{C_0} P_0,\ \mathcal{U}'' = \mathcal{U} \otimes_{C_0} P_1.$$

Note that $\mathcal{U}_n = \mathcal{U}'_n \oplus \mathcal{U}''_n$ and $\mathcal{U} = \mathcal{U}' \oplus \mathcal{U}''$. The summands $\mathcal{U}'_0$ and $\mathcal{U}''_0$ correspond to spinor representation and we shall use here the standard argument on dualization.

In the case of an even $d = 2m$, another property of φ and the quadratic form Q introduced in the proof of Lemma 49, may be verified by a direct computation:

Lemma 51. *In case of $d = 2m$*

i) *if m is even, then $P_i = (1 \pm \delta)C_0$ are orthogonal to each other, hence self-dual;*
ii) *if m is odd, then $P_i = (1 \pm \delta)C_0$ are totally isotropic, hence dual to each other;*
iii) $\varphi(1 \pm \delta) = (1 \mp \delta)\varphi$. □

Corollary 52. *In case of $d = 2m$,*

i) *$\mathcal{U}'\check{} \cong \mathcal{U}'(2d-1)$ and $\mathcal{U}''\hat{} \cong \mathcal{U}''(2d-1)$ for even m;*
ii) *$\mathcal{U}'\check{} \cong \mathcal{U}''(2d-1)$ and $\mathcal{U}''\hat{} \cong \mathcal{U}'(2d-1)$ for odd m;*

iii) $\mathrm{End}_X(\mathcal{U}') \cong \mathrm{End}_X(\mathcal{U}'') \cong M_{2^m}(F)$;
iv) *the exact sequence* (6.12) *splits into two exact sequences*

$$0 \to \mathcal{U}_0' \xrightarrow{\varphi\cdot} \mathcal{O}_X \otimes P_0 \to \mathcal{U}_0''(1) \to 0,$$
$$0 \to \mathcal{U}_0'' \xrightarrow{\varphi\cdot} \mathcal{O}_X \otimes P_1 \to \mathcal{U}_0'(1) \to 0. \quad \square$$

The trivial observation, that an automorphism of X induced by a reflection with respect to a nonsingular hyperplane interchanges $\mathcal{U}'$ with $\mathcal{U}''$, will be important in the following.

A standard way to determine indecomposable components is tensoring by the simple left module over an appropriate endomorphism algebra.

Definition 13.
i) *in case of* $d = 2m+1$: $\mathcal{V} = \mathcal{U} \otimes_{C_0} F^{2^{m+1}}$;
ii) *in case of* $d = 2m$: $\mathcal{V}_0 = \mathcal{U}' \otimes_{M_{2^m}(F)} F^{2^m}$, $\mathcal{V}_1 = \mathcal{U}'' \otimes_{M_{2^m}(F)} F^{2^m}$.

For convenience we will use mod 2 subscripts in $\mathcal{V}_i$. Since $M_n(F) = (F^n)^n$ as a left $M_n(F)$-module, indecomposable components inherit properties of the Swan bundle. We have:

Proposition 53.
a) *In case of* $d = 2m+1$:
i) $\mathcal{U} \cong \mathcal{V}^{2^{m+1}}$;
ii) $\mathcal{V}\check{\ } = \mathcal{V}(2d-1)$;
iii) $\mathrm{End}_X(\mathcal{V}) \cong F$ *and* $\mathrm{rank}\,\mathcal{V} = 2^m$;
iv) $[\mathcal{V}(d-1)] + [\mathcal{V}(d)] = 2^m$ *in* $K_0(X)$.
b) *In case of* $d = 2m$:
i) $\mathcal{U}' = \mathcal{V}_0^{2^m}$ *and* $\mathcal{U}'' = \mathcal{V}_1^{2^m}$;
ii) $\mathcal{V}_i\check{\ } = \mathcal{V}_{i+m}(2d-1)$;
iii) $\mathrm{End}_X(\mathcal{V}_i) \cong F$ *and* $\mathrm{rank}\,\mathcal{V}_i = 2^{m-1}$;
iv) $[\mathcal{V}_i(d-1)] + [\mathcal{V}_{i+1}(d)] = 2^m$ *in* $K_0(X)$. $\square$

In particular $\mathrm{Hom}_X(\mathcal{V}_i, \mathcal{V}_{i+1}) = 0$. Moreover, once again a reflection interchanges $\mathcal{V}_i$ with $\mathcal{V}_{i+1}$.

Example 35. *If* $d = 1$ *the conic* $X \overset{i}{\hookrightarrow} \mathbb{P}^2$ *given by* $x_0y_0 + {z_2}^2 = 0$ *is isomorphic to the projective line* $\mathbb{P}^1$:

$$\chi \;:\; \mathbb{P}^1 \xrightarrow{\sim} X,$$
$$\chi(t : u) = (t^2 : -u^2 : tu).$$

In this case

$$\mathcal{O}_X(-1) = i^*\left(\mathcal{O}_{\mathbb{P}^2}(-1)\right),$$

and the canonical truncated resolution of $\mathcal{O}_X(1)$ *is*

$$0 \to \mathcal{U} \to {\mathcal{O}_X}^3 \to \mathcal{O}_X(1) \to 0.$$

It follows that $\mathcal{U}$ has rank 2, and

$$\begin{aligned} \bigwedge^3 {\mathcal{O}_X}^3 &\cong \bigwedge^2 \mathcal{U} \otimes \mathcal{O}_X(1), \\ \bigwedge^2 \mathcal{U} &\cong \mathcal{O}_X(-1). \end{aligned}$$

Moreover,

$$\mathcal{V} = \chi_* \left(\mathcal{O}_{\mathbb{P}^1}(-1)\right) = \left(\chi^{-1}\right)^* \left(\mathcal{O}_{\mathbb{P}^1}(-1)\right),$$
$$\mathcal{V} \oplus \mathcal{V} = \mathcal{U},$$
$$\mathcal{V} \otimes \mathcal{V} = \bigwedge^2 \mathcal{U} = \mathcal{O}_X(-1).$$

□

Example 36. *In the case $d = 2$, the quadric surface $X \hookrightarrow \mathbb{P}^3$ given by $x_0y_0 + x_1y_1 = 0$ has two projections*

$$\begin{aligned} p_0 &: X \longrightarrow \mathbb{P}^1, \qquad p_0\left(x_0 : x_1 : y_0 : y_1\right) = \begin{cases} (x_0 : x_1) \\ (-y_1 : y_0) \end{cases}, \\ p_1 &: X \longrightarrow \mathbb{P}^1, \qquad p_1\left(x_0 : x_1 : y_0 : y_1\right) = \begin{cases} (x_0 : y_1) \\ (-x_1 : y_0) \end{cases} \end{aligned}$$

which define an isomorphism

$$\psi : X \longrightarrow \mathbb{P}^1 \times \mathbb{P}^1.$$

In this case one may check that

$$\mathcal{V}_0(1) = {p_0}^* \left(\mathcal{O}_{\mathbb{P}^1}(-1)\right), \qquad \mathcal{V}_1(1) = {p_1}^* \left(\mathcal{O}_{\mathbb{P}^1}(-1)\right)$$

or vice versa, and

$$\mathcal{O}_X(-1) = \mathcal{V}_0(1) \otimes \mathcal{V}_1(1).$$

□

Let $H = [\mathcal{O}_X] - [\mathcal{O}_X(-1)]$ be the class of a hyperplane section in $K_0(X)$.

Corollary 54. *In case of $d = 2m$ the following identities hold in $K_0(X)$:*

i) $([\mathcal{V}_0] - [\mathcal{V}_1]) \cdot H = 0$;
ii) $([\mathcal{V}_0] - [\mathcal{V}_1]) \cdot [\mathcal{O}_X(n)] = [\mathcal{V}_0] - [\mathcal{V}_1]$;
iii) $([\mathcal{V}_0] - [\mathcal{V}_1])\check{} = (-1)^m([\mathcal{V}_0] - [\mathcal{V}_1])$.

Proof. Proposition 53.b) iv) yields

$$[\mathcal{V}_0(d-1)] + [\mathcal{V}_1(d)] = [\mathcal{V}_1(d-1)] + [\mathcal{V}_0(d)].$$

Tensoring by $\mathcal{O}_X(-d)$ one obtains

$$[\mathcal{V}_0] - [\mathcal{V}_1] = ([\mathcal{V}_0] - [\mathcal{V}_1]) \cdot [\mathcal{O}_X(-1)],$$

hence i) and ii). Thus iii) results from 53. b) ii). □

Proposition 55. *$K_\bullet(X)$ is a free $K_\bullet(F)$-module of rank $2m + 2$; moreover*

i) *in the case of $d = 2m + 1$ the classes $[\mathcal{O}_X]$, $[\mathcal{O}_X(-1)]$, ..., $[\mathcal{O}_X(1 - d)]$, $[\mathcal{V}]$ form a basis of $K_\bullet(X)$;*

ii) *in the case of* $d = 2m$ *the classes* $[\mathcal{O}_X]$, $[\mathcal{O}_X(-1)]$, $\ldots$, $[\mathcal{O}_X(1-d)]$, $[\mathcal{V}_0]$, $[\mathcal{V}_1]$ *form a basis of* $K_\bullet X$.

Proof. Apply Theorem 46. □

In case of a (-1)-regular $\mathcal{F}$ to obtain an expression for $[\mathcal{F}] \in K_0(X)$ in terms of the basis from Proposition 55 one truncates the canonical resolution of $\mathcal{F}$:

$$0 \to \mathcal{Z}_{d-1} \to \mathcal{O}_X(1-d)^{k_{d-1}} \to \cdots \to \mathcal{O}_X(-1)^{k_1} \to \mathcal{O}_X{}^{k_0} \to \mathcal{F} \to 0$$

and replaces $\mathcal{Z}_{d-1}$ by $\mathcal{U} \otimes_{C_0} \operatorname{Hom}_X(\mathcal{U}, \mathcal{Z}_{d-1}) \cong \mathcal{Z}_{d-1}$. Then in $K_0(X)$

$$[\mathcal{F}] = \sum_{i=0}^{d-1} (-1)^i k_i [\mathcal{O}_X(-i)] + [\mathcal{U} \otimes_{C_0} \operatorname{Hom}_X(\mathcal{U}, \mathcal{Z}_{d-1})].$$

Depending on the parity of d we have either

$$[\mathcal{U} \otimes_{C_0} \operatorname{Hom}_X(\mathcal{U}, \mathcal{Z}_{d-1})] = a[\mathcal{V}]$$

or

$$[\mathcal{U} \otimes_{C_0} \operatorname{Hom}_X(\mathcal{U}, \mathcal{Z}_{d-1})] = a[\mathcal{V}_0] + b[\mathcal{V}_1],$$

where the integers a, b in turn depend on the decomposition of $\operatorname{Hom}_X(\mathcal{U}, \mathcal{Z}_{d-1})$ into a direct sum of simple left C_0-modules. Conversely, if for a given $\mathcal{F}$ the equality

$$[\mathcal{F}] = \sum_{i=0}^{d-1} (-1)^i k_i [\mathcal{O}_X(-i)] + W$$

holds, where W is either $a[\mathcal{V}]$ or $a[\mathcal{V}_0] + b[\mathcal{V}_1]$, then k_0 is the Euler characteristic $\sum (-1)^i \dim \mathrm{H}^i(X, \mathcal{F})$ of $\mathcal{F}$. So if $\mathcal{F}$ is regular, then $k_0 = \dim \Gamma(X, \mathcal{F})$. Next, $\mathcal{Z}_0(1) = \operatorname{Ker}\big(\mathcal{O}_X(1)^{k_0} \to \mathcal{F}(1)\big)$ is regular, and iterating this as in the recursive process of constructing the canonical resolution, one obtains that for a regular $\mathcal{F}$ the congruence

$$[\mathcal{F}] \equiv \sum_{i=0}^{d-1} (-1)^i k_i [\mathcal{O}_X(-i)] \qquad \operatorname{mod} \operatorname{Im}(K_0(C_0) \xrightarrow{u} K_0(X))$$

holds if and only if the integers the k_i satisfy (6.10). In case of $d = 2m+1$, in order to express class $[\mathcal{F}]$ of a regular sheaf $\mathcal{F}$ in terms of the basis of Proposition 55, it is enough to know the dimensions of $\Gamma(X, \mathcal{F}(i))$ for $i = 0, 1, 2, \ldots, d-1$ to determine the k_i's. Then the rank of $\mathcal{F}$ is sufficient to determine the coefficient a of $[\mathcal{V}]$. An analogous statement remains valid for an arbitrary sheaf $\mathcal{F}$ with the Euler characteristic of $\mathcal{F}(i)$ in place of $\dim \Gamma(X, \mathcal{F}(i))$. In case of $d = 2m$, in view of Corollary 54 ii) and Proposition 53 ii), the bundles $\mathcal{V}_0$ and $\mathcal{V}_1$ have the same Euler characteristic, rank and even the highest exterior power. Thus, without special considerations, one can express a class $[\mathcal{F}]$ in terms of the basis of Proposition 55 only up to a multiple of $[\mathcal{V}_0] - [\mathcal{V}_1]$.

7.1.3. Generating function for a truncated canonical resolution. Now we compute some generating functions for canonical resolutions.

Example 37. *For a linear section* $H^l = (1 - [\mathcal{O}_X(-1)])^l$ *of codimension* l *in* X

$$G_{H^l}(t) = (1+t)^l. \qquad \square$$

Example 38. *We follow the notation of 7.1.1. Since* X *splits, it contains linear subvarieties* $S_k = \operatorname{Proj} F[x_0, \dots, x_k]$ *given by the following equations:*

a) *in case of* $d = 2m$*:*

$$y_0 = \dots = y_m = x_{k+1} = \dots = x_m = 0 \text{ for } k < m \text{ and}$$
$$y_0 = \dots = y_m = 0 \text{ for } k = m;$$

b) *in case of* $d = 2m+1$*:*

$$y_0 = \dots = y_m = z_d = x_{k+1} = \dots = x_m = 0 \text{ for } k < m \text{ and}$$
$$y_0 = \dots = y_m = z_d = 0 \text{ for } k = m.$$

S_k *is isomorphic to* $\mathbb{P}^k_F$*, in particular its structural sheaf* $\mathcal{L}_k$ *is regular. Therefore*

$$G_{\mathcal{L}_k}(t) = \frac{P_k(-t)}{Q_d(-t)} = \frac{(1+t)^{-k-1}}{(1-t)\,/(1+t)^{d+1}} = \frac{(1+t)^{d-k}}{1-t}.$$

Lemma 56.
$$2G_{\mathcal{L}_k} - G_{\mathcal{L}_{k-1}} = (1+t)^{d-k}. \qquad \square$$

To obtain a compact formula for the truncated canonical resolution of a sheaf $\mathcal{F}$, we truncate $G_{\mathcal{F}}$, neglecting all terms of degree $\geq d$. Truncating generating function $G_{\mathcal{F}}$ one obtains a polynomial $T_{\mathcal{F}}$. For $l < d$ the canonical resolution for H^l is itself truncated:

$$T_{H^l}(t) = (1+t)^l \qquad \text{for} \qquad l < d.$$

The sequence (c_i) of coefficients of the canonical resolution of the sheaf $\mathcal{L}_k$ stabilizes starting from the degree $d-k$ onwards:

$$G_{\mathcal{L}_k}(t) = \frac{(1+t)^{d-k}}{1-t} = (1+t)^{d-k} \cdot \sum_{i=0}^{\infty} t^i = \sum_{i=0}^{\infty} c_i t^i$$

so

$$c_{d-k} = c_{d-k+1} = \dots = 2^{d-k}.$$

Thus

$$T_{\mathcal{L}_k}(t) = \frac{(1+t)^{d-k} - 2^{d-k} t^d}{1-t}.$$

Proposition 57. *If, for a fixed* k*,* $\mathcal{L}_k$ *is the structural sheaf of a linear subvariety* S_k *of dimensionk in* X*, then in* $K_0(X)$*:*

a) *in case of* $d = 2m+1$

$$[\mathcal{L}_k] = \sum_{i=0}^{d-1} \left(\sum_{p=0}^{i} \binom{d-k}{p} \right) (-1)^i [\mathcal{O}_X(-i)] + 2^{m-k}[\mathcal{V}];$$

b) *in case of* $d = 2m$ *for a suitable integer* a,

$$[\mathcal{L}_k] = \sum_{i=0}^{d-1} \left(\sum_{p=0}^{i} \binom{d-k}{p} \right) (-1)^i [\mathcal{O}_X(-i)] + a[\mathcal{V}_0] + (2^{m-k} - a)[\mathcal{V}_1];$$

c) $2[\mathcal{L}_k] - [\mathcal{L}_{k-1}] = H^{d-k}$.

Proof. Substituting $t = -[\mathcal{O}_X(-1)]$ into the expansion for $T_{\mathcal{L}_k}(t)$ yields, depending on the parity of d, the expressions

$$[\mathcal{L}_k] = \sum_{i=0}^{d-1} \left(\sum_{p=0}^{i} \binom{d-k}{p} \right) (-1)^i [\mathcal{O}_X(-i)] + a[\mathcal{V}];$$

$$[\mathcal{L}_k] = \sum_{i=0}^{d-1} \left(\sum_{p=0}^{i} \binom{d-k}{p} \right) (-1)^i [\mathcal{O}_X(-i)] + a[\mathcal{V}_0] + b[\mathcal{V}_1].$$

for suitable integers a, b. Thus

$$0 = \operatorname{rank}[\mathcal{L}_k] = \begin{cases} T_{\mathcal{L}_k}(-1) + (-1)^d a \cdot 2^m = \\ \quad = (-1)^d (2^m a - 2^{d-k-1}) & \text{for } d = 2m+1 \\ T_{\mathcal{L}_k}(-1) + (-1)^d (a+b) \cdot 2^{m-1} = \\ \quad = (-1)^d (2^{m-1}(a+b) - 2^{d-k-1}) & \text{for } d = 2m. \end{cases}$$

To prove c) it is enough to show that $2T_{\mathcal{L}_k}(t) - T_{\mathcal{L}_{k-1}}(t) = (1+t)^{d-k}$, which follows directly from Lemma 56. □

7.1.4. The topological filtration. Now we shall find a basis of $K_0(X)$ which is convenient for computations. Since the quadric X is regular, $K_0'(X) = K_0(X)$ and one may transfer the topological filtration

$$\mathrm{F}^p K_0'(X) = \text{ subgroup generated by}$$
$$\left\{ [\mathcal{F}] : \begin{array}{c} \text{the stalk } \mathcal{F}_x = 0 \text{ for all generic points} \\ x \text{ of subvarieties of codimension} < p \end{array} \right\}$$

of $K_0'(X)$ to $K_0(X)$. We know that for a split projective quadric X the Chow groups $CH^p(X)$ are isomorphic to the corresponding factors of the topological filtration:

$$CH^p(X) \cong \mathrm{F}^p K_0(X) / \mathrm{F}^{p+1} K_0(X).$$

We follow the notation of 7.1.1. The K-cohomology computation 26 yields Chow groups of the quadric X.

Proposition 58. *For a split projective quadric X of dimension d*

a) *in case of* $d = 2m$,

$$CH^p(X) \cong \mathbb{Z} \textit{ for } p \neq m,\ 0 \le p \le 2m \textit{ and } CH^m(X) \cong \mathbb{Z} \oplus \mathbb{Z};$$

b) *in case of* $d = 2m+1$,

$$CH^p(X) \cong \mathbb{Z} \textit{ for all } 0 \le p \le 2m+1.$$

Explicit generators are given as follows:

Case $d = 2m$*:*

i) *for* $p > m$*, the class of any linear subvariety of dimension* $d - p$*, e.g.,* S_{d-p} : $y_0 = \cdots = y_m = x_{d-p+1} = \cdots = x_m = 0$;

ii) *for* $p < m$*, the class* H^p *of a linear section of codimension* p*;*

iii) *for* $p = m$*,* $CH^m(X)$ *is generated by two classes of linear subvarieties* S'_m : $x_0 = \cdots = x_m = 0$ *and* S''_m : $y_0 = x_1 = \cdots = x_m = 0$; *the classes in* $CH^m(X)$ *remain unchanged if an even number of* x_i*'s is replaced by corresponding* y_i *in these equations.*

Case $d = 2m + 1$*:*

i) *for* $p > m$*, the class of any linear subvariety of dimension* $d - p$*, e.g.,* S_{d-p} : $y_0 = \cdots = y_m = z_{d+1} = x_{d-p+1} = \cdots = x_m = 0$;

ii) *for* $p \le m$*, a class* H^p *of a linear section of codimension* p*.*

Note that the reflection of Lemma 47 interchanges one x_i with y_i; so it interchanges S'_m with S''_m. Moreover, this reflection interchanges $[\mathcal{V}_0]$ with $[\mathcal{V}_1]$.

Now we can give an explicit description of the ring structure in $K_0(X)$. To do this denote $L_p = [\mathcal{L}_p]$ the class of the structural sheaf of the linear subvariety S_p of dimension p. Moreover, in case of an even $d = 2m$ denote by L'_m and L''_m the class of the structural sheaf of S'_m and S''_m respectively.

Theorem 59. *Let* X *be a split projective quadric of dimension* d*. Then*

i) *in the case of* $d = 2m+1$ *the classes* $1, H, H^2, \ldots, H^m, L_m, \ldots, L_0$ *form a basis of the free abelian group* $K_0(X)$*;*

ii) *in the case of* $d = 2m$ *the classes* $1, H, H^2, \ldots, H^{m-1}, L'_m, L''_m, L_{m-1}, \ldots, L_0$ *form a basis of the free abelian group* $K_0(X)$*;*

iii) *in the case of* $d = 2m$ *the classes of (co)dimension* m *may be chosen as follows:*

$$L'_m = \sum_{i=0}^{d-1}\left(\sum_{p=0}^{i}\binom{m}{p}\right)(-1)^i[\mathcal{O}_X(-i)] + [\mathcal{V}_0],$$

$$L''_m = \sum_{i=0}^{d-1}\left(\sum_{p=0}^{i}\binom{m}{p}\right)(-1)^i[\mathcal{O}_X(-i)] + [\mathcal{V}_1]$$

and for dimensions $k < m$

$$L_k = \sum_{i=0}^{d-1}\left(\sum_{p=0}^{i}\binom{d-k}{p}\right)(-1)^i[\mathcal{O}_X(-i)] + 2^{m-k-1}([\mathcal{V}_0] + [\mathcal{V}_1]);$$

iv) *if* $d = 2m$*, then* $H^m = L'_m + L''_m - L_{m-1}$*;*

v) $H \cdot L_p = L_{p-1}$, $H \cdot L'_m = H \cdot L''_m = L_{m-1}$*;*

vi) $H^{d-k} = 2L_k - L_{k-1}$ *for* $k \le \dfrac{d-1}{2}$*,* $H^d = 2L_0$*,* $H^{d+1} = 0$*;*

vii) $L_p \cdot L_q = L_p \cdot L'_m = L_p \cdot L''_m = 0$;
viii) *if* $d = 2m$ *and* m *is even, then* ${L'_m}^2 = {L''_m}^2 = L_0$, $L'_m \cdot L''_m = 0$, *if* $d = 2m$ *and* m *is odd, then* ${L'_m}^2 = {L''_m}^2 = 0$, $L'_m \cdot L''_m = L_0$.

Proof. Obviously H^i belongs to ith group $F^iK_0(X)$ of topological filtration. We first check that $H^{d+1} = 0$. If $\pi : X \to \mathbb{P}^d_F$ is a two-fold cover ramified along non-singular hyperplane section, then $\pi^* : K_0\left(\mathbb{P}^d_F\right) \to K_0(X)$ is a ring homomorphism and

$$\pi^*\left(\mathcal{O}_{\mathbb{P}^d_F}(n)\right) = \mathcal{O}_X(n)$$

so $\pi^*(H) = H$ and $H^{d+1} = 0$.

Next, $L_p \in F^{d-p}K_0(X)$, $L'_m, L''_m \in F^mK_0(X)$. To verify iii), recall that the reflection ρ_{v_1} fixes $v_0, v_2, \ldots, v_{d+1}$ and $\rho_{v_1}(v_1) = -v_1$ (7.1.2 above). Thus, this reflection induces an automorphism of the symmetric algebra $S(V\check{\ })$, which interchanges x_0 with y_0 and fixes other coordinates and q. Therefore it induces an automorphism of $S(V\check{\ })/(q)$, $X = \operatorname{Proj} S(V\check{\ })/(q)$, a semilinear automorphism of $\mathcal{O}_X(n)$ for all n, and an automorphism of $K_0(X)$. By Lemma 48 ii), the reflection ρ_{v_1} interchanges the P_i's. So the induced automorphism of $\mathcal{U}$ interchanges direct summands $\mathcal{U}' = \mathcal{U} \otimes_{C_0} P_0$ and $\mathcal{U}'' = \mathcal{U} \otimes_{C_0} P_1$ of $\mathcal{U}$ and their indecomposable components $\mathcal{V}_0, \mathcal{V}_1$. Therefore, the induced automorphism of $K_0(X)$ fixes the basic elements $[\mathcal{O}_X]$, $[\mathcal{O}_X(-1)]$, $\ldots$, $[\mathcal{O}_X(1-d)]$ and interchanges $[\mathcal{V}_0]$ with $[\mathcal{V}_1]$. This automorphism fixes $L_0, \ldots, L_{m-1}$. The explicit description given in Proposition 58 ii) shows that this automorphism interchanges L'_m with L''_m. Hence, by the explicit formula of Proposition 57 ii), for $k < m$, the integer a in the following formula

$$L_k = [\mathcal{L}_k] = \sum_{i=0}^{d-1}\left(\sum_{p=0}^{i}\binom{d-k}{p}\right)(-1)^i[\mathcal{O}_X(-i)] + a[\mathcal{V}_0] + (2^{m-k} - a)[\mathcal{V}_1]$$

must be equal to 2^{m-k-1}. This same argument for $k = m$ yields

$$L'_m = \sum_{i=0}^{d-1}\left(\sum_{p=0}^{i}\binom{d-k}{p}\right)(-1)^i[\mathcal{O}_X(-i)] + a[\mathcal{V}_0] + (1-a)[\mathcal{V}_1],$$

$$L''_m = \sum_{i=0}^{d-1}\left(\sum_{p=0}^{i}\binom{d-k}{p}\right)(-1)^i[\mathcal{O}_X(-i)] + (1-a)[\mathcal{V}_0] + a[\mathcal{V}_1].$$

From the K-cohomology computation, by induction on m, it follows that the classes

$$L'_m \bmod F^{m+1}K_0(X), \quad L''_m \bmod F^{m+1}K_0(X)$$

form a basis of $CH^m(X)$. Since statement ii) of the Theorem holds, the integer a must be 0 or 1 (this also follows from the regularity of the structural sheaves of S'_m and S''_m.) Statements i) and ii) follow from Proposition 58. Statements iv)–vii) are easy to see by explicit computations with generating functions for truncated canonical resolution, like in Proposition 57. To prove viii) assume, without loss of generality, that L'_m is the class of S'_m and L''_m is the class of S''_m. Consider the class

L_m of the subvariety $S_m : y_0 = \ldots = y_m = 0$. In the case of even m the classes L''_m and L_m coincide, and S_m, S'_m have no common points, so $L'_m \cdot L''_m = 0$. Moreover, S_m meets S''_m transversally at the rational point S_0, so ${L''_m}^2 = L_0$. Analogously, ${L'_m}^2 = L_0$.
In the case of odd m we have $L'_m = L_m$, so ${L'_m}^2 = {L''_m}^2 = 0$, $L'_m \cdot L''_m = L_0$. □

To obtain multiplicative rule in $CH^\bullet(X)$ it is enough to neglect the summand of lower dimension in iv) and vi):

$$\begin{aligned} \text{if } d &= 2m, \text{ then } H^m = L'_m + L''_m \text{ in } CH^m(X), \\ H^{d-k} &= 2L_k \text{ for } k \leq \frac{d-1}{2} \text{ in } CH^{d-k}(X). \end{aligned}$$

Exercise 6. *Refine the result to the form*

$$\begin{aligned} CH^\bullet(X) &\cong \mathbb{Z}[x,y]/\left(x^{m+1} - 2y, y^2\right) \text{ for } d = 2m+1, \\ CH^\bullet(X) &\cong \mathbb{Z}[x,y]/\left(x^{m+1} - 2xy, y^2\right) \text{ for } d = 2m,\ m \text{ odd}, \\ CH^\bullet(X) &\cong \mathbb{Z}[x,y]/\left(x^{m+1} - 2xy, y^2 - x^m y\right) \text{ for } d = 2m,\ m \text{ even}. \end{aligned}$$

Remark 2. *Chow ring of a quadric over an algebraically closed field was computed in* 1883 *by Segre (see* [26]*) as a cohomology ring of a quadric.*

7.2. Hilbert 90 for K_2 of fields

The celebrated Merkurjev Theorem from 1981, asserting that:

$$K_2(F)/2K_2(F) \cong H^2(F, \mu_2)$$

where the latter group is $H^2(Gal(F_s/F), \mu_2) = H^2_{ét}(\operatorname{Spec} F, \mu_2)$, and F_s is a separable closure of F, was a prototype of several Theorems proved by Merkurjev and Suslin. The main argument is "Hilbert 90 for K_2". We prove it for quadratic extensions; it remains valid for cyclic extensions of degree n provided F contains a primitive nth root of unity.

If $E = F[\sqrt{a}]$ is a quadratic extension, we denote by σ the nontrivial automorphism of E/F. The automorphism σ acts on $K_2(E)$ by

$$\sigma\{x, y\} = \{\sigma x, \sigma y\}.$$

There is the canonical homomorphism $r_{E/F} : K_2(F) \to K_2(E)$ induced by the inverse image functor $\mathcal{P}(\operatorname{Spec} F) \to \mathcal{P}(\operatorname{Spec} E)$:

$$r_{E/F}\{b, c\} = \{b, c\}.$$

The transfer map $N_{E/F} : K_2(E) \to K_2(F)$ is induced by direct image functor $\mathcal{P}(\operatorname{Spec} E) \to \mathcal{P}(\operatorname{Spec} F)$. It may be expressed by $\sigma : K_2(E) \to K_2(E)$ as follows.

Lemma 60. *The identity*

$$\{b, c\} = \left\{b + c, -\frac{b}{c}\right\}$$

holds in $K_2(F)$ *if* $b + c \neq 0$.

Proof. The identity $\dfrac{b}{b+c}+\dfrac{c}{b+c}=1$ implies

$$\begin{aligned}
\left\{\frac{b}{b+c},\frac{c}{b+c}\right\} &= 0 && \text{by (3.3)}\\
\{b,c\}-\{b+c,c\}-\{b,b+c\}+\{b+c,b+c\} &= 0 && \text{by (3.1)}\\
\{b,c\}-\{b+c,c\}+\{b+c,b\}+\{b+c,-1\} &= 0 && \text{by Exercise 3 (4) and (3)}\\
\{b,c\}+\{b+c,-\frac{b}{c}\} &= 0 && \text{by (3.1).} \quad \square
\end{aligned}$$

Lemma 61. *For a quadratic extension $E=F[\sqrt{a}]$, the map*

$$\begin{aligned}
K_1(F)\otimes K_1(E) &\to K_2(E),\\
\{b\}\otimes\{x\} &\mapsto \{b,x\}
\end{aligned}$$

is surjective.

Proof. Consider $\alpha=\{r+s\sqrt{a},t+u\sqrt{a}\}\in K_2(E)$. If $s=0$ or $u=0$, then α is the value of this map. The last case $\{s\sqrt{a},u\sqrt{a}\}$ is obvious by identity $\{x,x\}=\{-1,x\}$ (Exercise 3 (3)). In the remaining case by (3.1)

$$\begin{aligned}
\{r+s\sqrt{a},t+u\sqrt{a}\} &= \{ru+su\sqrt{a},t+u\sqrt{a}\}-\{u,t+u\sqrt{a}\}\\
&= \{ru+su\sqrt{a},-st-su\sqrt{a}\}-\{ru+su\sqrt{a},-s\}-\{u,t+u\sqrt{a}\}\\
&= \left\{ru-st,-\frac{ru+su\sqrt{a}}{-st-su\sqrt{a}}\right\}+\{-s,ru+su\sqrt{a}\}-\{u,t+u\sqrt{a}\}. \quad \square
\end{aligned}$$

Proposition 62. *For $b\in F^*$, $x\in E^*$*

$$N_{E/F}\{b,x\}=(1+\sigma)\{b,x\} \text{ and } N_{E/F}\{x,b\}=(1+\sigma)\{x,b\}.$$

Proof. It follows from the projection formula (Proposition 16) that:

$$\begin{aligned}
N_{E/F}\{b,x\} &= N_{E/F}\left(r_{E/F}\{b\}\cdot\{x\}\right)=\{b\}\cdot N_{E/F}\{x\}\\
&= \{b\}\cdot\{N_{E/F}x\}=\{b,(1+\sigma)x\}=(1+\sigma)\{b,x\}. \quad \square
\end{aligned}$$

Theorem 63 (Hilbert 90 for quadratic extensions). *For a quadratic extension $E=F[\sqrt{a}]$ the sequence*

$$K_2(E)\xrightarrow{1-\sigma}K_2(E)\xrightarrow{N_{E/F}}K_2(F)$$

is exact.

It is obvious that the sequence of the Theorem is a complex which functorially depends on F. Denote by $V_a(F)$ the homology group of this complex. To prove that $V_a(F)$ is trivial for all F, we first consider a particular case.

Proposition 64. *If the norm map $N_{E/F}:E^*\to F^*$ is surjective, then $V_a(F)=0$.*

Proof. A map

$$\begin{array}{rcl} F^* \otimes F^* & \to & K_2(E)/(1-\sigma)K_2(E) \\ b \otimes c & \mapsto & \{\beta, c\} \text{ if } N_{E/F}\beta = b \end{array}$$

is well defined by classical Hilbert 90. To show that it factors through $K_2(F)$ providing the inverse for $\overline{N_{E/F}} : K_2(E)/(1-\sigma)K_2(E) \to K_2(F)$, it is enough to show that it is a symbol, i.e., that

$$\text{if } N_{E/F}\beta + c = 1 \text{ then } \{\beta, c\} \in (1-\sigma)K_2(E).$$

- If $N_{E/F}\beta = b = 1 - c$ is a square in F, $b = d^2$, then

$$N_{E/F}\frac{\beta}{d} = \frac{b}{b} = 1,$$

so by classical Hilbert 90 there exists a $\gamma \in E^{\cdot}$ such that

$$\frac{\beta}{b} = \frac{\gamma}{\sigma\gamma},$$

and

$$\{\beta, c\} = \{\beta, 1-b\} = \left\{\frac{\beta}{b}, 1-b\right\}$$

since $\{b, 1-b\} = 0$,

$$\left\{\frac{\beta}{b}, 1-b\right\} = \left\{\frac{\gamma}{\sigma\gamma}, 1-b\right\} = (1-\sigma)\{\gamma, 1-b\} \in (1-\sigma)K_2(E).$$

- If $N_{E/F}\beta = b = 1 - c$ is not a square in F, then let $L = F(\sqrt{b})$, $M = E(\sqrt{b})$ and let τ be the nontrivial automorphism of M/E. We denote by σ the nontrivial automorphism of M/L, which is harmless, since its restriction to E is "the old" σ. We have

$$\{\beta, c\} = \{\beta, 1-b\} = \left\{\frac{\beta}{b}, 1-b\right\}$$

since $\{b, 1-b\} = 0$,

$$\left\{\frac{\beta}{b}, 1-b\right\} = 2\left\{\frac{\beta}{\sqrt{b}}, 1-b\right\}$$

by bimultiplicativity, and

$$2\left\{\frac{\beta}{\sqrt{b}}, 1-b\right\} = \left\{\frac{\beta}{\sqrt{b}}, 1-b\right\} + \left\{-\frac{\beta}{\sqrt{b}}, 1-b\right\} + \{-1, 1-b\}$$

by bimultiplicativity again. So

$$\begin{array}{rcl} \left\{-\frac{\beta}{\sqrt{b}}, 1-b\right\} & = & \tau\left\{\frac{\beta}{\sqrt{b}}, 1-b\right\} \\ \left\{\frac{\beta}{\sqrt{b}}, 1-b\right\} + \left\{-\frac{\beta}{\sqrt{b}}, 1-b\right\} & = & N_{M/E}\left\{\frac{\beta}{\sqrt{b}}, 1-b\right\} \end{array}$$

by Proposition 62, and finally

$$\{\beta, c\} = \{-1, 1-b\} + N_{M/E}\left\{\frac{\beta}{\sqrt{b}}, 1-b\right\}.$$

Now $N_{M/L}\frac{\beta}{\sqrt{b}} = 1$ and there exists $\gamma \in M^*$ such that

$$\frac{\beta}{\sqrt{b}} = \frac{\gamma}{\sigma\gamma}.$$

We have

$$\{-1, 1-b\} = (1-\sigma)\{\sqrt{a}, 1-b\} \in (1-\sigma)K_2(E),$$

and

$$\begin{aligned} N_{M/E}\left\{\frac{\beta}{\sqrt{b}}, 1-b\right\} &= N_{M/E}\left\{\frac{\gamma}{\sigma\gamma}, 1-b\right\} = N_{M/E}(1-\sigma)\{\gamma, 1-b\} \\ &= (1-\sigma)N_{M/E}\{\gamma, 1-b\} \in (1-\sigma)K_2(E). \end{aligned}$$

Thus there exists a homomorphism $K_2(F) \to K_2(E)/(1-\sigma)K_2(E)$ inverse to $\overline{N_{E/F}} : K_2(E)/(1-\sigma)K_2(E) \to K_2(F)$ and $V_a(F) = 0$. □

For a generalization to arbitrary cyclic extension of a field F containing appropriate roots of unity, see [27].

Next step shows how to enlarge the image of $N_{E/F} : K_1(E) = E^* \to F^* = K_1(F)$ not affecting $V_a(F)$. Since

$$N_{E/F}\left(x + y\sqrt{a}\right) = x^2 - ay^2,$$

any given $b \in F^*$ is a value of $N_{E(X)/F(X)}$ where X is a conic in $\mathbb{P}^2_F$ given by the equation $x^2 - ay^2 - bz^2 = 0$.

Proposition 65. *If for the projective conic* $X = \mathrm{Proj}F[x,y,z]/(x^2 - ay^2 - bz^2)$ *and* $X_E = X \times_{\mathrm{Spec}\,F} \mathrm{Spec}\,E$, *the induced map* $H^1(X, \mathcal{K}_2) \to H^1(X_E, \mathcal{K}_2)$ *is injective, then the map* $V_a(F) \to V_a(F(X))$ *is injective.*

Proof. Denote by $F(X)$ and $E(X)$ the function fields of conics X and X_E respectively. Since X_E has rational points, it is isomorphic to $\mathbb{P}^1_E$, which yields exactness of the sequence

$$\begin{aligned} K_2(E) = H^0(X_E, \mathcal{K}_2) \rightarrowtail K_2(E(X)) \xrightarrow{\partial} \coprod_{x \in (X_E)_1} K_1(E(x)) \twoheadrightarrow H^1(X_E, \mathcal{K}_2) \\ = K_1(E). \end{aligned}$$

A coset $\alpha + (1-\sigma)K_2(E)$ is in the kernel of $V_a(F) \to V_a(F(X))$ iff $N_{E/F}\alpha = 0$ and there exists $\beta \in K_2(E(X))$ such that $r_E(\alpha) = (1-\sigma)\beta$. A little chase in the commutative diagram:

$$\begin{array}{ccccccc}
H^1(X,\mathcal{K}_2) & \xrightarrow{\rho} & H^1(X_E,(\mathcal{K}_2) & & & & \\
\uparrow{\scriptstyle \kappa_F} & & \uparrow{\scriptstyle \kappa_E} & & & & \\
\coprod_{x\in X^1} K_1(F(x)) & \xrightarrow{\coprod r_x} & \coprod_{x\in {X_E}^1} K_1(E(x)) & \xrightarrow{1-\sigma} & \coprod_{x\in {X_E}^1} K_1(E(x)) & & \\
\uparrow{\scriptstyle \partial} & & \uparrow{\scriptstyle \partial} & & \uparrow{\scriptstyle \partial} & & \\
K_2(F(X)) & \xrightarrow{r_{F(X)}} & K_2(E(X)) & \xrightarrow{1-\sigma} & K_2(E(X)) & \xrightarrow{N} & K_2(F(X)) \\
 & & \uparrow{\scriptstyle r_E} & & \uparrow{\scriptstyle r_E} & & \uparrow{\scriptstyle r_F} \\
 & & K_2(E) & \xrightarrow[1-\sigma]{} & K_2(E) & \xrightarrow[N_{E/F}]{} & K_2(F)
\end{array}$$

- $\partial(1-\sigma)\beta = \partial r_E(\alpha) = 0$;
- $(1-\sigma)\partial\beta = \partial(1-\sigma)\beta = 0$;
- there exists $\gamma \in \coprod_{x\in X^1} K_1(F(x))$ such that $\coprod r_x(\gamma) = \partial\beta$;
- $\rho\kappa_F(\gamma) = \kappa_E\left(\coprod r_x(\gamma)\right) = \kappa_E\,(\partial\beta) = 0$;
- $\kappa_F(\gamma) = 0$;
- there exists $\delta \in K_2(F(X))$ such that $\partial\delta = \gamma$;
- $\partial\beta = \coprod r_x(\gamma) = \coprod r_x(\partial\delta) = \partial r_X(\delta)$;
- $\partial\,(\beta - r_X(\delta)) = 0$;
- there exists $\epsilon \in K_2(E)$ such that $\beta - r_X(\delta) = r_E(\epsilon)$;
- $r_E(\alpha) = (1-\sigma)\beta = (1-\sigma)\,(\beta - r_X(\delta)) = (1-\sigma)\,r_E(\epsilon) = r_E\,((1-\sigma)\,\epsilon)$;
- $\alpha = (1-\sigma)\,\epsilon$

shows that the coset $\alpha + (1-\sigma)K_2(E)$ is trivial. Thus the map $V_a(F) \to V_a(F(X))$ is injective. □

To show that the assumption of injectivity $H^1(X,\mathcal{K}_2) \rightarrowtail H^1(X_E,\mathcal{K}_2)$ is always valid, i.e., to compute $H^1(X,\mathcal{K}_2)$, we need several simple facts on K-theory of central simple algebras. First of all $K_\bullet(X) = K_\bullet(F) \oplus K_\bullet(C_0)$ and C_0 in this case is a quaternion algebra $\left(\frac{a,b}{F}\right)$. This algebra has basis $1, i, j, k$ such that

$$i^2 = a,\ j^2 = b,\ ij = -ji = k,\ k^2 = -ab,$$

and is a division algebra iff its reduced norm form $x^2 - ay^2 - bz^2 + abt^2$ has no nontrivial zeros (x, y, z, t), which is equivalent to the condition that the form $x^2 - ay^2 - bz^2$ has no nontrivial zeros, or that the conic X has no rational points.

Remark 3. *This equivalence is not so straightforward. It is so since the quadratic surface $x^2 - ay^2 - bz^2 + abt^2 = 0$ in $\mathbb{P}^3_F$ is isomorphic to $X \times X$. There is also an elementary proof in quadratic form theory, using Witt Cancellation Theorem*

([25, Chapter 1, Corollary 5.8]) *and basic property of Pfister forms* ([25, Chapter 4, Corollary 1.5]).

Any field extension L/F such that $\left(\frac{a,b}{F}\right) \otimes_F L = \left(\frac{a,b}{L}\right)$ splits (i.e., has zero divisors, so is a matrix algebra) is called a *splitting field* of $\left(\frac{a,b}{F}\right)$. Any maximal subfield $F(\alpha)$ of $\left(\frac{a,b}{F}\right)$, where α is a non-central element of the algebra, is its splitting field. The function field $F(X)$ is also a ("generic") splitting field of C_0.

Since modules over a division algebra D are classified up to isomorphism by their dimension,

$$K_0(D) = \mathbb{Z} \cdot [D].$$

The forgetful functor $\mathcal{P}\left(\left(\frac{a,b}{F}\right)\right) \to \mathcal{P}(F)$ is exact. The induced homomorphisms are norms $N_{C_0/F} : K_\bullet(C_0) \to K_\bullet(F)$; $N_{C_0/F}(K_0(C_0))$ has index 4 in $K_0(F) = \mathbb{Z}$, and $N_{C_0/F} : K_1(C_0) \to K_1(F) = F^*$ is a polynomial map of degree 4.

If L/F is a splitting field of C_0 which is finite over F, then consider the following composite functor:

$$\begin{array}{ccccc} \mathcal{P}(L) & \xrightarrow[\text{equivalence}]{\text{Morita}} & \mathcal{P}(M_2(L)) & = & \mathcal{P}(C_0 \otimes_F L) \\ & & & & \Big\downarrow \text{forgetful} \\ & & & & \mathcal{P}(C_0) \end{array}$$

which is obviously exact. Let $\theta_{L/C_0} : K_\bullet(L) \to K_\bullet(C_0)$ be the induced homomorphism. Another composition of functors for arbitrary splitting field M:

$$\mathcal{P}(C_0) \xrightarrow{-\otimes_F M} \mathcal{P}(C_0 \otimes_F M) = \mathcal{P}(M_2(M)) \xrightarrow[\text{equivalence}]{\text{Morita}} \mathcal{P}(M)$$

induces the homomorphism $p_{M/C_0} : K_\bullet(C_0) \to K_\bullet(L)$. Moreover, there is the following commutative diagram

$$\begin{array}{ccc} K_*(L) & \xrightarrow{N_{L/F}} & K_*(F) \\ {\scriptstyle \theta_{L/C_0}}\Big\downarrow & & \Big\downarrow \\ K_*(C_0) & \xrightarrow[p_{M/C_0}]{} & K_*(M) \end{array}.$$

The *reduced norm* $Nrd : K_p(C_0) \to K_p(F)$ is the homomorphism for which all diagrams

$$\begin{array}{ccc} K_p(L) & \xrightarrow{\theta_{L/C_0}} & K_p(C_0) \\ & {\scriptstyle N_{L/F}}\searrow \quad \swarrow{\scriptstyle Nrd} & \\ & K_p(F) & \end{array}$$

commute. It is known that the reduced norm exists for $p = 0, 1, 2$, and for division algebras C of square-free degree, $Nrd : K_1(C) \to K_1(F)$ is injective. In the case of quaternion algebra C, every nonzero element of $K_1(C) = C^*/[C^*, C^*]$ is in the image of $\theta_{L/C} : K_1(L) \to K_1(C)$ for some quadratic extension L of F. Thus Nrd for a quaternion algebra is a polynomial mapping of degree 2.

If L is a splitting field of C_0, then X_L has rational points, so it is isomorphic to the projective line. It is easy to check, on the level of functors, that for $h : X_L \to X$

$$h^* = \begin{pmatrix} r_{L/F} & 0 \\ 0 & p_{L/C_0} \end{pmatrix} : K_p(F) \oplus K_p(C_0) \to K_p(L) \oplus K_p(L).$$

If L/F is finite, then

$$h_* = \begin{pmatrix} N_{L/F} & 0 \\ 0 & \theta_{L/C_0} \end{pmatrix} : K_p(L) \oplus K_p(L) \to K_p(F) \oplus K_p(C_0).$$

Moreover, if $i_x : x \to X$ is a closed point, $\eta : \operatorname{Spec} F(X) \to X$ is the generic point, $f : X \to \operatorname{Spec} F$ is the structural map, then

$$i_{x*} = \begin{pmatrix} N_{F(x)/F} \\ -\theta_{F(x)/C_0} \end{pmatrix}, \qquad i^* = \left(r_{F(x)/F}, p_{F(x)/C_0}\right)$$

$$\eta^* = \left(r_{F(X)/F}, p_{F(X)/C_0}\right), \qquad f^* = \begin{pmatrix} 1 \\ 0 \end{pmatrix}.$$

The BGQ spectral sequence for the conic X has two columns, so it is the exact sequence

$$\cdots \xrightarrow{\partial} \coprod_{x \in X_1} K_i(F(x)) \xrightarrow{\sum_x i_{x*}} K_i(F) \oplus K_i(C_0) \xrightarrow{\eta^*} K_i(F(X)) \xrightarrow{\partial} \cdots .$$

Applying the automorphism $\begin{pmatrix} 1 & Nrd \\ 0 & 1 \end{pmatrix}$ to $K_i(F) \oplus K_i(C_0)$ for $i = 0, 1, 2$, one transforms it to

$$\cdots \xrightarrow{\partial} \coprod_{x \in X_1} K_i(F(x)) \xrightarrow{\sum \begin{pmatrix} 0 \\ -\theta_{F(x)/C_0} \end{pmatrix}} K_i(F) \oplus K_i(C_0) \xrightarrow{(r_{F(X)/F}, 0)} K_i(F(X)) \xrightarrow{\partial} \cdots .$$

It follows that

$$H^0(X, \mathcal{K}'_i) = \ker\left(K_i(F(X)) \xrightarrow{\partial} \coprod_{x \in X_1} K_{i-1}(F(x))\right) = \operatorname{im}(r_{F(X)/F}, 0) = K_i(F)$$

$$H^1(X, \mathcal{K}'_i) = \operatorname{co\,ker}\left(K_i(F(X)) \xrightarrow{\partial} \coprod_{x \in X_1} K_{i-1}(F(x))\right) \cong K_{i-1}(C_0).$$

Thus $H^1(X, \mathcal{K}'_2) = K_1(C_0)$ is identified with the subgroup $Nrd(K_1(C_0))$ of $K_1(F) = F^*$ and maps injectively into $H^1(X_E, \mathcal{K}'_2) = K_1(E) = E^*$.

Now we know that $V_a(F)$ injects into $V_a(F(X))$ and standard arguments show that $V_a(F) = 0$. Namely, put $F_0 = F$, F_{n+1} = compositum of $F_n(X_b)$ for all conics X_b given by $x^2 - ay^2 - bz^2 = 0$ for all $b \in F_n^*$. Thus

- $F_n^* \subset N_{F_{n+1}(\sqrt{a})/F_{n+1}}(F_{n+1}(\sqrt{a})^*)$
- $V_a(F_n) \hookrightarrow V_a(F_{n+1})$;

and therefore $V_a(F) \hookrightarrow V_a(F_{n+1})$. Next put $F_\infty = \bigcup_n F_n$; the map $N_{F_\infty(\sqrt{a})/F_\infty}$: $F_\infty(\sqrt{a})^* \to F_\infty^*$ is surjective and $V_a(F) \hookrightarrow V_a(F_\infty)$. But $V_a(F_\infty) = 0$ By Proposition 64. So $V_a(F) = 0$ and the Hilbert 90 for K_2 holds. □

References

[1] M. Artin, *Brauer Groups in Ring Theory and Algebraic Geometry*, in: *Brauer groups in ring theory and algebraic geometry*, Proc. Antwerp 1981, Lecture Notes in Math., **917** (1982), 194–210.

[2] H. Bass, *Algebraic K-theory*, NY, Benjamin 1968.

[3] P. Berthelot, A. Grothendieck, L. Illusie, *Théorie des intersections et théorème de Riemann-Roch* (SGA 6), Lecture Notes in Math. **225**, Springer, Berlin 1971.

[4] C. Chevalley, *The algebraic theory of spinors*, NY, Columbia Univ. Press 1954.

[5] J. Dieudonné, *La géométrie des groupes classiques*, 3rd ed. Springer, Berlin 1971.

[6] W. Fulton, *Intersection Theory*, MSM F. 3 B. 2, Springer, Berlin 1984.

[7] D.R. Grayson *Higher algebraic K-theory: II*, Lecture Notes in Math., **551** (1976) pp. 217–240.

[8] A. Grothendieck, *Classes de faisceaux et théorème de Riemann-Roch*, mimeographic notes, Princeton 1957 (reproduced in [3] as an appendix).

[9] A. Grothendieck, *Le groupe de Brauer I*, in: J. Giraud, A. Grothendieck, S.L. Kleiman, M. Raynaud, J. Tate, *Dix exposes sur la cohomologie des schemas*, North-Holland, Amsterdam 1968.

[10] R. Hartshorne, *Algebraic geometry*, GTM 52, Springer, Berlin 1977.

[11] B. Keller, *Chain complexes and stable categories*, Manuscripta Math. **67** (1990), pp. 379–417.

[12] R. Lee, R.H. Szczarba, *The group $K_3(\mathbb{Z})$ is cyclic of order 48*, Annals of Math. **104** (1976), pp. 31–60.

[13] S. Maclane, *Homology*, GMW b. 114, Springer, Berlin 1963.

[14] Y.I. Manin, *Lectures on algebraic K-functor in algebraic geometry* (in Russian), Usp. Mat. Nauk **24** No. 5 (1969), pp. 3–86.

[15] A.S. Merkurjev, A.A. Suslin, *K-cohomology of Severi-Brauer varieties and the norm residue homomorphism*, Izv. Akad. Nauk SSSR Ser. Matem. 46 (1982) No. 5 pp. 1011–1046 (in Russian); Math. USSR Izvestiya vol. **21** (1983) No. 2, pp. 307–340 (English translation).

[16] J. Milnor, *Algebraic K-theory and quadratic forms*, Inv. Math. **9** (1970), pp. 318–344.

[17] J. Milnor, *Introduction to algebraic K-theory*, Princeton Univ. Press, Princeton, New Jersey 1971.

[18] D. Mumford, *Lectures on curves on algebraic surface*, Annals of Math. Studies **59**, Princeton Univ. Press, Princeton, New Jersey 1966.

[19] A. Nenashev, *Double short exact sequences produce all elements of Quillens* K_1, Contemp. Math. **199**(1996), pp. 151–160.

[20] A. Nenashev, K_1 *by generators and relations*. J. Pure Appl. Alg. **131** (1998), 195–212.

[21] I.A. Panin, *K-theory of Grassmann fibre bundles and its twisted forms*, LOMI Preprints E-5-89, Leningrad 1989.

[22] I.A. Panin, *On algebraic K-theory of generalized flag fiber bundles and some their twisted forms*, Adv. in Soviet Math. **4** (1991), pp. 21–46.

[23] D. Quillen, *On the cohomology and K-theory of the general linear groups over a finite field*, Annals of Math. **96** (1972) No. 3, pp. 552–586.

[24] D. Quillen, *Higher algebraic K-theory I*, Lecture Notes in Math. **341**, Springer, Berlin 1973.

[25] W. Scharlau, *Quadratic and Hermitian forms*, GMW 270, Springer, Berlin 1985.

[26] C. Segre, *Studio sulle quadriche in uno spazio lineare ad un numero qualunque di dimensioni*, Mem. Della Reale Academia delle Scienze di Torino (2) **36** (1883), pp. 3–86

[27] A.A. Suslin, *Algebraic K-theory and norm residue homomorphism*, Results of science and technique, Contemp. problems of math. **25** (1984), pp. 115–209.

[28] R. G. Swan, *K-theory of quadric hypersurfaces*, Annals of Math. **122** (1985), 113–153.

[29] M. Szyjewski, *An invariant of quadratic forms over schemes*, Documenta Mathematica Journal DMV **1** (1996), 449–478.

[30] C. Weibel, *An Introduction to Algebraic K-theory*, `http://www.math.uiuc.edu/K-theory/0105/`.

Marek Szyjewski
Instytut Matematyki Uniwersytetu Śląskiego
Bankowa 14
PL-40007 Katowice, Poland
e-mail: `szyjewsk@gate.math.us.edu.pl`

Trends in Mathematics:
Topics in Cohomological Studies of Algebraic Varieties, 271–297

Gromov-Witten Invariants and Quantum Cohomology of Grassmannians

Harry Tamvakis

Abstract. This is the written version of my five lectures at the Banach Center mini-school on 'Schubert Varieties', in Warsaw, Poland, May 18–22, 2003.

Mathematics Subject Classification (2000). Primary 14N35; Secondary 14M15, 14N15, 05E15.

Keywords. Gromov-Witten invariants, Grassmannians, Flag varieties, Schubert varieties, Quantum cohomology, Littlewood-Richardson rule.

1. Lecture One

The aim of these lectures is to show that three-point genus zero Gromov-Witten invariants on Grassmannians are equal (or related) to classical triple intersection numbers on homogeneous spaces of the same Lie type, and to use this to understand the multiplicative structure of their (small) quantum cohomology rings. This theme will be explained in more detail as the lectures progress. Much of this research is part of a project with Anders S. Buch and Andrew Kresch, presented in the papers [Bu1], [KT1], [KT2], and [BKT1]. I will attempt to give the original references for each result as we discuss the theory.

1.1. The classical theory

We begin by reviewing the classical story for the type A Grassmannian. Let $E = \mathbb{C}^N$ and $X = G(m,E) = G(m,N)$ be the Grassmannian of m-dimensional complex linear subspaces of E. One knows that X is a smooth projective algebraic variety of complex dimension mn, where $n = N - m$.

The space X is stratified by Schubert cells; the closures of these cells are the *Schubert varieties* $X_\lambda(F_\bullet)$, where λ is a partition and

$$F_\bullet \;:\; 0 = F_0 \subset F_1 \subset \cdots \subset F_N = E$$

is a complete flag of subspaces of E, with $\dim F_i = i$ for each i. The partition $\lambda = (\lambda_1 \geqslant \lambda_2 \geqslant \cdots \geqslant \lambda_m \geqslant 0)$ is a decreasing sequence of nonnegative integers such that $\lambda_1 \leqslant n$. This means that the Young diagram of λ fits inside an $m \times n$ rectangle, which is the diagram of (n^m). We denote this containment relation of diagrams by $\lambda \subset (n^m)$. The diagram shown in Figure 1 corresponds to a Schubert variety in $G(4,10)$.

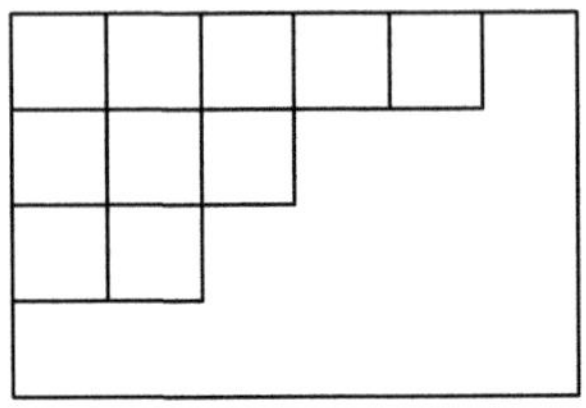

FIGURE 1. The rectangle (6^4) containing $\lambda = (5,3,2)$

The precise definition of $X_\lambda(F_\bullet)$ is

$$X_\lambda(F_\bullet) = \{\, V \in X \mid \dim(V \cap F_{n+i-\lambda_i}) \geqslant i, \ \forall\, 1 \leqslant i \leqslant m \,\}. \tag{1}$$

Each $X_\lambda(F_\bullet)$ is a closed subvariety of X of codimension equal to the weight $|\lambda| = \sum \lambda_i$ of λ. Using the Poincaré duality isomorphism between homology and cohomology, $X_\lambda(F_\bullet)$ defines a *Schubert class* $\sigma_\lambda = [X_\lambda(F_\bullet)]$ in $H^{2|\lambda|}(X,\mathbb{Z})$. The algebraic group $GL_N(\mathbb{C})$ acts transitively on X and on the flags in E. The action of an element $g \in GL_N(\mathbb{C})$ on the variety $X_\lambda(F_\bullet)$ is given by $g \cdot X_\lambda(F_\bullet) = X_\lambda(g \cdot F_\bullet)$. It follows that σ_λ does not depend on the choice of flag $F_\bullet$ used to define X_λ. As all cohomology classes in these lectures will occur in even degrees, we will adopt the convention that the degree of a class $\alpha \in H^{2k}(X,\mathbb{Z})$ is equal to k. The definition and basic properties of Schubert varieties in Grassmannians and flag manifolds are also discussed in Brion's lectures [Br, § 1].

We next review the classical facts about the cohomology of $X = G(m,N)$.

1) The additive structure of $H^*(X,\mathbb{Z})$ is given by

$$H^*(X,\mathbb{Z}) = \bigoplus_{\lambda \subset (n^m)} \mathbb{Z} \cdot \sigma_\lambda,$$

that is, $H^*(X,\mathbb{Z})$ is a free abelian group with basis given by the Schubert classes.

2) To describe the cup product in $H^*(X,\mathbb{Z})$, we will use Schubert's Duality Theorem. This states that for any λ and μ with $|\lambda|+|\mu| = mn$, we have $\sigma_\lambda \sigma_\mu = \delta_{\widehat{\lambda}\mu} \cdot [\mathrm{pt}]$, where $[\mathrm{pt}] = \sigma_{(n^m)}$ is the class of a point, and $\widehat{\lambda}$ is the dual partition to λ. The diagram of $\widehat{\lambda}$ is the complement of λ in the rectangle (n^m), rotated by 180°. This is illustrated in Figure 2.

For general products, we have an equation

$$\sigma_\lambda \, \sigma_\mu = \sum_{|\nu| = |\lambda| + |\mu|} c_{\lambda\mu}^{\nu} \, \sigma_\nu$$

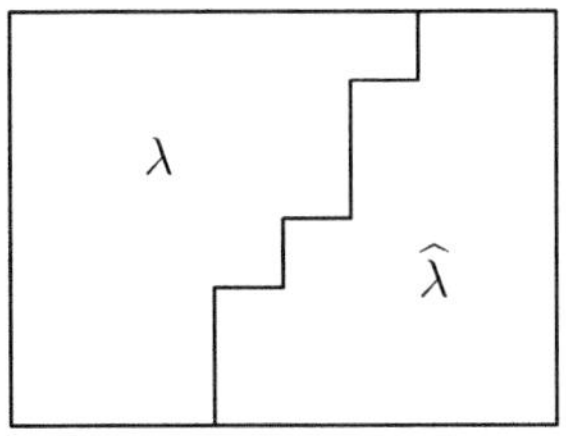

FIGURE 2. Dual Young diagrams

in $H^*(X,\mathbb{Z})$, and the *structure constants* $c^\nu_{\lambda\mu}$ are given by

$$c^\nu_{\lambda\mu} = \int_X \sigma_\lambda \sigma_\mu \sigma_{\widehat{\nu}} = \langle \sigma_\lambda, \sigma_\mu, \sigma_{\widehat{\nu}} \rangle_0 = \# X_\lambda(F_\bullet) \cap X_\mu(G_\bullet) \cap X_{\widehat{\nu}}(H_\bullet),$$

for general full flags $F_\bullet$, $G_\bullet$ and $H_\bullet$ in E. Later, we will discuss a combinatorial formula for these structure constants.

3) The classes $\sigma_1, \ldots, \sigma_n$ are called *special Schubert classes.* Observe that there is a unique Schubert class in codimension one: $H^2(X,\mathbb{Z}) = \mathbb{Z}\,\sigma_1$. If

$$0 \to S \to E_X \to Q \to 0 \tag{2}$$

is the tautological short exact sequence of vector bundles over X, with $E_X = X \times E$, then one can show that σ_i is equal to the ith Chern class $c_i(Q)$ of the quotient bundle Q, for $0 \leqslant i \leqslant n$.

Theorem 1 (Pieri rule, [Pi]**).** *For* $1 \leqslant p \leqslant n$ *we have* $\sigma_\lambda \, \sigma_p = \sum \sigma_\mu$*, where the sum is over all* $\mu \subset (n^m)$ *obtained from* λ *by adding* p *boxes, with no two in the same column.*

Example 1. Suppose $m = n = 2$ and we consider the Grassmannian $X = G(2,4)$ of 2-planes through the origin in $E = \mathbb{C}^4$. Note that X may be identified with the Grassmannian of all lines in projective 3-space $P(E) \cong \mathbb{P}^3$. The list of Schubert classes for X is

$$\sigma_0 = 1, \quad \sigma_1, \quad \sigma_2, \quad \sigma_{1,1}, \quad \sigma_{2,1}, \quad \sigma_{2,2} = [\mathrm{pt}].$$

Observe that the indices of these classes are exactly the six partitions whose diagrams fit inside a 2×2 rectangle. Using the Pieri rule, we compute that

$$\sigma_1^2 = \sigma_2 + \sigma_{1,1}, \; \sigma_1^3 = 2\,\sigma_{2,1}, \; \sigma_1^4 = 2\,\sigma_{2,2} = 2\,[\mathrm{pt}].$$

The last relation means that there are exactly 2 points in the intersection

$$X_1(F_\bullet) \cap X_1(G_\bullet) \cap X_1(H_\bullet) \cap X_1(I_\bullet),$$

for general flags $F_\bullet$, $G_\bullet$, $H_\bullet$, and $I_\bullet$. Since, e.g., $X_1(F_\bullet)$ may be identified with the locus of lines in $P(E)$ meeting the fixed line $P(F_2)$, this proves the enumerative fact that there are two lines in $\mathbb{P}^3$ which meet four given lines in general position.

4) Any Schubert class σ_λ may be expressed as a polynomial in the special classes in the following way. Let us agree here and in the sequel that $\sigma_p = 0$ if $p < 0$ or $p > n$.

Theorem 2 (Giambelli formula, [G]). *We have* $\sigma_\lambda = \det(\sigma_{\lambda_i+j-i})_{1\leqslant i,j\leqslant m}$, *that is,* σ_λ *is equal to a Schur determinant in the special classes.*

5) The ring $H^*(X,\mathbb{Z})$ is presented as a quotient of the polynomial ring $\mathbb{Z}[\sigma_1,\dots,\sigma_n]$ by the relations

$$D_{m+1} = \cdots = D_N = 0$$

where $D_k = \det(\sigma_{1+j-i})_{1\leqslant i,j\leqslant k}$. To understand where these relations come from, note that the Whitney sum formula applied to (2) says that $c_t(S)c_t(Q) = 1$, which implies, since $\sigma_i = c_i(Q)$, that $D_k = (-1)^k c_k(S) = c_k(S^*)$. In particular, we see that D_k vanishes for $k > m$, since S^* is a vector bundle of rank m.

1.2. Gromov-Witten invariants

Our starting point is the aforementioned fact that the classical structure constant $c_{\lambda\mu}^\nu$ in the cohomology of $X = G(m,N)$ can be realized as a triple intersection number $\#X_\lambda(F_\bullet)\cap X_\mu(G_\bullet)\cap X_{\widehat{\nu}}(H_\bullet)$ on X. The three-point, genus zero Gromov-Witten invariants on X extend these numbers to more general enumerative constants, which are furthermore used to define the 'small quantum cohomology ring' of X.

A *rational map of degree* d to X is a morphism $f\colon \mathbb{P}^1 \to X$ such that

$$\int_X f_*[\mathbb{P}^1]\cdot\sigma_1 = d,$$

i.e., d is the number of points in $f^{-1}(X_1(F_\bullet))$ when $F_\bullet$ is in general position.

Definition 1. Given a degree $d \geqslant 0$ and partitions λ, μ, and ν such that $|\lambda| + |\mu| + |\nu| = mn + dN$, we define the Gromov-Witten invariant $\langle\sigma_\lambda,\sigma_\mu,\sigma_\nu\rangle_d$ to be the number of rational maps $f\colon \mathbb{P}^1 \to X$ of degree d such that $f(0)\in X_\lambda(F_\bullet)$, $f(1)\in X_\mu(G_\bullet)$, and $f(\infty)\in X_\nu(H_\bullet)$, for given flags $F_\bullet$, $G_\bullet$, and $H_\bullet$ in general position.

We shall show later that $\langle\sigma_\lambda,\sigma_\mu,\sigma_\nu\rangle_d$ is a well-defined, finite integer. Notice that for the degree zero invariants, we have

$$\langle\sigma_\lambda,\sigma_\mu,\sigma_\nu\rangle_0 = \int_X \sigma_\lambda\sigma_\mu\sigma_\nu = \#X_\lambda(F_\bullet)\cap X_\mu(G_\bullet)\cap X_\nu(H_\bullet),$$

as a morphism of degree zero is just a constant map to X.

Key example. Consider the Grassmannian $G(d,2d)$ for any $d \geqslant 0$. We say that two points U, V of $G(d,2d)$ are in general position if the intersection $U\cap V$ of the corresponding subspaces is the zero subspace.

Proposition 1 ([BKT1]). *Let* U, V, *and* W *be three points of* $\mathcal{Z} = G(d,2d)$ *which are pairwise in general position. Then there is a unique morphism* $f\colon \mathbb{P}^1 \to \mathcal{Z}$ *of degree* d *such that* $f(0) = U$, $f(1) = V$, *and* $f(\infty) = W$. *In particular, the Gromov-Witten invariant which counts degree* d *maps to* $\mathcal{Z}$ *through three general points is equal to* 1.

Proof. Let U, V, and W be given, in pairwise general position. Choose a basis $(v_1, \ldots, v_d)$ of V. Then we can construct a morphism $f\colon \mathbb{P}^1 \to \mathcal{Z}$ of degree d such that $f(0) = U$, $f(1) = V$, and $f(\infty) = W$ as follows. For each i with $1 \leqslant i \leqslant d$, we let u_i and w_i be the projections of v_i onto U and W, respectively. If $(s{:}t)$ are the homogeneous coordinates on $\mathbb{P}^1$, then the morphism

$$f(s{:}t) = \operatorname{Span}\{su_1 + tw_1, \ldots, su_d + tw_d\}$$

satisfies the required conditions. Observe that f does not depend on the chosen basis for V. Indeed, if $v_i' = \sum a_{ij}v_j$, then $u_i' = \sum a_{ij}u_j$, $w_i' = \sum a_{ij}w_j$ and one checks easily that

$$\operatorname{Span}\{su_1 + tw_1, \ldots, su_d + tw_d\} = \operatorname{Span}\{su_1' + tw_1', \ldots, su_d' + tw_d'\}.$$

Exercise. Show that the map f is an embedding of $\mathbb{P}^1$ into $\mathcal{Z}$ such that $f(p_1)$ and $f(p_2)$ are in general position, for all points p_1, p_2 in $\mathbb{P}^1$ with $p_1 \neq p_2$. Show also that f has degree d.

Next, suppose that $f\colon \mathbb{P}^1 \to \mathcal{Z}$ is any morphism of degree d which sends 0, 1, ∞ to U, V, W, respectively. Let $S \subset \mathbb{C}^{2d} \otimes \mathcal{O}_{\mathcal{Z}}$ be the tautological rank d vector bundle over $\mathcal{Z}$, and consider the pullback $f^*S \to \mathbb{P}^1$. The morphism $f\colon \mathbb{P}^1 \to G(d, 2d)$ is determined by the inclusion of f^*S in $\mathbb{C}^{2d} \otimes \mathcal{O}_{\mathbb{P}^1}$, i.e., a point $p \in \mathbb{P}^1$ is mapped by f to the fiber over p of the image of this inclusion.

Every vector bundle over $\mathbb{P}^1$ splits as a direct sum of line bundles, so $f^*S \cong \oplus_{i=1}^d \mathcal{O}(a_i)$. Each $\mathcal{O}(a_i)$ is a subbundle of a trivial bundle, hence $a_i \leqslant 0$, and $\sum a_i = -d$ as f has degree d. We deduce that $a_i = -1$ for each i, since otherwise f^*S would have a trivial summand, and this contradicts the general position hypothesis. It follows that we can write $f(s{:}t) = \operatorname{Span}\{su_1 + tw_1, \ldots, su_d + tw_d\}$ for suitable vectors $u_i, w_i \in \mathbb{C}^{2d}$, which depend on the chosen identification of f^*S with $\oplus_{i=1}^d \mathcal{O}(-1)$. We conclude that f is the map constructed as above from the basis $(v_1, \ldots, v_d)$, where $v_i = u_i + w_i$. □

We now introduce the key definition upon which the subsequent analysis depends.

Definition 2 ([Bu1])**.** For any morphism $f\colon \mathbb{P}^1 \to G(m, N)$, define the *kernel* of f to be the intersection of all the subspaces $V \subset E$ corresponding to image points of f. Similarly, the *span* of f is the linear span of these subspaces.

$$\operatorname{Ker}(f) = \bigcap_{p \in \mathbb{P}^1} f(p); \qquad \operatorname{Span}(f) = \sum_{p \in \mathbb{P}^1} f(p).$$

Note that for each $f\colon \mathbb{P}^1 \to X$, we have $\operatorname{Ker}(f) \subset \operatorname{Span}(f) \subset E$.

Lemma 1 ([Bu1])**.** *If $f\colon \mathbb{P}^1 \to G(m, N)$ is a morphism of degree d, then* $\dim \operatorname{Ker}(f) \geqslant m - d$ *and* $\dim \operatorname{Span}(f) \leqslant m + d$.

Proof. Let $S \to X$ be the rank m tautological bundle over $X = G(m,N)$. Given any morphism $f\colon \mathbb{P}^1 \to X$ of degree d, we have that $f^*S \cong \oplus_{i=1}^m \mathcal{O}(a_i)$, where $a_i \leqslant 0$ and $\sum a_i = -d$. Moreover, the map f is induced by the inclusion $f^*S \subset E \otimes \mathcal{O}_{\mathbb{P}^1}$. There are at least $m-d$ zeroes among the integers a_i, hence f^*S contains a trivial summand of rank at least $m-d$. But this corresponds to a fixed subspace of E of the same dimension which is contained in $\mathrm{Ker}(f)$, hence $\dim \mathrm{Ker}(f) \geqslant m-d$.

Similar reasoning shows that if $Q \to X$ is the rank n universal quotient bundle over X, then f^*Q has a trivial summand of rank at least $n-d$. It follows that the image of the map $f^*S \to E \otimes \mathcal{O}_{\mathbb{P}^1}$ factors through a subspace of E of codimension at least $n-d$, and hence of dimension at most $m+d$. □

In the next lecture, we will see that for those maps f which are counted by a degree d Gromov-Witten invariant for X, we have $\dim \mathrm{Ker}(f) = m-d$ and $\dim \mathrm{Span}(f) = m+d$. In fact, it will turn out that the pair $(\mathrm{Ker}(f), \mathrm{Span}(f))$ determines f completely!

2. Lecture Two

2.1. The main theorem

Given integers a and b, we let $F(a,b;E) = F(a,b;N)$ denote the two-step flag variety parametrizing pairs of subspaces (A,B) with $A \subset B \subset E$, $\dim A = a$ and $\dim B = b$. We agree that $F(a,b;N)$ is empty unless $0 \leqslant a \leqslant b \leqslant N$; when the latter condition holds then $F(a,b;N)$ is a projective complex manifold of dimension $(N-b)b+(b-a)a$. For any non-negative integer d we set $Y_d = F(m-d, m+d; E)$; this will be the parameter space of the pairs $(\mathrm{Ker}(f), \mathrm{Span}(f))$ for the relevant morphisms $f\colon \mathbb{P}^1 \to X$. Our main theorem will be used to identify Gromov-Witten invariants on $X = G(m,E)$ with classical triple intersection numbers on the flag varieties Y_d.

To any subvariety $\mathcal{W} \subset X$ we associate the subvariety $W^{(d)}$ in Y_d defined by

$$\mathcal{W}^{(d)} = \{\, (A,B) \in Y_d \mid \exists\, V \in \mathcal{W} : A \subset V \subset B \,\}. \tag{3}$$

Let $F(m-d, m, m+d; E)$ denote the variety of three-step flags in E of dimensions $m-d$, m, and $m+d$. There are natural projection maps

$$\pi_1\colon F(m-d, m, m+d; E) \to X \quad \text{and} \quad \pi_2\colon F(m-d, m, m+d; E) \to Y_d.$$

We then have $\mathcal{W}^{(d)} = \pi_2(\pi_1^{-1}(\mathcal{W}))$. Moreover, as the maps π_i are GL_N-equivariant, if $\mathcal{W} = X_\lambda(F_\bullet)$ is a Schubert variety in X, then $\mathcal{W}^{(d)} = X_\lambda^{(d)}(F_\bullet)$ is a Schubert variety in Y_d. We will describe this Schubert variety in more detail after we prove the main theorem.

Remarks. 1) One computes that $\dim Y_d = mn + dN - 3d^2$.

2) Since the fibers of π_2 are isomorphic to $G(d,2d)$, the codimension of $X_\lambda^{(d)}(F_\bullet)$ in Y_d is at least $|\lambda| - d^2$.

Theorem 3 ([BKT1]). *Let λ, μ, and ν be partitions and d be an integer such that $|\lambda| + |\mu| + |\nu| = mn + dN$, and let $F_\bullet$, $G_\bullet$, and $H_\bullet$ be complete flags of $E = \mathbb{C}^N$ in general position. Then the map $f \mapsto (\mathrm{Ker}(f), \mathrm{Span}(f))$ gives a bijection of the set of rational maps $f\colon \mathbb{P}^1 \to G(m,N)$ of degree d satisfying $f(0) \in X_\lambda(F_\bullet)$, $f(1) \in X_\mu(G_\bullet)$, and $f(\infty) \in X_\nu(H_\bullet)$, with the set of points in the intersection $X_\lambda^{(d)}(F_\bullet) \cap X_\mu^{(d)}(G_\bullet) \cap X_\nu^{(d)}(H_\bullet)$ in $Y_d = F(m-d, m+d; N)$.*

It follows from Theorem 3 that we can express any Gromov-Witten invariant of degree d on $G(m,N)$ as a classical intersection number on Y_d. Let $[X_\lambda^{(d)}]$ denote the cohomology class of $X_\lambda^{(d)}(F_\bullet)$ in $H^*(Y_d, \mathbb{Z})$.

Corollary 1. *Let λ, μ, and ν be partitions and $d \geqslant 0$ an integer such that $|\lambda| + |\mu| + |\nu| = mn + dN$. We then have*

$$\langle \sigma_\lambda, \sigma_\mu, \sigma_\nu \rangle_d = \int_{F(m-d,m+d;N)} [X_\lambda^{(d)}] \cdot [X_\mu^{(d)}] \cdot [X_\nu^{(d)}].$$

Proof of Theorem 3. Let $f\colon \mathbb{P}^1 \to X$ be a rational map as in the statement of the theorem.

Claim 1. We have $d \leqslant \min(m,n)$, $\dim \mathrm{Ker}(f) = m - d$ and $\dim \mathrm{Span}(f) = m + d$.

Indeed, let $a = \dim \mathrm{Ker}(f)$ and $b = \dim \mathrm{Span}(f)$. In the two-step flag variety $Y' = F(a, b; E)$ there are associated Schubert varieties $X'_\lambda(F_\bullet)$, $X'_\mu(G_\bullet)$, and $X'_\nu(H_\bullet)$, defined as in (3). Writing $e_1 = m - a$ and $e_2 = b - m$, we see that the codimension of $X'_\lambda(F_\bullet)$ in Y' is at least $|\lambda| - e_1e_2$, and similar inequalities hold with μ and ν in place of λ. Since $(\mathrm{Ker}(f), \mathrm{Span}(f)) \in X'_\lambda(F_\bullet) \cap X'_\mu(G_\bullet) \cap X'_\nu(H_\bullet)$ and the three flags $F_\bullet$, $G_\bullet$ and $H_\bullet$ are in general position, we obtain

$$mn + dN - 3e_1e_2 \leqslant \dim F(a, b; E) = (N-b)(m+e_2) + (e_1+e_2)a,$$

and hence, by a short computation,

$$dN \leqslant 2e_1e_2 + e_2(N-b) + ae_1. \tag{4}$$

Lemma 1 says that $e_1 \leqslant d$ and $e_2 \leqslant d$, and therefore that the right-hand side of (4) is at most $2e_1e_2 + d(N - b + a)$. Since $b - a = e_1 + e_2$, it follows that

$$(e_1+e_2)^2 \leqslant 2d(e_1+e_2) \leqslant 4e_1e_2,$$

and hence $e_1 = e_2 = d$. This proves Claim 1.

Let $\mathcal{M}$ denote the set of rational maps in the statement of the theorem, and set $\mathcal{I} = X_\lambda^{(d)}(F_\bullet) \cap X_\mu^{(d)}(G_\bullet) \cap X_\nu^{(d)}(H_\bullet)$. If $f \in \mathcal{M}$ then Claim 1 shows that $(\mathrm{Ker}(f), \mathrm{Span}(f)) \in \mathcal{I}$. We next describe the inverse of the resulting map $\mathcal{M} \to \mathcal{I}$.

Given $(A, B) \in \mathcal{I}$, we let $\mathcal{Z} = G(d, B/A) \subset X$ be the set of m-dimensional subspaces of E between A and B. Observe that $\mathcal{Z} \cong G(d, 2d)$, and that $X_\lambda(F_\bullet) \cap \mathcal{Z}$, $X_\mu(G_\bullet) \cap \mathcal{Z}$, and $X_\nu(H_\bullet) \cap \mathcal{Z}$ are non-empty Schubert varieties in $\mathcal{Z}$. (Indeed, e.g., $X_\lambda(F_\bullet) \cap \mathcal{Z}$ is defined by the attitude of V/A with respect to the flag $\widetilde{F}_\bullet$ in B/A with $\widetilde{F}_i = ((F_i + A) \cap B)/A$ for each i.) We assert that each of $X_\lambda(F_\bullet) \cap \mathcal{Z}$, $X_\mu(G_\bullet) \cap \mathcal{Z}$, and $X_\nu(H_\bullet) \cap \mathcal{Z}$ must be a single point, and that these three points are subspaces of B/A in pairwise general position. Proposition 1 then provides the unique $f\colon \mathbb{P}^1 \to X$ in $\mathcal{M}$ with $\mathrm{Ker}(f) = A$ and $\mathrm{Span}(f) = B$.

Claim 2. Let U, V, and W be three points in $\mathcal{Z}$, one from each intersection. Then the subspaces U, V and W are in pairwise general position.

Assuming this claim, we can finish the proof as follows. Observe that any positive-dimensional Schubert variety in $\mathcal{Z}$ must contain a point U' which meets U non-trivially, and similarly for V and W. Indeed, on $G(d, 2d)$, the locus of d-dimensional subspaces Σ with $\Sigma \cap U \neq \{0\}$ is, up to a general translate, the unique Schubert variety in codimension 1. It follows that this locus must meet any other Schubert variety non-trivially, unless the latter is zero-dimensional, in other words, a point. Therefore Claim 2 implies that $X_\lambda(F_\bullet) \cap \mathcal{Z}$, $X_\mu(G_\bullet) \cap \mathcal{Z}$, and $X_\nu(H_\bullet) \cap \mathcal{Z}$ are three points on $\mathcal{Z}$ in pairwise general position.

To prove Claim 2, we again use a dimension counting argument to show that if the three reference flags are chosen generically, no two subspaces among U, V, W can have non-trivial intersection. Consider the three-step flag variety $Y'' = F(m-d, m-d+1, m+d; E)$ and the projection $\pi\colon Y'' \to Y_d$. Note that $\dim Y'' = \dim Y_d + 2d - 1$, as Y'' is a $\mathbb{P}^{2d-1}$-bundle over Y_d.

To each subvariety $\mathcal{W} \subset G(m, E)$ we associate $\mathcal{W}'' \subset Y''$ defined by

$$\mathcal{W}'' = \{\, (A, A', B) \in Y'' \mid \exists\, V \in \mathcal{W} : A' \subset V \subset B \,\}.$$

We find that the codimension of $X''_\mu(G_\bullet)$ in Y'' is at least $|\mu| - d^2 + d$, and similarly for $X''_\nu(H_\bullet)$. Since the three flags are in general position, and $\pi^{-1}(X^{(d)}_\lambda(F_\bullet))$ has codimension at least $|\lambda| - d^2$ in Y'', we must have

$$\pi^{-1}(X^{(d)}_\lambda(F_\bullet)) \cap X''_\mu(G_\bullet) \cap X''_\nu(H_\bullet) = \emptyset,$$

and the same is true for the other two analogous triple intersections. This completes the proof of Claim 2, and of the theorem. □

It is worth pointing out that we may rephrase Theorem 3 using rational curves in X, instead of rational maps to X. For this, recall from the Exercise given in the first lecture that every rational map f that is counted in Theorem 3 is an embedding of $\mathbb{P}^1$ into X of degree equal to the degree of the curve $\mathrm{Im}(f)$. Moreover, the bijection of the theorem shows that all of these maps have different images.

2.2. Parametrizations of Schubert varieties

We now describe an alternative way to parametrize the Schubert varieties on $G(m, N)$, by replacing each partition λ by a 01-string $I(\lambda)$ of length N, with m zeroes (compare with [Br, § 1.1]). Begin by drawing the Young diagram of the partition λ in the upper-left corner of an $m \times n$ rectangle. We then put a label on each step of the path from the lower-left to the upper-right corner of this rectangle which follows the border of λ. Each vertical step is labeled "0", while the remaining n horizontal steps are labeled "1". The string $I(\lambda)$ then consists of these labels in lower-left to upper-right order.

Example 2. On the Grassmannian $G(4,9)$, the 01-string of the partition $\lambda = (4,4,3,1)$ is $I(\lambda) = 101101001$. This is illustrated below.

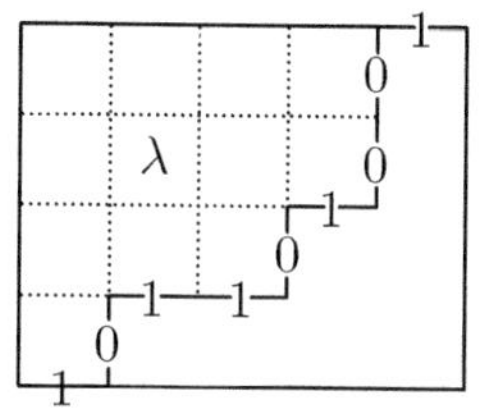

Alternatively, each partition $\lambda \subset (n^m)$ corresponds to a Grassmannian permutation w_λ in the symmetric group S_N, which is a minimal length representative in the coset space $S_N/(S_m \times S_n)$. The element $w = w_\lambda$ is such that the positions of the "0"s (respectively, the "1"s) in the 01-string $I(\lambda)$ are given by $w(1), \ldots, w(m)$ (respectively, by $w(m+1), \ldots, w(N)$). In Example 2, we have $w_\lambda = 257813469 \in S_9$.

In a similar fashion, the Schubert varieties on the two-step flag variety $F(a,b;N)$ are parametrized by permutations $w \in S_N$ with $w(i) < w(i+1)$ for $i \notin \{a,b\}$. For each such permutation w and fixed full flag $F_\bullet$ in E, the Schubert variety $X_w(F_\bullet) \subset F(a,b;N)$ is defined as the locus of flags $A \subset B \subset E$ such that

$$\dim(A\cap F_i) \geqslant \#\{p \leqslant a \mid w(p) > N-i\} \text{ and } \dim(B\cap F_i) \geqslant \#\{p \leqslant b \mid w(p) > N-i\}$$

for each i. The codimension of $X_w(F_\bullet)$ in $F(a,b;N)$ is equal to the *length* $\ell(w)$ of the permutation w. Furthermore, these indexing permutations w correspond to 012-strings $J(w)$ of length N with a "0"s and $b-a$ "1"s. The positions of the "0"s (respectively, the "1"s) in $J(w)$ are recorded by $w(1), \ldots, w(a)$ (respectively, by $w(a+1), \ldots, w(b)$).

Finally, we describe the 012-string $J^d(\lambda)$ associated to the modified Schubert variety $X_\lambda^{(d)}(F_\bullet)$ in $Y_d = F(m-d, m+d; N)$. This string is obtained by first multiplying each number in the 01-string $I(\lambda)$ by 2, to get a 02-string $2I(\lambda)$. We then get the 012-string $J^d(\lambda)$ by changing the first d "2"s and the last d "0"s of $2I(\lambda)$ to "1"s. Taking $d = 2$ in Example 2, we get $J^2(4,4,3,1) = 101202112$, which corresponds to a Schubert variety in $F(2,6;9)$.

Corollary 2 ([Y]). *Let λ, μ, and ν be partitions and $d \geqslant 0$ be such that $|\lambda| + |\mu| + |\nu| = mn + dN$. If any of λ_d, μ_d, and ν_d is less than d, then $\langle \sigma_\lambda, \sigma_\mu, \sigma_\nu \rangle_d = 0$.*

Proof. By computing the length of the permutation corresponding to $J^d(\lambda)$, one checks easily that when $\lambda_d < d$, the codimension of $X_\lambda^{(d)}(F_\bullet)$ in Y_d is strictly greater than $|\lambda| - d^2$. Therefore, when any of λ_d, μ_d, or ν_d is less than d, the sum of the codimensions of the three Schubert varieties $X_\lambda^{(d)}(F_\bullet)$, $X_\mu^{(d)}(G_\bullet)$, and

$X_\nu^{(d)}(H_\bullet)$ which appear in the statement of Theorem 3 is strictly greater than the dimension of $Y_d = F(m-d, m+d; N)$. We deduce that

$$\langle \sigma_\lambda, \sigma_\mu, \sigma_\nu \rangle_d = \int_{Y_d} [X_\lambda^{(d)}] \cdot [X_\mu^{(d)}] \cdot [X_\nu^{(d)}] = 0. \qquad \square$$

3. Lecture Three

3.1. Classical and quantum Littlewood-Richardson rules

The problem we turn to now is that of finding a positive combinatorial formula for the Gromov-Witten invariants $\langle \sigma_\lambda, \sigma_\mu, \sigma_\nu \rangle_d$. For the classical structure constants (the case $d = 0$), this problem was solved in the 1930's by Littlewood and Richardson, although complete proofs of their rule only appeared in the early 1970's. In the past few years, there has been a resurgence of interest in this question (see, for example, [F2]), which has led to a new formulation of the rule in terms of 'puzzles' (due to Knutson, Tao, and Woodward). The lectures of Buch [Bu4] discuss an extension of the Littlewood-Richardson rule to the K-theory of Grassmannians.

Define a *puzzle* to be a triangle decomposed into *puzzle pieces* of the three types displayed below.

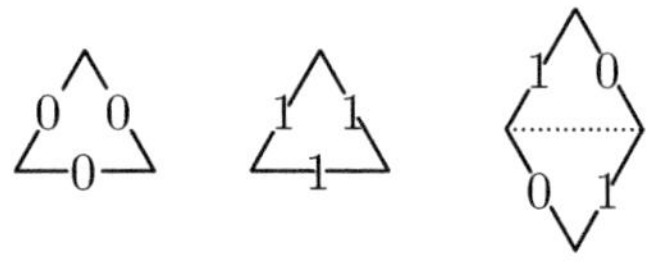

A puzzle piece may be rotated but not reflected when used in a puzzle. Furthermore, the common edges of two puzzle pieces next to each other must have the same labels. Recall from the last lecture that a Schubert class σ_λ in $H^*(G(m,N),\mathbb{Z})$ may also be indexed by a 01-string $I(\lambda)$ with m "0"s and n "1"s.

Theorem 4 ([KTW]). *For any three Schubert classes σ_λ, σ_μ, and σ_ν in the cohomology of $X = G(m,N)$, the integral $\int_X \sigma_\lambda \sigma_\mu \sigma_\nu$ is equal to the number of puzzles such that $I(\lambda)$, $I(\mu)$, and $I(\nu)$ are the labels on the north-west, north-east, and south sides when read in clockwise order.*

The formula in Theorem 4 is bijectively equivalent to the classical Littlewood-Richardson rule, which describes the same numbers as the cardinality of a certain set of Young tableaux (see [V, § 4.1]).

Example 3. In the projective plane $\mathbb{P}^2 = G(1,3)$, two general lines intersect in a single point. This corresponds to the structure constant

$$\langle \sigma_1, \sigma_1, \sigma_0 \rangle_0 = \int_{\mathbb{P}^2} \sigma_1^2 = 1.$$

The figure below displays the unique puzzle with the corresponding three strings 101, 101, and 011 on its north-west, north-east, and south sides.

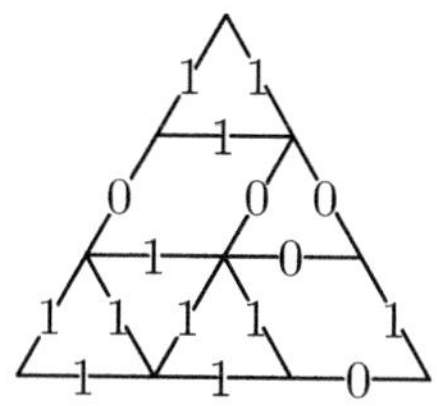

We suggest that the reader works out the puzzle which corresponds to the intersection $\sigma_k \sigma_\ell = \sigma_{k+\ell}$ on $\mathbb{P}^n$, for $k + \ell \leqslant n$.

It is certainly tempting to try to generalize Theorem 4 to a result that would hold for the flag variety SL_N/B. This time the three sides of the puzzle would be labeled by permutations, and one has to specify the correct set of puzzle pieces to make the rule work. In the fall of 1999, Knutson proposed such a general conjecture for the Schubert structure constants on all partial flag varieties, which specialized to Theorem 4 in the Grassmannian case. However, he soon discovered counterexamples to this conjecture (in fact, it fails for the three-step flag variety $F(1, 3, 4; 5)$).

Motivated by Theorem 3, Buch, Kresch and the author were especially interested in a combinatorial rule for the structure constants on two-step flag varieties. Surprisingly, there is extensive computer evidence which suggests that Knutson's conjecture is true in this special case. Recall from the last lecture that the Schubert classes on two-step flag varieties are indexed by the 012-strings $J(w)$, for permutations $w \in S_N$. In this setting we have the following six different types of puzzle pieces.

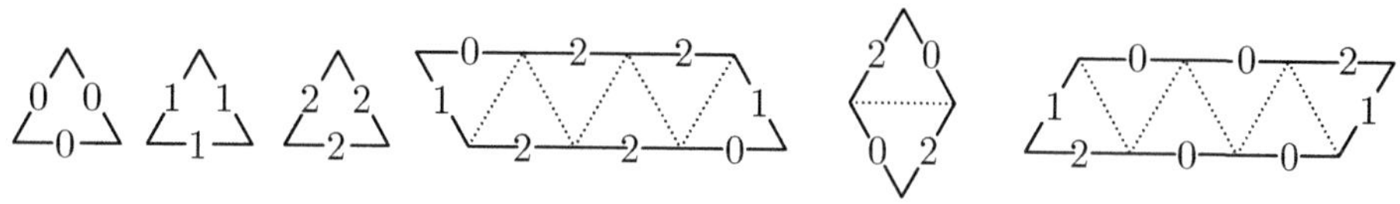

The length of the fourth and of the sixth piece above may vary. The fourth piece can have any number of "2"s (including none) to the right of the "0" on the top edge and equally many to the left of the "0" on the bottom edge. Similarly the sixth piece can have an arbitrary number of "0"s on the top and bottom edges. Again each puzzle piece may be rotated but not reflected. Figure 3 shows two examples of such puzzles.

We can now state Knutson's conjecture in the case of two-step flag varieties. This conjecture has been verified by computer for all two-step flag varieties $F(a, b; N)$ for which $N \leqslant 16$.

Conjecture 1 (Knutson). *For any three Schubert varieties X_u, X_v, and X_w in the flag variety $F(a, b; N)$, the integral $\int_{F(a,b;N)} [X_u] \cdot [X_v] \cdot [X_w]$ is equal to the*

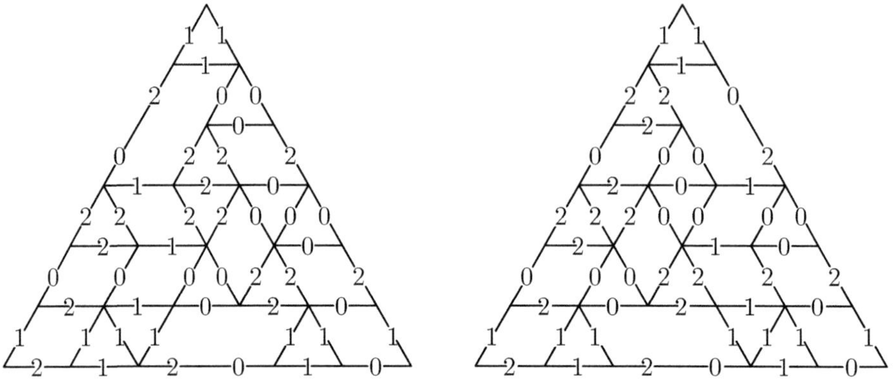

FIGURE 3. Two puzzles with the same boundary labels

number of puzzles such that $J(u)$, $J(v)$, and $J(w)$ are the labels on the north-west, north-east, and south sides when read in clockwise order.

By combining Theorem 3 with Conjecture 1, we arrive at a conjectural 'quantum Littlewood-Richardson rule' for the Gromov-Witten invariants $\langle \sigma_\lambda, \sigma_\mu, \sigma_\nu \rangle_d$. This time we use the 012-string $J^d(\lambda)$ associated to the Schubert variety $X_\lambda^{(d)}(F_\bullet)$ in $F(m-d, m+d; N)$.

Conjecture 2 ([BKT1]). *For partitions λ, μ, ν such that $|\lambda| + |\mu| + |\nu| = mn + dN$ the Gromov-Witten invariant $\langle \sigma_\lambda, \sigma_\mu, \sigma_\nu \rangle_d$ is equal to the number of puzzles such that $J^d(\lambda)$, $J^d(\mu)$, and $J^d(\nu)$ are the labels on the north-west, north-east, and south sides when read in clockwise order.*

The verified cases of Conjecture 1 imply that Conjecture 2 holds for all Grassmannians $G(m, N)$ for which $N \leqslant 16$. It has also been proved in some special cases including when λ has length at most 2 or when m is at most 3.

Example 4. On the Grassmannian $G(3, 6)$, the Gromov-Witten invariant

$$\langle \sigma_{3,2,1},\ \sigma_{3,2,1},\ \sigma_{2,1} \rangle_1$$

is equal to 2. We have $J^1(3,2,1) = 102021$ and $J^1(2,1) = 010212$. Figure 3 displays the two puzzles with the labels $J^1(3,2,1)$, $J^1(3,2,1)$, and $J^1(2,1)$ on their sides.

3.2. Quantum cohomology of $G(m, N)$

As was alluded to in the first lecture and also by the phrase 'quantum Littlewood-Richardson rule', the above Gromov-Witten invariants are the structure constants in a deformation of the cohomology ring of $X = G(m, N)$. This (small) *quantum cohomology ring* $QH^*(X)$ was introduced by string theorists, and is a $\mathbb{Z}[q]$-algebra which is isomorphic to $H^*(X, \mathbb{Z}) \otimes_{\mathbb{Z}} \mathbb{Z}[q]$ as a module over $\mathbb{Z}[q]$. Here q is a formal variable of degree $N = m + n$. The ring structure on $QH^*(X)$ is determined by the relation

$$\sigma_\lambda \cdot \sigma_\mu = \sum \langle \sigma_\lambda, \sigma_\mu, \sigma_{\widehat{\nu}} \rangle_d \, \sigma_\nu \, q^d, \tag{5}$$

the sum over $d \geqslant 0$ and partitions ν with $|\nu| = |\lambda| + |\mu| - dN$. Note that the terms corresponding to $d = 0$ just give the classical cup product in $H^*(X, \mathbb{Z})$. We will need to use the hard fact that equation (5) defines an *associative* product, which turns $QH^*(X)$ into a commutative ring with unit. The reader can find a proof of this basic result in the expository paper [FP].

We now prove, following [Bu1], analogues of the basic structure theorems about $H^*(X, \mathbb{Z})$ for the quantum cohomology ring $QH^*(X)$. For any Young diagram $\lambda \subset (n^m)$, let $\overline{\lambda}$ denote the diagram obtained by removing the leftmost d columns of λ. In terms of partitions, we have $\overline{\lambda}_i = \max\{\lambda_i - d, 0\}$. For any Schubert variety $X_\lambda(F_\bullet)$ in $G(m, E)$, we consider an associated Schubert variety $X_{\overline{\lambda}}(F_\bullet)$ in $G(m+d, E)$. It is easy to see that if $\pi : F(m-d, m+d; E) \to G(m+d, E)$ is the projection map, then $\pi(X_\lambda^{(d)}(F_\bullet)) = X_{\overline{\lambda}}(F_\bullet)$.

Corollary 3. *If* $\langle \sigma_\lambda, \sigma_\mu, \sigma_\nu \rangle_d \neq 0$, *then* $[X_{\overline{\lambda}}] \cdot [X_{\overline{\mu}}] \cdot [X_{\overline{\nu}}] \neq 0$ *in* $H^*(G(m+d, E), \mathbb{Z})$.

Corollary 4. *If* $\langle \sigma_\lambda, \sigma_\mu, \sigma_\nu \rangle_d \neq 0$ *and* $\ell(\lambda) + \ell(\mu) \leqslant m$, *then* $d = 0$.

Proof. We know a priori that $|\lambda| + |\mu| + |\nu| = mn + dN$. The assumption on the lengths of λ and μ implies that

$$|\overline{\lambda}| + |\overline{\mu}| + |\overline{\nu}| \geqslant |\lambda| + |\mu| + |\nu| - 2md = \dim G(m+d, E) + d^2.$$

By Corollary 3, we must have $d = 0$. □

Corollary 4 implies that if $\ell(\lambda) + \ell(\mu) \leqslant m$, then

$$\sigma_\lambda \cdot \sigma_\mu = \sum_{d,\nu} \langle \sigma_\lambda, \sigma_\mu, \sigma_{\widehat{\nu}} \rangle_d \, \sigma_\nu \, q^d = \sum_{|\nu| = |\lambda| + |\mu|} \langle \sigma_\lambda, \sigma_\mu, \sigma_{\widehat{\nu}} \rangle_0 \, \sigma_\nu,$$

that is, there are no quantum correction terms in the product $\sigma_\lambda \cdot \sigma_\mu$.

Theorem 5 (Quantum Giambelli, [Be]**).** *We have* $\sigma_\lambda = \det(\sigma_{\lambda_i + j - i})_{1 \leqslant i,j \leqslant m}$ *in* $QH^*(X)$. *That is, the classical Giambelli and quantum Giambelli formulas coincide for* $G(m, N)$.

Proof. Define a linear map $\phi : H^*(X, \mathbb{Z}) \to QH^*(X)$ by $\phi([X_\lambda]) = \sigma_\lambda$. It follows from Corollary 4 that $\sigma_p \cdot \sigma_\mu = \phi([X_p][X_\mu])$ whenever $\ell(\mu) \leqslant m - 1$. Using the classical Pieri rule and induction, we see that

$$\sigma_{p_1} \cdots \sigma_{p_m} = \phi([X_{p_1}] \cdots [X_{p_m}]),$$

for any m special Schubert classes $\sigma_{p_1}, \ldots, \sigma_{p_m}$. This implies that

$$\det(\sigma_{\lambda_i + j - i})_{1 \leqslant i,j \leqslant m} = \phi(\det([X_{\lambda_i + j - i}])_{1 \leqslant i,j \leqslant m}) = \phi([X_\lambda]) = \sigma_\lambda. \qquad \square$$

Theorem 6 (Quantum Pieri, [Be]). *For $1 \leqslant p \leqslant n$, we have*

$$\sigma_\lambda \cdot \sigma_p = \sum_\mu \sigma_\mu + \sum_\nu \sigma_\nu \, q, \tag{6}$$

where the first sum is over diagrams μ obtained from λ by adding p boxes, no two in the same column, and the second sum is over all ν obtained from λ by removing $N-p$ boxes from the 'rim' of λ, at least one from each row.

Here the 'rim' of a diagram λ is the rim hook (or 'border strip') contained in λ whose south-east border follows the path we used earlier to define the 01-string corresponding to λ.

Example 5. For the Grassmannian $G(3,6)$, we have

$$\sigma_{3,2,1} \cdot \sigma_2 = \sigma_{3,3,2} + (\sigma_2 + \sigma_{1,1})\, q$$

in $QH^*(G(3,6))$. The rule for obtaining the two q-terms is illustrated below.

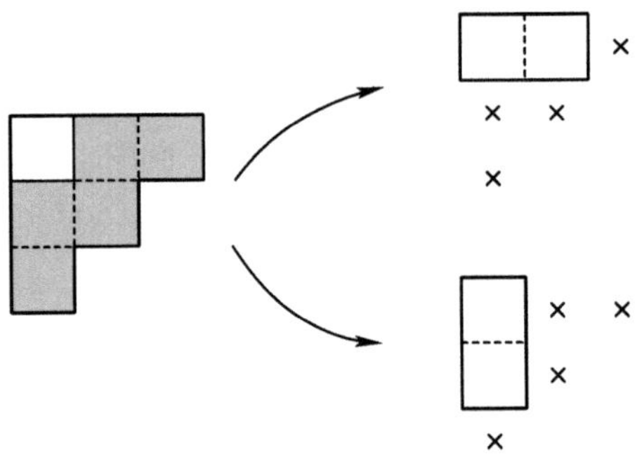

Proof of Theorem 6. By applying the vanishing Corollary 2, we see that it will suffice to check that the line numbers (that is, the Gromov-Witten invariants for $d=1$) agree with the second sum in (6). We will sketch the steps in this argument, and leave the omitted details as an exercise for the reader.

Let $\sigma_{\overline{\lambda}} = [X_{\overline{\lambda}}]$ denote the cohomology class in $H^*(G(m+1,E),\mathbb{Z})$ associated to σ_λ for $d=1$, and define $\sigma_{\overline{\mu}}$ and $\sigma_{\overline{p}}$ in a similar way. One then uses the classical Pieri rule to show that the prescription for the line numbers in $\sigma_\lambda \cdot \sigma_p$ given in (6) is equivalent to the identity

$$\langle \sigma_\lambda, \sigma_\mu, \sigma_p \rangle_1 = \langle \sigma_{\overline{\lambda}}, \sigma_{\overline{\mu}}, \sigma_{\overline{p}} \rangle_0, \tag{7}$$

where the right-hand side of (7) is a classical intersection number on $G(m+1,E)$.

To prove (7), observe that the right-hand side is given by the classical Pieri rule on $G(m+1,N)$, and so equals 0 or 1. If $\langle \sigma_{\overline{\lambda}}, \sigma_{\overline{\mu}}, \sigma_{\overline{p}} \rangle_0 = 0$, then Corollary 3 shows that $\langle \sigma_\lambda, \sigma_\mu, \sigma_p \rangle_1 = 0$ as well.

Next, assume that $\langle \sigma_{\overline{\lambda}}, \sigma_{\overline{\mu}}, \sigma_{\overline{p}} \rangle_0 = 1$, so that there is a unique $(m+1)$-dimensional subspace B in the intersection $X_{\overline{\lambda}}(F_\bullet) \cap X_{\overline{\mu}}(G_\bullet) \cap X_{\overline{p}}(H_\bullet)$, for generally chosen reference flags. Note that the construction of B ensures that it lies in the intersection of the corresponding three Schubert *cells* in $G(m+1, E)$, where the defining inequalities in (1) are all equalities. It follows that the two subspaces

$$V_m = B \cap F_{N-\lambda_m} \quad \text{and} \quad V'_m = B \cap G_{N-\mu_m}$$

each have dimension m, and in fact $V_m \in X_\lambda(F_\bullet)$ and $V'_m \in X_\mu(G_\bullet)$. Since

$$|\lambda| + |\mu| = mn + N - p > \dim G(m, N),$$

we see that $X_\lambda(F_\bullet) \cap X_\mu(G_\bullet) = \emptyset$, and hence $V_m \neq V'_m$. As V_m and V'_m are both codimension one subspaces of B, this proves that $A = V_m \cap V'_m$ has dimension $m-1$. We deduce that the only line (corresponding to the required map $f : \mathbb{P}^1 \to X$ of degree one) meeting the three Schubert varieties $X_\lambda(F_\bullet)$, $X_\mu(G_\bullet)$, and $X_p(H_\bullet)$ is the locus $\{V \in X \mid A \subset V \subset B\}$. □

We conclude with Siebert and Tian's presentation of $QH^*(G(m, N))$ in terms of generators and relations.

Theorem 7 (Ring presentation, [ST]**).** *The ring $QH^*(X)$ is presented as a quotient of the polynomial ring $\mathbb{Z}[\sigma_1, \dots, \sigma_n, q]$ by the relations*

$$D_{m+1} = \cdots = D_{N-1} = 0 \quad \text{and} \quad D_N + (-1)^n q = 0,$$

where $D_k = \det(\sigma_{1+j-i})_{1 \leqslant i,j \leqslant k}$ for each k.

Proof. We will justify why the above relations hold in $QH^*(X)$, and then sketch the rest of the argument. Since the degree of q is N, the relations $D_k = 0$ for $k < N$, which hold in $H^*(X, \mathbb{Z})$, remain true in $QH^*(X)$. For the last relation we use the formal identity of Schur determinants

$$D_N - \sigma_1 D_{N-1} + \sigma_2 D_{N-2} - \cdots + (-1)^n \sigma_n D_m = 0$$

to deduce that $D_N = (-1)^n \sigma_n D_m = (-1)^n \sigma_n \sigma_{(1^m)}$. Therefore it will suffice to show that $\sigma_n \sigma_{(1^m)} = q$; but this is a consequence of Theorem 6.

With a bit more work, one can show that the quotient ring in the theorem is in fact isomorphic to $QH^*(X)$ (see, e.g., [Bu1]). Alternatively, one may use an algebraic result of Siebert and Tian [ST]. This states that for a homogeneous space X, given a presentation

$$H^*(X, \mathbb{Z}) = \mathbb{Z}[u_1, \dots, u_r]/(f_1, \dots, f_t)$$

of $H^*(X, \mathbb{Z})$ in terms of homogeneous generators and relations, if $f'_1, \dots, f'_t$ are homogeneous elements in $\mathbb{Z}[u_1, \dots, u_r, q]$ such that $f'_i(u_1, \dots, u_r, 0) = f_i(u_1, \dots, u_r)$ in $\mathbb{Z}[u_1, \dots, u_r, q]$ and $f'_i(u_1, \dots, u_r, q) = 0$ in $QH^*(X)$, then the canonical map

$$\mathbb{Z}[u_1, \dots, u_r, q]/(f'_1, \dots, f'_t) \to QH^*(X)$$

is an isomorphism. For a proof of this, see [FP, Prop. 11]. □

Remark. The proofs of Theorems 5, 6, and 7 do not require the full force of our main Theorem 3. Indeed, the notion of the kernel and span of a map to X together with Lemma 1 suffice to obtain the simple proofs presented here. For instance, to prove Corollary 3 one can check directly that the span of a rational map which contributes to the Gromov-Witten invariant $\langle \sigma_\lambda, \sigma_\mu, \sigma_\nu \rangle_d$ must lie in the intersection $X_{\overline{\lambda}}(F_\bullet) \cap X_{\overline{\mu}}(G_\bullet) \cap X_{\overline{\nu}}(H_\bullet)$ in $G(m+d, E)$. This was the original approach in [Bu1].

The proof of Theorem 5 used the surprising fact that in the expansion of the Schur determinant in the quantum Giambelli formula, each individual monomial is purely classical, that is, has no q correction terms. This was also observed and generalized to partial flag varieties by Ciocan-Fontanine [C-F, Thm. 3.14].

4. Lecture Four

4.1. Schur polynomials

People have known for a long time about the relation between the product of Schubert classes in the cohomology ring of $G(m, N)$ and the multiplication of *Schur polynomials*, which are the characters of irreducible polynomial representations of GL_n. Recall that if Q denotes the universal (or tautological) quotient bundle of rank n over X, then the special Schubert class σ_i is just the ith Chern class $c_i(Q)$. If the variables $x_1, \ldots, x_n$ are the Chern roots of Q, then the Giambelli formula implies that for any partition λ,

$$\sigma_\lambda = \det(c_{\lambda_i+j-i}(Q)) = \det(e_{\lambda_i+j-i}(x_1, \ldots, x_n)) = s_{\lambda'}(x_1, \ldots, x_n), \qquad (8)$$

where λ' is the conjugate partition to λ (whose diagram is the transpose of the diagram of λ), and $s_{\lambda'}(x_1, \ldots, x_n)$ is a Schur S-polynomial in the variables $x_1, \ldots, x_n$. The Schur polynomials $s_\lambda(x_1, \ldots, x_n)$ for λ of length at most n form a $\mathbb{Z}$-basis for the ring $\Lambda_n = \mathbb{Z}[x_1, \ldots, x_n]^{S_n}$ of symmetric polynomials in n variables. It follows that the structure constants $N_{\lambda\mu}^\nu$ for Schur polynomials

$$s_\lambda s_\mu = \sum_\nu N_{\lambda\mu}^\nu s_\nu$$

agree with the Schubert structure constants $c_{\lambda\mu}^\nu$ in $H^*(X, \mathbb{Z})$.

Our main theorem implies that the Gromov-Witten invariants $\langle \sigma_\lambda, \sigma_\mu, \sigma_\nu \rangle_d$ are structure constants in the product of the two *Schubert polynomials* indexed by the permutations for the modified Schubert varieties $X_\lambda^{(d)}$ and $X_\mu^{(d)}$. Postnikov [P] has shown how one may obtain the same numbers as the coefficients when certain 'toric Schur polynomials' are expanded in the basis of the regular Schur polynomials.

For the rest of these lectures, we will present the analogue of the theory developed thus far in the other classical Lie types. To save time, there will be very little discussion of proofs, but only an exposition of the main results. The arguments are often analogous to the ones in type A, but there are also significant differences. For instance, we shall see that in the case of maximal isotropic Grassmannians, the equation directly analogous to (8) defines a family of '$\widetilde{Q}$-polynomials'. The latter polynomials have the property that the structure constants in their product expansions contain both the classical and quantum invariants for these varieties.

4.2. The Lagrangian Grassmannian $LG(n, 2n)$

We begin with the symplectic case and work with the Lagrangian Grassmannian $LG = LG(n, 2n)$ parametrizing Lagrangian subspaces of $E = \mathbb{C}^{2n}$ equipped with a symplectic form $\langle\ ,\ \rangle$. Recall that a subspace V of E is *isotropic* if the restriction of the form to V vanishes. The maximal possible dimension of an isotropic subspace is n, and in this case V is called a *Lagrangian* subspace. The variety LG is the projective complex manifold of dimension $n(n+1)/2$ which parametrizes Lagrangian subspaces in E.

The Schubert varieties $X_\lambda(F_\bullet)$ in $LG(n, 2n)$ now depend on a strict partition $\lambda = (\lambda_1 > \lambda_2 > \cdots > \lambda_\ell > 0)$ with $\lambda_1 \leqslant n$; we let $\mathcal{D}_n$ denote the parameter space of all such λ (a partition is *strict* if all its parts are distinct). We also require a *complete isotropic flag* of subspaces of E:

$$0 = F_0 \subset F_1 \subset \cdots \subset F_n \subset E$$

where $\dim(F_i) = i$ for each i, and F_n is Lagrangian. The codimension $|\lambda|$ Schubert variety $X_\lambda(F_\bullet) \subset LG$ is defined as the locus of $V \in LG$ such that

$$\dim(V \cap F_{n+1-\lambda_i}) \geqslant i, \quad \text{for} \quad i = 1, \ldots, \ell(\lambda). \tag{9}$$

Let σ_λ be the class of $X_\lambda(F_\bullet)$ in the cohomology group $H^{2|\lambda|}(LG, \mathbb{Z})$. We then have a similar list of classical facts, analogous to those for the type A Grassmannian. However, these results were obtained much more recently than the theorems of Pieri and Giambelli.

1) We have $H^*(LG, \mathbb{Z}) \cong \bigoplus_{\lambda \in \mathcal{D}_n} \mathbb{Z}\,\sigma_\lambda$, that is, the cohomology group of LG is free abelian with basis given by the Schubert classes σ_λ.

2) There is an equation $\sigma_\lambda \sigma_\mu = \sum_\nu e_{\lambda\mu}^\nu \sigma_\nu$ in $H^*(LG, \mathbb{Z})$, with

$$e_{\lambda\mu}^\nu = \int_{LG} \sigma_\lambda \sigma_\mu \sigma_{\nu^\vee} = \# X_\lambda(F_\bullet) \cap X_\mu(G_\bullet) \cap X_{\nu^\vee}(H_\bullet), \tag{10}$$

for general complete isotropic flags $F_\bullet$, $G_\bullet$ and $H_\bullet$ in E. Here the 'dual' partition $\nu^\vee$ is again defined so that $\int_{LG} \sigma_\lambda \sigma_\mu = \delta_{\lambda^\vee \mu}$, and it has the property that the parts of $\nu^\vee$ are the complement of the parts of ν in the set $\{1, \ldots, n\}$. For example, the partitions $(4, 2, 1)$ and $(5, 3)$ form a dual pair in $\mathcal{D}_5$.

Stembridge [Ste] has given a combinatorial rule similar to the classical Littlewood-Richardson rule, which expresses the structure constants $e_{\lambda\mu}^\nu$ in terms of certain sets of shifted Young tableaux. It would be interesting to find an analogue of the 'puzzle rule' of Theorem 4 that works in this setting.

3) The classes $\sigma_1, \ldots, \sigma_n$ are called *special Schubert classes*, and again we have $H^2(LG, \mathbb{Z}) = \mathbb{Z}\,\sigma_1$. If

$$0 \to S \to E_X \to Q \to 0$$

denotes the tautological short exact sequence of vector bundles over LG, then we can use the symplectic form on E to identify Q with the dual of the vector bundle S, and we have $\sigma_i = c_i(S^*)$, for $0 \leqslant i \leqslant n$.

Let us say that two boxes in a (skew) diagram α are connected if they share a vertex or an edge; this defines the connected components of α. We now have the following Pieri rule for LG, due to Hiller and Boe.

Theorem 8 (Pieri rule for LG, [HB]). *For any $\lambda \in \mathcal{D}_n$ and $p \geqslant 0$ we have*

$$\sigma_\lambda \, \sigma_p = \sum_\mu 2^{N(\lambda,\mu)} \sigma_\mu \tag{11}$$

in $H^*(LG,\mathbb{Z})$, where the sum is over all strict partitions μ obtained from λ by adding p boxes, with no two in the same column, and $N(\lambda,\mu)$ is the number of connected components of μ/λ which do not meet the first column.

4) The Pieri rule (11) agrees with the analogous product of Schur Q-functions. This was used by Pragacz to obtain a Giambelli formula for LG, which expresses each Schubert class as a polynomial in the special Schubert classes.

Theorem 9 (Giambelli formula for LG, [P]). *For $i > j > 0$, we have*

$$\sigma_{i,j} = \sigma_i \sigma_j + 2 \sum_{k=1}^{n-i} (-1)^k \sigma_{i+k} \sigma_{j-k},$$

while for λ of length greater than two,

$$\sigma_\lambda = \mathrm{Pfaffian}[\sigma_{\lambda_i,\lambda_j}]_{1 \leqslant i < j \leqslant r}, \tag{12}$$

where r is the smallest even integer such that $r \geqslant \ell(\lambda)$.

For those who are not so familiar with Pfaffians, we recall that they are analogous to (and in fact, square roots of) determinants; see, e.g., [FPr, Appendix D] for more information. The Pfaffian formula (12) is equivalent to the Laplace-type expansion for Pfaffians

$$\sigma_\lambda = \sum_{j=1}^{r-1} (-1)^{j-1} \sigma_{\lambda_j,\lambda_r} \sigma_{\lambda \smallsetminus \{\lambda_j,\lambda_r\}}.$$

5) The ring $H^*(LG,\mathbb{Z})$ is presented as a quotient of the polynomial ring $\mathbb{Z}[c(S^*)] = \mathbb{Z}[\sigma_1,\dots,\sigma_n]$ modulo the relations coming from the Whitney sum formula

$$c_t(S)c_t(S^*) = (1 - \sigma_1 t + \sigma_2 t^2 - \cdots)(1 + \sigma_1 t + \sigma_2 t^2 + \cdots) = 1. \tag{13}$$

By equating the coefficients of like powers of t in (13), we see that the relations are given by

$$\sigma_i^2 + 2 \sum_{k=1}^{n-i} (-1)^k \sigma_{i+k} \sigma_{i-k} = 0$$

for $1 \leqslant i \leqslant n$, where it is understood that $\sigma_0 = 1$ and $\sigma_j = 0$ for $j < 0$. In terms of the Chern roots $x_1,\dots,x_n$ of S^*, the equations (13) may be written as

$$\prod_i (1 - x_i t) \prod_i (1 + x_i t) = \prod_i (1 - x_i^2 t^2) = 1.$$

We thus see that $H^*(LG, \mathbb{Z})$ is isomorphic to the ring $\Lambda_n = \mathbb{Z}[x_1, \ldots, x_n]^{S_n}$ modulo the relations $e_i(x_1^2, \ldots, x_n^2) = 0$, for $1 \leqslant i \leqslant n$.

4.3. $\widetilde{Q}$-polynomials

We turn now to the analogue of Schur's S-polynomials in type C, as suggested by the discussion in § 4.1. These are a family of polynomials symmetric in the variables $X = (x_1, \ldots, x_n)$, which are modelled on Schur's Q-polynomials. They were defined by Pragacz and Ratajski [PR] in the course of their work on degeneracy loci.

For strict partitions $\lambda \in \mathcal{D}_n$, the polynomials $\widetilde{Q}_\lambda(X)$ are obtained by writing

$$\sigma_\lambda = \widetilde{Q}_\lambda(S^*) = \widetilde{Q}_\lambda(x_1, \ldots, x_n)$$

as a polynomial in the Chern roots of S^*, as we did in (8). So $\widetilde{Q}_i(X) = e_i(X)$ for $0 \leqslant i \leqslant n$,

$$\widetilde{Q}_{i,j}(X) = \widetilde{Q}_i(X)\widetilde{Q}_j(X) + 2\sum_{k=1}^{n-i}(-1)^k \widetilde{Q}_{i+k}(X)\widetilde{Q}_{j-k}(X),$$

for $i > j > 0$, and for $\ell(\lambda) \geqslant 3$,

$$\widetilde{Q}_\lambda(X) = \text{Pfaffian}[\widetilde{Q}_{\lambda_i,\lambda_j}(X)]_{1 \leqslant i < j \leqslant r}.$$

Pragacz and Ratajski noticed that this definition also makes sense for *non-strict* partitions λ. For $\lambda_1 > n$, one checks easily that $\widetilde{Q}_\lambda(X) = 0$. Let $\mathcal{E}_n$ denote the parameter space of all partitions λ with $\lambda_1 \leqslant n$. We then obtain polynomials $\widetilde{Q}_\lambda(X)$ for $\lambda \in \mathcal{E}_n$ with the following properties:

a) The set $\{\widetilde{Q}_\lambda(X) \mid \lambda \in \mathcal{E}_n\}$ is a free $\mathbb{Z}$-basis for Λ_n.

b) $\widetilde{Q}_{i,i}(X) = e_i(x_1^2, \ldots, x_n^2)$, for $1 \leqslant i \leqslant n$.

c) (Factorization Property) If $\lambda = (\lambda_1, \ldots, \lambda_\ell)$ and λ^+ is defined by $\lambda^+ = \lambda \cup (i,i) = (\lambda_1, \ldots, i, i, \ldots, \lambda_\ell)$, then

$$\widetilde{Q}_{\lambda^+}(X) = \widetilde{Q}_\lambda(X) \cdot \widetilde{Q}_{i,i}(X).$$

d) For strict λ, the $\widetilde{Q}_\lambda(X)$ enjoy the same Pieri rule as in (11)

$$\widetilde{Q}_\lambda(X) \cdot \widetilde{Q}_p(X) = \sum_\mu 2^{N(\lambda,\mu)} \widetilde{Q}_\mu(X), \tag{14}$$

only this time the sum in (14) is over all partitions $\mu \in \mathcal{E}_n$ (possibly not strict) obtained from λ by adding p boxes, with no two in the same column. In particular, it follows that

$$\widetilde{Q}_n(X) \cdot \widetilde{Q}_\lambda(X) = \widetilde{Q}_{(n,\lambda)}(X)$$

for all $\lambda \in \mathcal{E}_n$.

e) There are structure constants $e_{\lambda\mu}^\nu$ such that

$$\widetilde{Q}_\lambda(X) \cdot \widetilde{Q}_\mu(X) = \sum_\nu e_{\lambda\mu}^\nu \widetilde{Q}_\nu(X),$$

defined for $\lambda,\mu,\nu \in \mathcal{E}_n$ with $|\nu| = |\lambda| + |\mu|$. These agree with the integers in (10) if λ, μ, and ν are strict. In general, however, these integers can be negative, for example

$$e^{(4,4,2,2)}_{(3,2,1),(3,2,1)} = -4.$$

In the next lecture, we will see that some of the constants $e^{\nu}_{\lambda\mu}$ for non-strict ν must be positive, as they are equal to three-point Gromov-Witten invariants, up to a power of 2.

Finally, observe that the above properties allow us to present the cohomology ring of $LG(n,2n)$ as the quotient of the ring $\Lambda_n = \mathbb{Z}[X]^{S_n}$ of $\widetilde{Q}$-polynomials in X modulo the relations $\widetilde{Q}_{i,i}(X) = 0$, for $1 \leqslant i \leqslant n$.

5. Lecture Five

5.1. Gromov-Witten invariants on LG

As in the first lecture, by a rational map to LG we mean a morphism $f\colon \mathbb{P}^1 \to LG$, and its degree is the degree of $f_*[\mathbb{P}^1]\cdot\sigma_1$. The Gromov-Witten invariant $\langle\sigma_\lambda,\sigma_\mu,\sigma_\nu\rangle_d$ is defined for $|\lambda|+|\mu|+|\nu| = n(n+1)/2 + d(n+1)$ and counts the number of rational maps $f\colon \mathbb{P}^1 \to LG(n,2n)$ of degree d such that $f(0) \in X_\lambda(F_\bullet)$, $f(1) \in X_\mu(G_\bullet)$, and $f(\infty) \in X_\nu(H_\bullet)$, for given flags $F_\bullet$, $G_\bullet$, and $H_\bullet$ in general position.

We also define the kernel of a map $f:\mathbb{P}^1 \to LG$ as the intersection of the subspaces $f(p)$ for all $p \in \mathbb{P}^1$. In the symplectic case it happens that the span of f is the orthogonal complement of the kernel of f, and hence is not necessary. Therefore the relevant parameter space of kernels that replaces the two-step flag variety is the isotropic Grassmannian $IG(n-d,2n)$, whose points correspond to isotropic subspaces of E of dimension $n-d$.

If $d \geqslant 0$ is an integer, λ, μ, $\nu \in \mathcal{D}_n$ are such that $|\lambda|+|\mu|+|\nu| = n(n+1)/2 + d(n+1)$, and $F_\bullet$, $G_\bullet$, and $H_\bullet$ are complete isotropic flags of $E = \mathbb{C}^{2n}$ in general position, then similar arguments to the ones discussed earlier show that the map $f \mapsto \mathrm{Ker}(f)$ gives a bijection of the set of rational maps $f\colon \mathbb{P}^1 \to LG$ of degree d satisfying $f(0) \in X_\lambda(F_\bullet)$, $f(1) \in X_\mu(G_\bullet)$, and $f(\infty) \in X_\nu(H_\bullet)$, with the set of points in the intersection $X^{(d)}_\lambda(F_\bullet) \cap X^{(d)}_\mu(G_\bullet) \cap X^{(d)}_\nu(H_\bullet)$ in $Y_d = IG(n-d,2n)$. We therefore get

Corollary 5 ([BKT1]). *Let $d \geqslant 0$ and λ, μ, $\nu \in \mathcal{D}_n$ be chosen as above. Then*

$$\langle\sigma_\lambda,\sigma_\mu,\sigma_\nu\rangle_d = \int_{IG(n-d,2n)} [X^{(d)}_\lambda]\cdot[X^{(d)}_\mu]\cdot[X^{(d)}_\nu].$$

The *line numbers* $\langle\sigma_\lambda,\sigma_\mu,\sigma_\nu\rangle_1$ satisfy an additional relation, which is an extra ingredient needed to complete the analysis for $LG(n,2n)$.

Proposition 2 ([KT1]). *For λ, μ, $\nu \in \mathcal{D}_n$ we have*

$$\langle \sigma_\lambda, \sigma_\mu, \sigma_\nu \rangle_1 = \frac{1}{2} \int_{LG(n+1,2n+2)} [X_\lambda^+] \cdot [X_\mu^+] \cdot [X_\nu^+],$$

where X_λ^+, X_μ^+, X_ν^+ denote Schubert varieties in $LG(n+1, 2n+2)$.

The proof of Proposition 2 in [KT1] proceeds geometrically, by using a correspondence between lines on $LG(n, 2n)$ (which are parametrized by points of $IG(n-1, 2n)$) and points on $LG(n+1, 2n+2)$.

5.2. Quantum cohomology of $LG(n, 2n)$

The quantum cohomology ring of LG is a $\mathbb{Z}[q]$-algebra isomorphic to $H^*(LG, \mathbb{Z}) \otimes_{\mathbb{Z}} \mathbb{Z}[q]$ as a module over $\mathbb{Z}[q]$, but here q is a formal variable of degree $n+1$. The product in $QH^*(LG)$ is defined by the same equation (5) as before, but as $\deg(q) = n+1 < 2n$, we expect different behavior than what we have seen for $G(m, N)$. The previous results allow one to prove the following theorem (the original proofs in [KT1] were more involved).

Theorem 10 (Ring presentation and quantum Giambelli, [KT1]**).** *The ring $QH^*(LG)$ is presented as a quotient of the polynomial ring $\mathbb{Z}[\sigma_1, \ldots, \sigma_n, q]$ by the relations*

$$\sigma_i^2 + 2\sum_{k=1}^{n-i} (-1)^k \sigma_{i+k}\sigma_{i-k} = (-1)^{n-i} \sigma_{2i-n-1}\, q$$

for $1 \leqslant i \leqslant n$. The Schubert class σ_λ in this presentation is given by the Giambelli formulas

$$\sigma_{i,j} = \sigma_i \sigma_j + 2\sum_{k=1}^{n-i} (-1)^k \sigma_{i+k}\sigma_{j-k} + (-1)^{n+1-i} \sigma_{i+j-n-1}\, q$$

for $i > j > 0$, and for $\ell(\lambda) \geqslant 3$,

$$\sigma_\lambda = \text{Pfaffian}[\sigma_{\lambda_i, \lambda_j}]_{1 \leqslant i < j \leqslant r}. \tag{15}$$

The key observation here is that the quantum Giambelli formulas for $LG(n, 2n)$ coincide with the classical Giambelli formulas for $LG(n+1, 2n+2)$, when the class $2\sigma_{n+1}$ is identified with q. Note that this does not imply that the cohomology ring of $LG(n+1, 2n+2)$ is isomorphic to $QH^*(LG(n, 2n))$, because the relation $\sigma_{n+1}^2 = 0$, which holds in the former ring, does not hold in the latter (as $q^2 \neq 0$).

Using the $\widetilde{Q}$-polynomials, we can write the presentation of $QH^*(LG)$ as follows. Let $X^+ = (x_1, \ldots, x_{n+1})$ and let $\widetilde{\Lambda}_{n+1}$ be the subring of Λ_{n+1} generated by the polynomials $\widetilde{Q}_i(X^+)$ for $1 \leqslant i \leqslant n$ together with $2\,\widetilde{Q}_{n+1}(X^+)$. Then the map $\widetilde{\Lambda}_{n+1} \to QH^*(LG)$ which sends $\widetilde{Q}_\lambda(X^+)$ to σ_λ for $\lambda \in \mathcal{D}_n$ and $2\,\widetilde{Q}_{n+1}(X^+)$ to q extends to a surjective ring homomorphism, whose kernel is generated by the relations $\widetilde{Q}_{i,i}(X^+) = 0$ for $1 \leqslant i \leqslant n$.

Theorem 10 therefore implies that the algebra in $QH^*(LG)$ is controlled by the multiplication of $\widetilde{Q}$-polynomials. In particular, the quantum Pieri rule for LG is a specialization of the Pieri rule for $\widetilde{Q}$-polynomials.

Theorem 11 (Quantum Pieri rule for LG, [KT1]). *For any $\lambda \in \mathcal{D}_n$ and $p \geqslant 0$ we have*

$$\sigma_\lambda \cdot \sigma_p = \sum_\mu 2^{N(\lambda,\mu)} \sigma_\mu + \sum_\nu 2^{N'(\nu,\lambda)} \sigma_\nu \, q \tag{16}$$

in $QH^*(LG(n,2n))$, where the first sum is classical, as in (11), while the second is over all strict ν obtained from λ by subtracting $n+1-p$ boxes, no two in the same column, and $N'(\nu,\lambda)$ is one less than the number of connected components of λ/ν.

The informed reader will notice that the exponents $N'(\nu,\lambda)$ in the multiplicities of the quantum correction terms in (16) are of the kind encountered in the classical Pieri rule for *orthogonal* Grassmannians. This was the first indication of a more general phenomenon, which we will discuss at the end of this lecture.

Example 6. For the Lagrangian Grassmannian $LG(4,8)$, the relation

$$\sigma_{4,1} \cdot \sigma_2 = 2\,\sigma_{4,3} + \sigma_{4,2,1} + 2\,\sigma_2\, q$$

holds in the quantum cohomology ring $QH^*(LG)$.

For arbitrary products in $QH^*(LG)$, we have

Corollary 6. *In the relation*

$$\sigma_\lambda \cdot \sigma_\mu = \sum_{\substack{d \geqslant 0 \\ |\nu| = |\lambda| + |\mu| - d(n+1)}} \langle \sigma_\lambda, \sigma_\mu, \sigma_{\nu^\vee} \rangle_d \, \sigma_\nu \, q^d,$$

the quantum structure constant $\langle \sigma_\lambda, \sigma_\mu, \sigma_{\nu^\vee} \rangle_d$ is equal to $2^{-d} e_{\lambda,\mu}^{((n+1)^d,\nu)}$.

Corollary 6 follows immediately from Theorem 10 together with the identity

$$\widetilde{Q}_{((n+1)^d,\nu)}(X^+) = \widetilde{Q}_{n+1}(X^+)^d \cdot \widetilde{Q}_\lambda(X^+)$$

of $\widetilde{Q}$-polynomials. We deduce that the $\widetilde{Q}$-polynomial structure constants of the form $e_{\lambda,\mu}^{(n^d,\nu)}$ (for $\lambda, \mu, \nu \in \mathcal{D}_{n-1}$) are nonnegative integers, divisible by 2^d. A combinatorial rule for these numbers would give a quantum Littlewood-Richardson rule for LG.

The proofs of Theorems 10 and 11, as compared to those for the type A Grassmannian $G(m,N)$, are complicated by two facts. First, a different argument is needed to establish the quantum Pieri rule, which is related by Proposition 2 to the classical Pieri rule on $LG(n+1,2n+2)$. Second, in the quantum Giambelli Pfaffian expansion $\sigma_\lambda = \text{Pfaffian}[\sigma_{\lambda_i,\lambda_j}]_{1 \leqslant i < j \leqslant r}$, there are terms which do involve q-corrections, and these extra q-terms cancel each other out in the end. Thus more combinatorial work is required to prove that the Pfaffian formula (15) holds in $QH^*(LG)$.

5.3. The orthogonal Grassmannian $OG(n+1,2n+2)$

We now turn to the analogue of the above theory in the orthogonal Lie types. We will work with the even orthogonal Grassmannian $OG = OG(n+1,2n+2) = SO_{2n+2}/P_{n+1}$. This variety parametrizes (one component of) the locus of maximal isotropic subspaces of a $(2n+2)$-dimensional vector space E, equipped with a nondegenerate symmetric form. Note that there are two families of such subspaces; by convention, given a fixed isotropic flag $F_\bullet$ in E, we consider only those isotropic V in E such that $V \cap F_{n+1}$ has even codimension in F_{n+1}. We remark that OG is isomorphic to the odd orthogonal Grassmannian $OG(n,2n+1) = SO_{2n+1}/P_n$, hence our analysis (for the maximal isotropic case) will include both the Lie types B and D.

The Schubert varieties $X_\lambda(F_\bullet)$ in OG are again parametrized by partitions $\lambda \in \mathcal{D}_n$ and are defined by the same equations (9) as before, with respect to a complete isotropic flag $F_\bullet$ in E. Let τ_λ be the cohomology class of $X_\lambda(F_\bullet)$; the set $\{\tau_\lambda \,|\, \lambda \in \mathcal{D}_n\}$ is a $\mathbb{Z}$-basis of $H^*(OG,\mathbb{Z})$. Now much of the theory for OG is similar to that for $LG(n,2n)$. To save time, we will pass immediately to the results about the quantum cohomology ring of OG. We again have an isomorphism of $\mathbb{Z}[q]$-modules $QH^*(OG) \cong H^*(OG,\mathbb{Z}) \otimes \mathbb{Z}[q]$, but this time the variable q has degree $2n$.

Another difference between the symplectic and orthogonal case is that the natural embedding of $OG(n+1,2n+2)$ into the type A Grassmannian $G(n+1,2n+2)$ is degree doubling. This means that for every degree d map $f : \mathbb{P}^1 \to OG$, the pullback of the tautological quotient bundle over OG has degree $2d$. It follows that the relevant parameter space of kernels of the maps counted by a Gromov-Witten invariant is the sub-maximal isotropic Grassmannian $OG(n+1-2d,2n+2)$. We pass directly to the corollary of the corresponding 'main theorem':

Corollary 7 ([BKT1]). *Let $d \geqslant 0$ and λ, μ, $\nu \in \mathcal{D}_n$ be such that $|\lambda|+|\mu|+|\nu| = n(n+1)/2+2nd$. Then*

$$\langle \tau_\lambda, \tau_\mu, \tau_\nu \rangle_d = \int_{OG(n+1-2d,2n+2)} [X_\lambda^{(d)}] \cdot [X_\mu^{(d)}] \cdot [X_\nu^{(d)}].$$

This result may be used to obtain analogous structure theorems for $QH^*(OG)$.

Theorem 12 (Ring presentation and quantum Giambelli, [KT2]**).** *The ring $QH^*(OG)$ is presented as a quotient of the polynomial ring $\mathbb{Z}[\tau_1,\ldots,\tau_n,q]$ modulo the relations*

$$\tau_i^2 + 2\sum_{k=1}^{i-1}(-1)^k \tau_{i+k}\tau_{i-k} + (-1)^i \tau_{2i} = 0$$

for all $i < n$, together with the quantum relation

$$\tau_n^2 = q.$$

The Schubert class τ_λ *in this presentation is given by the Giambelli formulas*

$$\tau_{i,j} = \tau_i\tau_j + 2\sum_{k=1}^{j-1}(-1)^k\tau_{i+k}\tau_{j-k} + (-1)^j\tau_{i+j}$$

for $i > j > 0$*, and for* $\ell(\lambda) \geqslant 3$*,*

$$\tau_\lambda = \text{Pfaffian}[\tau_{\lambda_i,\lambda_j}]_{1\leqslant i<j\leqslant r}.$$

It follows from this that the quantum Giambelli formula for OG coincides with the classical Giambelli formula, and indeed the results in the orthogonal case are formally closer to those in type A.

One can use the $\widetilde{P}$-polynomials, defined by $\widetilde{P}_\lambda = 2^{-\ell(\lambda)}\widetilde{Q}_\lambda$ for each λ, to describe the multiplicative structure of $QH^*(OG)$. Let Λ'_n denote the $\mathbb{Z}$-algebra generated by the polynomials $\widetilde{P}_\lambda(X)$, for $\lambda \in \mathcal{E}_n$, where $X = (x_1, \dots, x_n)$. Then the map which sends $\widetilde{P}_\lambda(X)$ to τ_λ for all $\lambda \in \mathcal{D}_n$ and $\widetilde{P}_{n,n}(X)$ to q extends to a surjective ring homomorphism $\Lambda'_n \to QH^*(OG)$ with kernel generated by the relations $\widetilde{P}_{i,i}(X) = 0$, for all $i < n$. Define the structure constants $f_{\lambda\mu}^\nu$ by the relation

$$\widetilde{P}_\lambda(X)\cdot\widetilde{P}_\mu(X) = \sum_{|\nu|=|\lambda|+|\mu|} f_{\lambda\mu}^\nu\,\widetilde{P}_\nu(X).$$

Corollary 8. *The Gromov-Witten invariant (and quantum structure constant)* $\langle\tau_\lambda, \tau_\mu, \tau_{\nu^\vee}\rangle_d$ *is equal to* $f_{\lambda,\mu}^{(n^{2d},\nu)}$.

Theorem 13 (Quantum Pieri rule for OG**, [KT2]).** *For any* $\lambda \in \mathcal{D}_n$ *and* $p \geqslant 0$ *we have*

$$\tau_\lambda\cdot\tau_p = \sum_\mu 2^{N'(\lambda,\mu)}\tau_\mu + \sum_\nu 2^{N'(\lambda,\nu)}\tau_{\nu\smallsetminus(n,n)}\,q,$$

where the first sum is over strict μ *and the second over partitions* $\nu = (n, n, \overline{\nu})$ *with* $\overline{\nu}$ *strict, such that both* μ *and* ν *are obtained from* λ *by adding* p *boxes, with no two in the same column.*

The above quantum Pieri rule implies that

$$\tau_\lambda\cdot\tau_n = \begin{cases}\tau_{(n,\lambda)} & \text{if } \lambda_1 < n,\\ \tau_{\lambda\smallsetminus(n)}\,q & \text{if } \lambda_1 = n\end{cases}$$

in the quantum cohomology ring of $OG(n+1, 2n+2)$. We thus see that multiplication by τ_n is straightforward; it follows that to compute all the Gromov-Witten invariants for OG, it suffices to evaluate the $\langle\tau_\lambda, \tau_\mu, \tau_\nu\rangle_d$ for $\mu, \nu \in \mathcal{D}_{n-1}$. Define a map $\widehat{\ }: \mathcal{D}_n \to \mathcal{D}_{n-1}$ by setting $\widehat{\lambda} = (n - \lambda_\ell, \dots, n - \lambda_1)$ for any partition λ of length ℓ. Notice that $\widehat{\ }$ is essentially a type A Poincaré duality map (for the type A Grassmannian $G(\ell, n+\ell)$).

Partitions in $\mathcal{D}_{n-1}$ also parametrize the Schubert classes σ_λ in the cohomology of $LG(n-1, 2n-2)$. The two spaces $OG(n+1, 2n+2)$ and $LG(n-1, 2n-2)$ have different dimensions and seemingly little in common besides perhaps the fact the the degree of q in the quantum cohomology of the former is twice the degree of q in the latter. However, if $^\vee : \mathcal{D}_{n-1} \to \mathcal{D}_{n-1}$ denotes the Poincaré duality involution on $\mathcal{D}_{n-1}$, we have the following result.

Theorem 14 ([KT2]). *Suppose that $\lambda \in \mathcal{D}_n$ is a non-zero partition with $\ell(\lambda) = 2d + e + 1$ for some nonnegative integers d and e. For any $\mu, \nu \in \mathcal{D}_{n-1}$, we have an equality*

$$\langle \tau_\lambda, \tau_\mu, \tau_\nu \rangle_d = \langle \sigma_{\widehat{\lambda}}, \sigma_{\mu^\vee}, \sigma_{\nu^\vee} \rangle_e \tag{17}$$

of Gromov-Witten invariants for $OG(n+1, 2n+2)$ and $LG(n-1, 2n-2)$, respectively. If λ is zero or $\ell(\lambda) < 2d + 1$, then $\langle \tau_\lambda, \tau_\mu, \tau_\nu \rangle_d = 0$.

We remark that the left-hand side of (17) is symmetric in λ, μ, and ν, unlike the right-hand side. This reflects a $(\mathbb{Z}/2\mathbb{Z})^3$-symmetry shared by the Gromov-Witten invariants for both LG and OG. In fact, Theorem 14 is essentially equivalent to this symmetry. The proof in [KT1][KT2] proceeds by first establishing the symmetry by a clever use of the quantum Pieri rule, and then using the relation between the structure constants of $\widetilde{Q}$- and $\widetilde{P}$-polynomials to put everything together. As of this writing, we lack a purely geometric result that would explain Theorem 14.

5.4. Concluding remarks

The kernel and span ideas in these notes have been used by Buch in [Bu2] and [Bu3] to obtain simple proofs of the main structure theorems regarding the quantum cohomology of any partial flag manifold SL_N/P, where P is a parabolic subgroup of SL_N (the arguments assume the associativity of the quantum product). However, in [BKT1] it is shown that there is no direct analogue of Theorem 3 in this generality, at least not for the complete flag manifold SL_N/B.

In recent work with Buch and Kresch [BKT2], we present similar results to the ones described here (including an analogue of Theorem 3) for any homogeneous space of the form G/P, were G is a classical Lie group and P is a maximal parabolic subgroup of G. These manifolds include the Grassmannians parametrizing non-maximal isotropic subspaces which were mentioned earlier.

For the reader who is interested in learning more about the classical and quantum cohomology of homogeneous spaces, we recommend the texts by Fulton [F1] and Manivel [M] and the expository article [FP]. The latter reference features the general approach to Gromov-Witten theory using Kontsevich's moduli space of stable maps.

Acknowledgements. It is a pleasure to thank Piotr Pragacz and Andrzej Weber for their efforts in organizing the mini-school on 'Schubert Varieties' at the Banach Center in Warsaw. I am also grateful to Marc Levine and the Universität Essen for their hospitality and the Wolfgang Paul program of the Humbolt Foundation for support during the period when these notes were written. Finally, I thank my collaborators Anders Buch and Andrew Kresch for many enlightening discussions, and for all the fun we have had working together on this project.

References

[Be] A. Bertram: *Quantum Schubert calculus*, Adv. Math. **128** (1997), no. 2, 289–305.

[Br] M. Brion: *Lectures on the geometry of flag varieties*, to appear in this Birkhäuser volume.

[Bu1] A.S. Buch: *Quantum cohomology of Grassmannians*, Compositio Math. **137** (2003), no. 2, 227–235.

[Bu2] A.S. Buch: *A direct proof of the quantum version of Monk's formula*, Proc. Amer. Math. Soc. **131** (2003), no. 7, 2037–2042.

[Bu3] A.S. Buch: *Quantum cohomology of partial flag manifolds*, to appear in Trans. Amer. Math. Soc.

[Bu4] A.S. Buch: *Combinatorial K-theory*, to appear in this Birkhäuser volume.

[BKT1] A.S. Buch, A. Kresch, and H. Tamvakis: *Gromov-Witten invariants on Grassmannians*, J. Amer. Math. Soc. **16** (2003), 901–915.

[BKT2] A.S. Buch, A. Kresch, and H. Tamvakis: *Quantum Pieri rules for isotropic Grassmannians*, in preparation.

[C-F] I. Ciocan-Fontanine: *On quantum cohomology rings of partial flag varieties*, Duke Math. J. **98** (1999), no. 3, 485–524.

[F1] W. Fulton: *Young tableaux*, L.M.S. Student Texts **35**, Cambridge Univ. Press, Cambridge, 1997.

[F2] W. Fulton: *Eigenvalues, invariant factors, highest weights, and Schubert calculus*, Bull. Amer. Math. Soc. (N.S.) **37** (2000), no. 3, 209–249.

[FP] W. Fulton and R. Pandharipande: *Notes on stable maps and quantum cohomology*, in: Algebraic geometry (Santa Cruz, 1995), Proc. Sympos. Pure. Math. **62**, Part 2, Amer. Math. Soc., Providence (1997), pp. 45–96.

[FPr] W. Fulton and P. Pragacz: *Schubert varieties and degeneracy loci*, Lecture Notes in Math. **1689**, Springer-Verlag, Berlin, 1998.

[G] G.Z. Giambelli: *Risoluzione del problema degli spazi secanti*, Mem. R. Accad. Sci. Torino (2) **52** (1902), 171–211.

[HB] H. Hiller and B. Boe: *Pieri formula for SO_{2n+1}/U_n and Sp_n/U_n*, Adv. Math. **62** (1986), 49–67.

[KTW] A. Knutson, T. Tao and C. Woodward: *The honeycomb model of $GL(n)$ tensor products II: Puzzles determine facets of the Littlewood-Richardson cone*, J. Amer. Math. Soc. **17** (2004), 19–48.

[KT1] A. Kresch and H. Tamvakis: *Quantum cohomology of the Lagrangian Grassmannian*, J. Algebraic Geom. **12** (2003), 777–810.

[KT2] A. Kresch and H. Tamvakis: *Quantum cohomology of orthogonal Grassmannians*, Compositio Math. **140** (2004), 482–500.

[M] L. Manivel: *Fonctions symétriques, polynômes de Schubert et lieux de dégénérescence*, Cours Spécialisés **3**, Société Mathématique de France, Paris, 1998.

[Pi] M. Pieri: *Sul problema degli spazi secanti. Nota 1^a*, Rend. Ist. Lombardo (2) **26** (1893), 534–546.

[Po] A. Postnikov: *Affine approach to quantum Schubert calculus*, preprint (2002), math.CO/0205165.

[P] P. Pragacz: *Algebro-geometric applications of Schur S- and Q-polynomials*, Séminaire d'Algèbre Dubreil-Malliavin 1989–1990, Lecture Notes in Math. **1478**, 130–191, Springer-Verlag, Berlin, 1991.

[PR] P. Pragacz and J. Ratajski: *Formulas for Lagrangian and orthogonal degeneracy loci; $\widetilde{Q}$-polynomial approach*, Compositio Math. **107** (1997), no. 1, 11–87.

[ST] B. Siebert and G. Tian: *On quantum cohomology rings of Fano manifolds and a formula of Vafa and Intriligator*, Asian J. Math. **1** (1997), no. 4, 679–695.

[Ste] J.R. Stembridge: *Shifted tableaux and the projective representations of symmetric groups*, Adv. Math. **74** (1989), 87–134.

[V] R. Vakil: *A geometric Littlewood-Richardson rule*, preprint (2003), math.AG/0302294.

[Y] A. Yong: *Degree bounds in quantum Schubert calculus*, Proc. Amer. Math. Soc. **131** (2003), 2649–2655.

Harry Tamvakis
Department of Mathematics
Brandeis University – MS 050
P.O. Box 9110
Waltham, MA 02454-9110, USA
e-mail: `harryt@brandeis.edu`

GPSR Compliance

The European Union's (EU) General Product Safety Regulation (GPSR) is a set of rules that requires consumer products to be safe and our obligations to ensure this.

If you have any concerns about our products, you can contact us on ProductSafety@springernature.com

In case Publisher is established outside the EU, the EU authorized representative is:

Springer Nature Customer Service Center GmbH
Europaplatz 3
69115 Heidelberg, Germany

Batch number: 10732470

Printed by Printforce, the Netherlands